# IMPROVING UPPER CANADA

## Agricultural Societies and State Formation, 1791–1852

# Improving Upper Canada

## *Agricultural Societies and State Formation, 1791–1852*

ROSS FAIR

UNIVERSITY OF TORONTO PRESS
Toronto Buffalo London

Toronto Buffalo London
utorontopress.com

ISBN 978-1-4875-5353-1 (cloth)
ISBN 978-1-4875-5355-5 (EPUB)
ISBN 978-1-4875-5354-8 (PDF)

---

**Library and Archives Canada Cataloguing in Publication**

Title: Improving Upper Canada : agricultural societies and state formation, 1791–1852 / Ross Fair.
Names: Fair, Ross, author.
Description: Includes bibliographical references and index.
Identifiers: Canadiana (print) 20240280156 | Canadiana (ebook) 20240280245 | ISBN 9781487553531 (hardcover) | ISBN 9781487553555 (EPUB) | ISBN 9781487553548 (PDF)
Subjects: LCSH: Agriculture – Ontario – Societies, etc. – History – 18th century. | LCSH: Agriculture – Ontario – Societies, etc. – History – 19th century. | LCSH: Agriculture and state – Ontario – History – 18th century. | LCSH: Agriculture and state – Ontario – History – 19th century.
Classification: LCC HD1486.C3 F32 2024 | DDC 334/.6830971309033 – dc23

---

Cover design: Rebecca Lown
Cover image: "Moss Park," the estate of William Allan (c. 1770–1853), viewed from the southwest, 1834. John George Howard (Courtesy of the Royal Ontario Museum, 995.35.2.)

We wish to acknowledge the land on which the University of Toronto Press operates. This land is the traditional territory of the Wendat, the Anishnaabeg, the Haudenosaunee, the Métis, and the Mississaugas of the Credit First Nation.

University of Toronto Press acknowledges the financial support of the Government of Canada, the Canada Council for the Arts, and the Ontario Arts Council, an agency of the Government of Ontario, for its publishing activities.

Canada Council for the Arts
Conseil des Arts du Canada

Funded by the Government of Canada
Financé par le gouvernement du Canada

*To Dad,*
*To Mom,*
*To Green Park Farm*

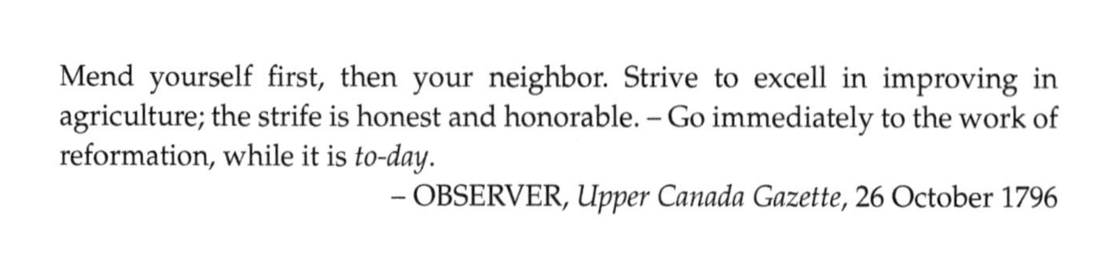

Mend yourself first, then your neighbor. Strive to excell in improving in agriculture; the strife is honest and honorable. – Go immediately to the work of reformation, while it is *to-day*.

– OBSERVER, *Upper Canada Gazette*, 26 October 1796

# Contents

# Illustrations and Maps

**Illustrations**

## Maps

# Acknowledgments

There are several individuals who I must acknowledge for their guidance and assistance along the (long) path this study has taken from an initial idea to it appearing in print. Thanks to the late Gerald Stortz and to Ken McLaughlin for allowing an ambitious fourth-year undergraduate student to discover for himself that "Ontario agricultural societies in the nineteenth century" was a bit too ambitious of a project to undertake and for also seeing the merit in me pursuing a more focused study on select societies across a reduced period of time. I cannot thank Jane Errington enough for her wise supervision of the dissertation on which this book is based. Thanks also to the many archivists, librarians, and others who assisted me in locating the sources of historical information necessary to produce this book. Thanks to the anonymous reviewers who provided excellent advice, in particular the reviewer who recommended extending the analysis beyond my dissertation's initial timeframe and reframing it as a study of improvement and state formation. I confess that undertaking the additional research and revisions took *much* longer than I originally envisioned but it has, I believe, produced a more substantial contribution to the study of Canada's colonial past.

My sincerest thanks to Grant and Carol Bridge. Those early days spent writing a first draft of this manuscript in their home were critical to advancing this project forward. It's unfortunate that Grant is not alive to see this book in print. My thanks, too, to Robert Teigrob and Suzanne Zerger for graciously offering me use of their cabin for a week of summer writing on several occasions. Those were brief but valuable opportunities to write and revise, blessedly absent of the fog of teaching and administrative duties.

Funding from the Department of History at Toronto Metropolitan University (formerly Ryerson University) helped secure many of the book's illustrations and its maps. Thanks to Julie Witmer for producing those maps. Thanks also to Len Husband at the University of Toronto Press for his assistance in shepherding this manuscript to print. Thanks to Sathya Shree Kumar for copyediting the manuscript, and thanks to Stefan Fergus for producing the index.

For completely different reasons, I must thank The Gang. Distractions are important for those of us whose brains are always ON as we think through ideas and arguments. I could not ask for a better group of friends, who are always prepared to help provide an OFF switch.

Finally, to Jennifer Bridge, I extend special thanks for offering much needed support and encouragement. Your gentle and infrequent question, "When will the book be done?" can now be answered definitively. It is finished.

# Abbreviations

| | |
|---|---|
| AAUC | Agricultural Association of Upper Canada |
| ASMD | Agricultural Society of the Midland District |
| FCAS | Frontenac County Agricultural Society |
| HDAS | Home District Agricultural Study |
| MDAS | Midland District Agricultural Society |
| NAS | Niagara Agricultural Society |
| NDAS | Niagara District Agricultural Society |
| UCACS | Upper Canada Agricultural and Commercial Society |
| UCAS | Upper Canada Agricultural Society |
| | |
| RASE | Royal Agricultural Society of England |
| RSA | Royal Society of Arts, Manufactures, and Commerce |
| | |
| NYSAS | New York State Agricultural Society |
| | |
| MHA | member of the House of Assembly |
| MPP | member of the Provincial Parliament |

# IMPROVING UPPER CANADA

## Agricultural Societies and State Formation, 1791–1852

# Introduction

William Hutton stood and read his gold-medal-winning essay to an assembled crowd at the annual provincial agricultural exhibition in Brockville, Upper Canada.[1] On that September 1851 evening, he spoke "On Agriculture and its Advantages as a Pursuit," suggesting practical improvements that his fellow colonists might introduce to their farming practices for beneficial results. Hutton, a schoolteacher and local office holder who farmed near Belleville, declared: "The idea 'that any fool can farm' is now antiquated and an unjust stigma on our noble profession, one of the first advantages of which, as a pursuit, is, that *it requires enlightenment*, that it demands never ceasing improvement of our mental capacities, which tend to raise us in the scale of intellectual, therefore, happy beings." To amplify the importance of pursuing agricultural improvements to his fellow colonists, he repeated the motto of Dr. James F.W. Johnston, professor of agricultural chemistry at Britain's University of Durham, a prominent agricultural improver of the day: "He that causes two blades of grass to grow where only one grew before is a benefactor of his country."[2]

Hutton's selection of Dr. Johnston's "two blades of grass" motto is instructive for situating the Upper Canadian experience within the longevity, the adaptability, and the transatlantic nature of an improving ideology first spawned during the Enlightenment and central to the idea of progress by the nineteenth century. Johnston's quote was far from original, however. Variants of it had appeared in print for decades.[3] While Johnston and Hutton employed the motto for their contemporary mid-nineteenth-century world of agricultural improvement and commerce, the motto's sentiment did not originate as a statement on the progress of a modern capitalist state – greater yields from improved farmland and improved breeds of livestock, which contributed to increased commerce, all combining to strengthen the progress of an increasingly prosperous nation. Rather, it appears that Jonathan Swift wrote the original form of the motto as satire in his 1726 *Gulliver's Travels*. Swift has Gulliver attempting to impress the King of Brobdingnag by explaining the superiority of England's government. To Gulliver's surprise, the king was far from impressed and "gave it for his opinion, 'that whoever could make two ears of corn, or two blades of grass, to grow upon a spot of ground where only one grew before, would deserve better of mankind, and do more essential service to his country, than the whole race of politicians put together.'" In this fantasy clash of cultures, Gulliver finds the Brobdingnagians' learning "defective" for it consisted "only in morality, history, poetry, and mathematics … But the last of these is wholly applied to what may be useful in life, to the improvement of agriculture, and all mechanical arts; so that among us, it would be little esteemed."[4] Here, the Anglo-Irish Swift employed satire to assert his belief that the English government of his day was betraying its duty to encourage its own citizens to improve the economy and society by growing those two ears of corn or two blades of grass where only one grew before. Instead, it was waging ever larger wars and engaging in exploitative trade with ever more distant lands.[5] The Brobdingnagian king's declaration, later detached from its original context as satire, became the agricultural improvers' motto and continued to have currency in that sense a century and a quarter on.

Hutton's essay was the sort of verbose composition that one might expect a Victorian-era colonist to write, but the treatise and its author are instructive to understanding the ideology of improvement's role, specifically agricultural improvement, in the pursuit of progress in Upper Canada and the formation of the Upper Canadian state. Hutton was a member of the agricultural society in his home district of Victoria, while the Johnstown District Agricultural Society, centred in Brockville, sponsored the competition to which Hutton had submitted his essay. Those who listened to Hutton read his essay were members of the Agricultural Association of Upper Canada (AAUC) that coordinated the provincial exhibition at Brockville. Within a few months, the recently established Board of Agriculture for Upper Canada would publish Hutton's essay as front-page content in its monthly publication, *The Canadian Agriculturist and Transactions of the Board of Agriculture of Upper Canada*.[6] By then, Upper Canada's numerous agricultural societies, the provincial agricultural association, and the board were all operating under the authority of a single provincial statute that set forth roles, relationships, and permanent annual funding. Meantime, improvers were busily lobbying for the provincial government to create a Bureau of Agriculture. Some drafted the legislation. Once analysed in this way, Hutton's moment in the limelight emphasizes the fact that agricultural improvement played a significant role in Upper Canadian state formation. Agriculture was, after all, the core economic activity in the colony and farming was the common labour that constructed rural communities. To fully appreciate that role, this study analyses Upper Canada's agricultural societies, the colony's institutional manifestation of the improving ideology, and the improvers who founded and led them. It reveals how these individuals employed their social, economic, and political influence to secure state support and leadership of agricultural improvement as an accepted function of government.

British North American settlers began weaving together the threads of improvement, agricultural societies, and colonial state formation in the months prior to the Constitutional Act of 1791 that created Upper Canada as a British North American colony. They did so by establishing an agricultural society at the Niagara settlement. It was a colonial variant of similarly named institutions founded by contemporaries among Britain's landed gentry to promote agricultural improvement among their class and, by extension, the nation at large. The attempt by Upper Canadian improvers to emulate the British gentry and their institutions was an ambitious effort to extend their pretended paternalistic authority throughout the scattered settlements of the expansive province. The colony's political and social elite were not equal to Britain's landed gentry, however. They did not live on and manage long-cultivated estates and they did not possess the gentry's leisure time and disposable wealth to conduct and report on agricultural experiments while supporting a local agricultural society through annual membership fees and donations. Establishing paternalistic leadership on a provincial scale proved an overly ambitious task during Upper Canada's first forty years due to the distances between settlements and the arrival of new settlers who saw no direct reason to be subservient to a provincial oligarchy located at a capital town elsewhere in the province.

In this study, an 1830 Upper Canadian statute that provided government support and public funding to establish an agricultural society in each district of the province serves as a fulcrum for my analysis of the role of agricultural improvement in Upper Canadian state formation during the 1791–1852 period. The legislation was passed by a colonial legislature divided between an emerging faction of reformers who had secured its first majority of seats in the House of Assembly. Yet this divided legislature found common values in the promotion of agricultural improvement, despite most tories looking to British examples of agricultural societies and other leadership of improvement, while many reformers looked to the United States as a model. In the end, a majority were able to see merits in each approach as the act provided public funds to support the establishment of an agricultural society in each district of the province while permitting the local elites to do the work of organizing and leading them. In the United States, improvement's adaptability from its origins among Britain's landed gentry had permitted elites of the republic to establish agricultural societies to promote improvement among the independent yeoman farmer, whom they praised as the model citizen and on whose shoulders, proclaimed the agrarian myth,[7] the strength of the republic relied. But American improvers had also faced problems sustaining an agricultural society financially and exerting influence over settlers. In working through such problems American agricultural improvers, particularly in New York state, were a step ahead of their Upper Canadian counterparts in convincing a state government that funding and leading agricultural improvement was an additional role it must undertake for the progress of the state's economy and society. That Upper Canadian improvers looked to the United States for a model to adopt was a prime example of the province's "dual heritage," identified by Jane Errington, who claims that colonists of this period "attempted to use the best of both worlds – the old and the new – in laying the foundations of their new society."[8]

The 1830 legislative debate and resulting agricultural statute permit a charting of the assumptions and processes of colonial state formation. It demonstrates that the Georgian, tory-dominated world of early Upper Canadian public life was changing in significant ways. Subsequent renewals of the agricultural societies' legislation during the 1830s and 1840s provide a developing set of semi-public institutions across the province that serve as a lens for us to chart the emergence of the Victorian liberal order. During the 1840s, agricultural societies became umbrella organizations for township-level agricultural societies founded under their authority. Leaders of district agricultural societies would also form a provincial association to represent all such institutions, and by the 1850s, the association and the network of local societies provided the foundation on which a board of agriculture and a bureau of agriculture were established. In drafting the establishing legislation for each, Upper Canadian agricultural improvers ensured the bureau and its minister of agriculture were reliant on the network of local publicly funded agricultural societies as a volunteer bureaucracy for the promotion of agricultural improvement throughout the colony.

The subject of this study is not merely a curiosity of a developing colonial government. Significantly, the provincial government established its new Bureau of Agriculture with a mandate that encompassed not only Upper Canada but also the Lower Canadian portion of the United Province of Canada. The appointment of a minister of agriculture was the first expansion of cabinet for the Province of Canada since Lord Sydenham had initiated his administrative reforms following the union of Upper and Lower Canada in 1841. Moreover, the ideology of improvement that informed the bureau's creation continues, to some degree, in form and function. By Confederation, the Bureau of Agriculture for the United Province of Canada had become a Department of Agriculture. After Confederation, it became the Dominion of Canada's Department of Agriculture, today's Agriculture and Agri-Food Canada. The Ontario Ministry of Agriculture, Food and Rural Affairs, another descendant of the colonial bureau, continues to govern the province's agriculturalists as well as the Ontario Association of Agricultural Societies, itself descended from the AAUC and the agricultural societies legislation of 1830. The association continues to provide a collective voice for the numerous local agricultural societies operating across Ontario, each best known for hosting an annual "Fall Fair." Completing the trifurcation of the pre-Confederation department was a provincial bureaucracy established to oversee agricultural improvements in Quebec, today's Ministère de l'Agriculture, des Pêcheries et de l'Alimentation du Québec.

To better understand the purpose and function of Upper Canadian agricultural societies and the mid-century bureaucracy that colonial agricultural improvers created, it is necessary to introduce the Enlightenment ideology of improvement upon which they were based and chart agricultural improvement's transatlantic connections between Britain and North America (and beyond). Also required is a brief assessment of the existing understanding of agricultural societies' contributions to Upper Canadian society and state formation in colonial British North America.

## Improvers, Improvement, and Progress

"To improve" appeared as a concept in England as early as 1302, meaning to reclaim waste or unoccupied land by enclosure and cultivation. By the sixteenth century, "to improve" had come to mean "make a profit, increase the value of" and, subsequently, "make greater in amount or degree." To do so effectively required private property; therefore, "to improve" became tied to the enclosure of common or "waste" land. By the seventeenth century, *improver* began to describe someone "who cultivated their own character." Over time, improvement's meaning continued to accumulate non-agricultural variants so that it became synonymous with *change* and, by the early 1700s, "had come to embrace the cultivation of the body or the mind."[9] "Improvement," which contributed to "progress," developed within Enlightenment undertakings through numerous means to establish a less religious, more secular intellectual foundation for society and governance.[10] Careful observations of the natural world provided a considerable amount of empirical evidence that could be applied to agriculture and utilized for the benefit of humankind. In the British Isles, it was the leaders of Scotland who, during the mid-eighteenth-century years of the Scottish Enlightenment, first began to "forge that alliance between science and agriculture which was to provide the basis for the improvers' optimism in Britain more generally." There, notes John Gascoigne, the landed aristocracy's awareness of their backwardness "strengthened their determination to promote schemes which would improve the productivity of Scotland's agriculture and put to better use both its land and its peoples."[11]

Enlightenment thought became "the ideological mortar of late-eighteenth century British society."[12] By then, gentlemen of science and agriculture had joined together in a variety of improving and scientific societies to share their knowledge, and agricultural societies and scientific institutions promised important means by which progress could be attained more quickly through the "cumulatively progressive effect of cooperative enterprise." *Progress*, at that time in Britain, was "the belief in the movement over time of some aspect or aspects of human existence, within a social setting, toward a better condition." It was so interwoven into British intellectual thought and trends, David Spadafora argues, that thought from that era cannot be understood without appreciating it. "Visible progress in the present gave a promise of still better things to come."[13] Enlightenment thinkers of the last half of the eighteenth century exuded a self-confidence, Gascoigne argues, "for they saw themselves as in the vanguard of change which, increasingly was enlisting the support of the privileged elites who governed society."[14] Elites were enclosing their estates and experimenting with improved forms of cultivation and livestock husbandry and breeding, and they were members of local agricultural societies and scientific organizations. By century's end, the improving ideology was ubiquitous throughout the British world, for "the confidence that the human condition could be improved rippled out from agriculture, the traditional centre of the nation's economic and social order, to most other areas of society."[15] Belief in improvement and progress was, of course, by no means limited to the British Empire. Landed elites of continental Europe were also busily implementing and recording agricultural advances; improvement was at the core of Spanish, French, and Russian expansions of their empires.[16]

While E.A. Heaman's recent study, *Civilization*, provides a significant contribution to our understanding of Enlightenment thought's role in shaping the Canadian nation, Upper Canada's historiography has changed only somewhat since Jeffrey McNairn observed that the colony is often framed as one set apart from the Enlightenment and the intellectual contributions it provided to society and governance.[17] Nevertheless, there is much we can learn from numerous other studies focused on Britain, other parts of the empire, as well as the United States, regarding the ways in which the pursuit of improvement affected agriculture, economy, commerce, society, imperialism and colonial settlement. In his economic history of Britain from 1700–1850, Joel Mokyr argues, "Enlightened farming was deeply interested in and committed to economic progress and innovation, and was hell bent on improvements and rationalization based on a better understanding of the natural processes at work." While he dismisses the notion that improvements brought about any "agricultural revolution" or that most enclosures were undertaken for any higher purpose than increased wealth, Mokyr notes that enclosed lands were "generally more versatile and capable of adopting improved cultivation techniques." Such changes in practice may not have revolutionized agriculture, because there were limits to what eighteenth-century improvements could do in the absence of precise knowledge of botany and soil chemistry, the ability to control insects and plant diseases, and little capability to produce machinery to reduce field labour. Nevertheless, the pursuit of improvement and its main effect of enclosure "changed the geography of the land, the organization of agricultural production, and the social composition of the agricultural labor force."[18]

Increased interest in agricultural improvement within the British Isles was concomitant with Britain's colonial expansion and led to its entwining with the functions of the state. Richard Drayton argues that the political and economic values of Britain's Second Empire (1783–1815) focused on the management of land and people, for imperial authorities valued science's potential for agriculture to make estate management at home more efficient and productive. It also provided hope for increased commerce with colonies and other distant territories. Particular attention was paid to the cultivation of new plants and crops as potential commodities.[19] Individuals such as Sir Joseph Banks, president of the Royal Society from 1778–1820, became a central figure in Britain's planning for the exploitation of its colonial resources. He directed scientific discoveries in newly explored territories and introduced new agricultural commodities to be grown in Britain's colonies around the globe.[20] As chapter 2 explains, improvers deemed European hemp varieties to be well suited to the soil and climate of Upper Canada. Drayton claims that, by the late eighteenth century, "'Improvement', once a slogan for the local activity of the gentry, became a new criterion for responsible authority, and a mission towards which government might legitimately expand its powers." George III, with his keen interest in sheep breeding, became "the paradigm of an 'improver,'" and in the face of ongoing pressures from the loss of its American colonies following the American Revolution and, then, war with France, there was a drive to use Britain's resources more efficiently. By the turn of the nineteenth century, agriculture had become so central to Britain's destiny, that it required state assistance. Improvers argued that "only the Crown could ensure that, across the nation, private virtue turned Nature into public abundance." As a result, parliament soon established a board of agriculture (1793–1822) to study and report on the state of agriculture in the counties and in 1801 it passed the General Enclosure Act to ease the process of enclosure.[21] Improvers in Upper Canadian took note of this development, and some wondered how a board might be replicated in the colony to guide the progress of its agricultural development.

Across the colonies, British officials, merchants, and settlers viewed their new territories through preconceived ideals of English landscapes, no matter the difference in local topography and flora and fauna. Thus, elite landowners erected English-style country houses in the colonies with views of English gardens and field landscapes to re-create a properly improved landscape for European uses.[22] Upper Canadian improvers with sufficient wealth attempted to clear their farms and build their buildings to look like an estate of the British gentry, too, setting a local example for other propertied gentlemen to attain. Some improvers were merchants, such as Robert Hamilton, who resided at his home at Queenston and who was president of the Niagara Agricultural Society (NAS). Thus, one must think broadly as to who was an agricultural "improver." As this study demonstrates, several Upper Canadian gentlemen who owned properties they farmed using improved methods might not have identified themselves as a "farmer." Likewise, historians may not have identified them as farmers due to the prominence they had gained through other social, political, or economic roles.[23]

Merchants, as David Hancock informs us, were among those who spread the ideology of improvement as they attempted to improve themselves. His analysis of London merchants and their world from the 1730s to the 1780s reveals that improvement "touched most aspects of everyday life, and it manifested itself in programs that were at once polite, industrious, and moral." Improvement was a preoccupation that charted their path to becoming gentlemen. Yes, it led to increased business and profits, but as Hancock asserts, to gain the respect of the peerage and landed gentry, to which merchants had not been born, they had to insinuate themselves into gentlemanly circles. Thus, improvement included the "domestic, noncommercial activities as necessary steps along the road to becoming a gentleman: estate building, house building, art collecting, gardening, farming, and charitable endeavours – road, bridge, and canal building, manufacturing and philanthropy." As they improved their status to that of gentlemen, their activities helped improve the lives of others, "since they believed society as a whole was advancing from barbarism toward civility."[24]

Tamara Plakins Thornton identifies similar trends that persisted among similar classes of improving gentlemen in antebellum New England. She notes: "It is a curious fact of history that the same men directly responsible for changing the Massachusetts economy from a farming to a commercial and industrial one – merchants, financiers, manufacturers, and their legal and political advocates – should have endeavored so assiduously to identify themselves with things rural and agrarian." Although they were mercantile in outlook and had rejected the landed aristocracy of Britain, they set about establishing themselves as agrarian gentlemen of Massachusetts by building estate homes, planting gardens and orchards, and improving their farms all while encouraging others to do the same via the agricultural and horticultural societies they established and led. The elite of Massachusetts knew "through a kind of cultural intuition," she argues, that being a gentleman farmer was appropriate to their "station and pretensions" and that interest in scientific agriculture was what defined a "proper gentleman."[25]

It was within this transatlantic milieu of late eighteenth-century improvers and improving ideology that British and Anglo-American settlement began in earnest within the Great Lakes territory that became Upper Canada in 1791. While clearing forests for farmland, not enclosure, was the Upper Canadian settlers' task, the improved forms of agriculture employed on enclosed lands were, according to British improvers of the era, highly transferrable in turning North American "wasteland" into farmland. Such agricultural improvement ideals informed the way the new colony should be organized and surveyed for settlement (straight boundaries, little matter the topography), what farms should aspire to look like and produce once cleared for cultivation, and what role an agricultural society might play in colonial development. At any point during the 1791–1852 era covered by this study, immigrants to Upper Canada who departed from various parts of the

British Isles possessed some level of awareness of improvement and progress, even if that awareness came by way of their eviction as a tenant due to enclosure. The Highland Clearances of the eighteenth and nineteenth centuries were a consequence of the Scottish Enlightenment and its pursuit of agricultural improvement to transform rural society and economy through state-sanctioned land enclosure, tenant evictions, and the introduction of large-scale sheep grazing. As rural areas were forcibly depopulated in Scotland, England, and Ireland, many of those displaced would immigrate to British North America. They had been "cleared" from farmland as impediments to improvement, but once they arrived in Upper Canada, they heard colonial improvers declare that the colony's progress rested on the improvements that they were to undertake in clearing their piece of forest into farmland.[26] In taking up this challenge of improvement, those thousands of immigrants themselves displaced Indigenous peoples of the Great Lakes region and erased their cultivated landscapes and settlements.[27] The tireless call of the colonial improver was directed at European and Anglo-American settlers; however, this study identifies rare moments when improvers attempted to engage Indigenous people in their pursuit of improvement and progress.

Agricultural improvement was largely about achieving potential outcomes on the farm, for the local community, and for the colony. In writing about the eighteenth-century American farmer, Richard Bushman notes: "Farming had less to do with scale than with an idea." It was the promise that a farm family with land could provide sufficient labour to produce enough food, first, to subsist and, then, produce some amount of surplus to trade for increased self-sufficiency.[28] In the frontier settlements of Upper Canada, where most settlers were Anglo-Americans during its first decades, thousands already familiar with the North American landscape and environment pursued this promise with considerable daily toil. They well knew that the idea of farming was fickle, given the forces of weather, disease, markets, and others. Improvement was another powerful idea associated with farming, for it too promised self-sufficiency, often without increased scale – "two blades of grass to grow where only one grew before." What if a farmer planted improved varieties of wheat to reap a larger return per acre? What if a farmer employed improved ploughing techniques using an improved plough designed for the soil type of his fields? What if that farmer employed new tools and machinery in farming the land to not only increase yields but also free up labour and time to permit the addition of livestock husbandry to the farm's economy and, perhaps, also an on-farm industry such as wool and dairy? What if a farmer also used the manure produced by that livestock on his farmland for even greater returns? What if a farmer raised a better breed of livestock and employed improved methods of animal husbandry to produce more meat, wool, or milk?

Improvement of soil and tillage were critical to not only maximizing crop yields but also for sustaining rural communities, argues Steven Stoll in his examination of soil and society in nineteenth-century America. "People are anchored in place only as securely as the ground they till," he notes. With an ever-present and shifting frontier, American farmers had the option of leaving a long-settled farm with its soil exhausted from a yearly cycle of crops for western land and its virgin soil. Observers of Upper Canadian agriculture in the decades that followed the War of 1812 complained that these same poor farming practices left exhausted farmland that could be improved with proper husbandry. Stoll notes that, at the same time in New England, fertility of soil became a concern and improvement began to be defined as "the changes that enabled land to be cultivated in the most prosperous possible way over the longest possible time." Improvement represented an ethic: "Those who took up this 'good system' of land use no matter what they grew or where they grew it, expressed a desire to endure, to persist, to cultivate as an expression of their stake in local society ... Permanence of society, landscape, and home was the paramount value of improvement." Here, Stoll connects early nineteenth century efforts at agricultural improvement to state formation in that it slowed a perceived rush towards a United States characterized by "unfettered democracy and unfettered expansion." The work of improvement on the farm, manuring fields, ploughing them properly, and rotating appropriate crops preserved the local landscape, culture, and community. Famers of this era, Stoll emphasizes, "thought through what we might today call land-use economics, or the interaction of cultivation and the market aimed at the best possible adaptation of both to the environment."[29] One needs to be cautious of drawing firm conclusions about the income earned from the choices farmers made at any given time, however. In her study of New York state improvers, Emily Pawley reminds us that we tend to sort success and failure retrospectively. Improvers were not certain their visions of improvement "were real or illusory, if their plans would work, their science would last, or the forms of the consumer desire on which they based their farms would prove durable ... Many entirely possible agricultural futures did not become concrete reality."[30]

Improvement, which first flourished in the mercantilist era, was adaptable to a capitalist economy in which the improvement of agricultural commerce relied on not only improved crop yields but also the improvement of infrastructure to increase ease and volumes of exchange. During the first half of the nineteenth century, improvement incorporated new scientific discoveries to advance agricultural production: chemistry, botany, pure-bred livestock, and labour-saving tools and machinery. Hence, in 1851, the celebrated agricultural chemist Professor Johnston was the improver Hutton admired and quoted in his essay. But contemporary Upper Canadian improvers also promoted additional canal construction and steamboats to increase capacity and speed along the St. Lawrence River–Great Lakes shipping routes for agricultural produce to Britain, and to secure access to American markets in consequence of Britain's shift to free trade in the mid-1840s. Improvers encouraged the expansion of its nascent industries, anticipating their invention and manufacturing of labour-saving machinery to ease farmers' toil and reduce the need for hired labour, a costly expense. They began to promote railways, too, which would be constructed across the colony in the coming years. Once again, Upper Canadian improvers looked to New York State as a model. There, as Pawley relates, improvers were busily applying science to agriculture to produce more goods and to open commercial markets for new types of agricultural products.[31] In advancing this vision of Upper Canadian progress, the colony's improvers were acting like contemporaries elsewhere in British North America. In his study of Nova Scotia, Daniel Samson analyses the discourse of improvement, "an inherently positive term that denoted not only progress and betterment for individuals and the nation but also the legitimation of capitalist consolidation and practice."[32]

Similar to the debates of 1830, improvers at mid-century had to again find consensus within a fractured political landscape populated by a new generation of political leaders from across the entire United Province of Canada whose power was derived from non-farm capitalist enterprises, including canal and railway promotion and construction. By then, these politicians were also adjusting to a form of responsible government. Vituperative debates in the arena of the provincial parliament were amplified by editorials published in newspapers supportive of each political faction. Yet, within the first years of responsible government, the ministry of Francis Hincks and Augustin-Norbert Morin appointed a minister of agriculture and improvers secured sufficient support across the spectrum of political allegiances to secure passage of legislation to establish a bureau of agriculture for the Canadas.

Lastly, regarding improvers, improvement, and progress, this study aims to identify a range of improvers, some who can be found in an online search for "Agriculture – Improvers and developers" in the

*Dictionary of Canadian Biography*,[33] as well as others who have remained unrecognized for their role as an Upper Canadian improver. The study might also be considered additional context for Suzanne Zeller's study of the culture of Victorian science in Canada,[34] as it provides new information about the pursuit of agricultural improvement and how Georgian and early Victorian science were used to understand the Upper Canadian environment in order to exploit it for agricultural production. Likewise, the study provides broader context to the subject of improvement and agricultural education examined by Douglas Lawr and Tom Nesmith, who sought in their studies to understand the intellectual nature of agricultural education in the Victorian era, particularly that which led to the founding of the Ontario Agricultural College in 1874.[35] *Improving Upper Canada* provides an important contribution to that discussion by analysing the improvers' successful lobbying for an agricultural professorship at the University of Toronto, their involvement in the candidate selection for the position, and the operation of an experimental farm on the university grounds. The study also provides some early insights into the subject of pure-bred cattle breeders in Ontario, a subject Margaret Derry has analysed, by identifying some of the earliest efforts to import improved breeds of cattle in Upper Canada.[36] Each of these issues emerge from the study's detailed analysis of Upper Canadian agricultural societies.

## Agricultural Societies of Upper Canada/Ontario

*Improving Upper Canada* builds on two sorts of histories related to Upper Canadian agricultural societies. The first category aims to identify the earliest agricultural societies formed in the province or the agricultural society that held the first recorded exhibition. Often, these connect Upper Canadian agricultural exhibitions to European medieval fairs to situate the colonial agricultural exhibitions within a long continuum of this cultural form. As chapter 1 explains, that bridge to the medieval past is less direct than it might appear. Such histories also tend to advance a chronology that connects the earliest identified Upper Canadian exhibition to later, larger provincial exhibitions and today's Canadian National Exhibition held each summer in Toronto.[37] Their collective focus on exhibitions obfuscates the important role they played in shaping the colonial state.

A second category of more recent scholarly works places Upper Canada's agricultural societies within a variety of contexts. Jodey Nurse explores the world of rural Ontario women from the mid-nineteenth century to the mid-twentieth century through the lens of their involvement and leadership in Ontario's agricultural societies. To the extent possible in the colonial context, this study highlights examples of the growing involvement of women in the work of improvement, particularly in domestic manufacturing, as well as floriculture and horticulture. E.A. Heaman analyses exhibitions and exhibition culture in nineteenth-century Canadian society and views Upper Canada's agricultural societies as "'societies' of citizens, to be founded on liberal principles of self-government." She identifies emulation as a key component of improvement. Agricultural societies hosted exhibitions with the goal of visitors viewing prize-winning submissions to see visible evidence of progress. Improvers hoped attendees would return home to emulate that quality in their own livestock, agricultural produce, or domestic manufactures, thereby contributing to the progress of society. Jeffery McNairn contextualizes agricultural societies as part of the developing associational life of Upper Canada's liberalizing colonial society. He places agricultural societies among other contemporary, voluntary organizations that to him identify a provincial experiment in "democratic sociability." In particular, he notes, agricultural societies, as part of the larger group of voluntary societies, reinforced the belief that the best solutions for society could be generated through organized public discussion rather than the actions of "solitary individuals or a self-selected few." Darren Ferry also examines Upper Canada's agricultural societies among a range of numerous other mid- to late nineteenth-century voluntary associations that represented central Canada's liberalizing society.[38] *Improving Upper Canada* draws extensively on reports of agricultural societies and their activities published in Upper Canadian newspapers, and, later, agricultural journals, to provide much more depth of analysis to "the agricultural society" explored in the previously mentioned studies. It does so to identify specific activities undertaken to improve agriculture by specific agricultural societies and demonstrate important variations in approach from society to society.

## Upper Canadian State Formation

It is a curious fact of Canadian history that the creation of a Bureau of Agriculture has been so little analysed, considering agriculture was the predominant economic force in the Canadas of the 1850s and 1860s. Part of the blame must be laid at the feet of the emergence-of-responsible-government canon of Canadian history, whose authors barely identify agriculture or the bureau as a subject of investigation. Rather, they highlight the political squabbles of the day to relate common themes of the era: political fragmentation, corruption, sectarianism, and the birth of railway construction and related financing schemes.[39]

In fact, few people have considered an agricultural aspect in Canadian state formation. In 1902, C.C. James, deputy minister of agriculture for Ontario, observed that "apart from the Legislature and town meetings ... agricultural societies are the oldest forms of organizations of this Province for mutual improvement," and in 1955, political scientist J.E. Hodgetts claimed in his study the Province of Canada's Department of Agriculture during the years prior to Confederation: "Here, surely, on the harsh Canadian soil, we see the last full-flowering of the Age of Enlightenment."[40] Aside from those two brief insights, most historians have reinforced Robert L. Jones's 1946 assessment that Upper Canada's pre-1850 agricultural societies "were failures, [although] the work they attempted seemed potentially useful." Their progress, Jones concludes, "was a disappointment to their promoters and friends," and it is "easier to over-estimate the importance of the early agricultural societies than to under-estimate it." Writing in the same year, Vernon Fowke claimed the government funding offered to Upper Canada's agricultural societies "was trifling in amount, formal (even ritualistic) in conception, and indicates that agriculture was no essential part of the real interests in government." Even Paul Romney, a biographer of Charles Fothergill, the Upper Canadian assemblyman who shepherded the 1830 agricultural societies bill to approval, dismissed that achievement, claiming that Fothergill's act is "nowadays thought to have done little to encourage agricultural progress in the province."[41]

Also impeding our understanding of the Upper Canadian state is the temporal divide produced by Britain's 1841 union of Upper and Lower Canada within a single colonial parliament following the 1837–8 rebellions. As previously mentioned, Sydenham, governor general of the Province of Canada, initiated a series of administrative reforms shortly after Union that established a liberal foundation for the operation of the colonial state. With the evolution of responsible government occurring in the years that followed, a historiographical consequence has been that most analyses of the developing colonial state are bookended by Union in 1841 and Confederation in 1867.[42] While there remain fundamental reasons to continue the analysis of this period, *Improving Upper Canada* employs a temporal framework of 1791–1852 that centres on 1830.

To establish its framework, this study expands on Daniel Samson's discussion in *Spirit of Industry and Improvement*, in which he connects Bruce Curtis's study of school inspectors in Upper Canada with the nature and role of state-supported agricultural societies in Nova Scotia.[43]

Curtis begins his study with William Hutton as a school inspector (not as an agricultural improver) to demonstrate how Hutton and other school inspectors were a new form of "state servant" that emerged in the 1840s and 1850s, a period Curtis describes as one of "political centralization and state formation." This was a process characterized by new departments of the state undertaking "systematic social policies, often using local government bodies as management agencies." But political centralization was paradoxical, he notes, for it relied on local agents or the creation of local government bodies to execute policy. While representative government permitted more propertied men to speak on behalf of the entire colony, the execution of state policies required the expansion of bureaucracy. As a consequence of this need, there remained "important continuities between bureaucratic and non-bureaucratic government forms in the colony, in personnel and in the class basis of political professionals." Curtis's proof of these exigencies is the appointment of prominent, respected local men to serve as school inspectors to ensure that standards of education issued by the central administration were being upheld in local schools. Such men also had to be trusted to report back from their locality with accurate information about the local conditions. By these means and the curriculum taught, the government aimed for a social consensus, all the while legitimizing the dominance of "a select group of propertied men and their allies."[44] William Hutton also operated within a world of contemporary agricultural improvers whose network of agricultural societies would come to provide a "volunteer bureaucracy" for the Bureau of Agriculture.

State formation in the 1791–1841 period of Upper Canada's development is more difficult to trace, although E.A. Heaman's recent study of the state in Canada offers a pathway of sorts for the analysis of agricultural societies' role in Upper Canadian state formation. She opens with the succinct observation: "There was not 'a' Canadian state through history but, rather, a chained series of institutions, across various territories constituting contemporary Canada, that people recognized as the state at any given time." On one hand, there were those who were "very obviously servants of the state" and deeds clearly undertaken on an official state basis. On the other hand, there were individuals whose actions were not widely recognized as a function of state as well as actions taken on behalf of the state that "were never perfectly distinguishable from other kinds of deeds."[45] Curtis's study of school inspectors is instructive for this earlier era, too, for he notes the shortcomings of physical and administrative connections between the government centre and colonial localities to develop and sustain a centralized authority such as what was possible after 1841. In the pre-1841 era, he suggests, the pattern of governance was to appoint a local overseer or commissioner to supervise publicly funded projects, such as the construction of roads, legislated during the 1830s.[46] Agricultural societies established under the 1830 statute fit this pattern for it entrusted a local oligarchy to employ annual public grants for local agricultural improvements that it deemed appropriate via the activities of the district agricultural society that it had formed and led.

In his study of the individual colonist's relations with the small number of state offices that governed the colony to 1841, J.K. Johnson concludes there to be "little evidence that during most of the Upper Canadian period there was much interest in finding means to improve the way in which the government operated or to make the relationship between the state and the people more congenial."[47] But perhaps searching for provincial-level examples of the state in this era misses important local developments. For example, Kari Dehli's analysis of state formation at the local level in Upper Canada offers useful insights. Her focus is the incorporation of the Town of York into the City of Toronto in 1834, an evolutionary development undertaken, she argues, because "the earlier machinery of appointed officials could no longer manage the beginning transition to an urban market centre and capitalist social relations." She makes a strong case for studying the sub-national state, for "it draws attention to the hows – the rituals, practices, mechanisms and techniques of representation, governance, surveillance and administration, in short, again, who regulates whom and how."[48] This study employs such insights by examining three agricultural societies across a twenty-year period. Doing so illustrates that each took different approaches to the establishment and operation of a district agricultural society despite each being enabled and funded by the same 1830 provincial statute. Moreover, no subsequent agricultural society statute required inter-district cooperation or communication, and a continuing local independence was constructed into the operation of the Bureau of Agriculture, even as its mandate extended to the entire Province of Canada.

*Improving Upper Canada's* first set of chapters examines the 1791–1830 era, presenting several failed attempts at establishing a single agricultural society with a province-wide mandate as well as debates surrounding the best forms of agricultural improvement suited to Upper Canada and the best methods for making them widely known. During this era an initial population of some 10,000 inhabitants in 1791 swelled to 77,000 by 1811, most Anglo-American settlers. Following the destruction of many agricultural settlements during the War of 1812, the province recovered slowly in the decade after the peace of 1814. Upper Canada's population of 150,000 by 1824 expanded to more than 213,000 by 1830, with much of that increase occurring in the last years of the decade as Britain enacted legislation providing inexpensive passage for immigrants on cargo ships crossing the Atlantic Ocean and began to promote emigration to British North America (compare areas of settlement shown in map 1.1 with map 4.1).[49] Throughout this period – and the entire period under study – wheat remained the major crop sold on local markets and shipped to British markets as supplies increased. Livestock was raised for home consumption and for local trade but provided no significant source of farm income.[50] Chapter 1 explores the history of the NAS in the context of its United Empire Loyalist-settler founders at the birth of Upper Canada. It explores the ambitions of its members to assert the society's place within the transatlantic world of improvers, as well as itself as the acknowledged authority in agricultural improvement locally and for Upper Canada in general. Chapters 2 and 3 shed new light on two significant political imbroglios in the province's history, one before the War of 1812 and one in its aftermath and recovery. To be properly understood, the crisis caused by Robert Thorpe soon after his arrival to the province in 1805 requires both the context of failed efforts by imperial officials and the colonial government to encourage hemp cultivation in the colony as well as Thorpe's efforts to establish an agricultural society to control this new initiative. Likewise, agricultural improvement provides important context for understanding the vicious response that Robert Gourlay received from Upper Canadian officials shortly after his arrival to the colony in 1817. A recognized Scottish improver, Gourlay asked settlers questions that differed little from those he had posed in England on behalf of the British Board of Agriculture. In Upper Canada, such questions, the public meetings that Gourlay hosted, and his pointed criticisms of those who governed contributed to his imprisonment and expulsion from the colony. At the same time, as this study reveals, those government leaders whom he criticized established an agricultural society at the colony's capital to re-assert leadership of agricultural improvement across Upper Canada.

The book's second set of chapters examines the district agricultural societies established by the legislation of 1830 and their operation through to 1850. During that first decade, the colony's population expanded by nearly 200,000, with increased and sustained immigration from the British Isles contributing to its total of over 410,000 by 1840. By 1852, Upper Canada would more than double in population to more than 952,000 individuals, some living in the City of Toronto (30,700), and other large urban centres, such as Hamilton (14,100) and Kingston (11,600); (compare limits of settlement shown in map 4.1 and map 4.3).[51] Chapter 4 examines calls for agricultural improvement expressed in the colonial legislature and newspapers during the 1820s to understand how the agricultural societies legislation came to be passed during Upper Canada's first reform-dominated legislature. It then charts the development of the legislation sponsoring district agricultural societies through to mid-century to identify the sustained pursuit of government support of agricultural improvement by both tories and reformers across an era that included rebellion and the consequent union of Upper and Lower Canada. Chapters 5 and 6 analyse three district agricultural societies to highlight the variations in form and approach within the Home, Niagara, and Midland Districts. Common among them, however, was the promotion of a "mixed farming" ideal. Guided by British and American sources, this ideal necessitated livestock husbandry to be part of a farmers' labour, in part, to produce the manure to spread on the fields to increase crop yields. But this mixed-farming ideal was possible only for more established farmers, particularly those with off-farm income and the land and labour to grow fodder crops to feed livestock and construct and maintain barns to house them.[52] Employing the examples of these three district agricultural societies, chapter 7 draws on the limited available data to answer the question: What was Upper Canada's return on investment from its annual public funding to the province's district agricultural societies between 1830 and 1850? In doing so, it highlights each agricultural society's interest in livestock, particularly "improved" breeds. With limited data from the records related to agricultural societies, this study attempts no conclusions about the levels of acceptance at the individual farm level; it analyses only the extent to which and the manners by which agricultural societies and their improving members promoted improved forms of agriculture.

The final section of chapters begins with chapter 8's examination of the AAUC, founded in 1846, and its role as host of ever larger annual provincial exhibitions. By hosting these events, the association's leaders supported efforts to improve agriculture for the expansion of commerce and industry and, thus, the improvement of the wider Upper Canadian economy. The final chapter examines the agricultural association as a political lobby. Not only did its sustained political efforts secure legislation for a board of agriculture, and then a bureau of agriculture with a minister of agriculture to lead it, but similar efforts also secured a chair of agriculture and an experimental farm as part of the new provincial university at Toronto.

One final point: the book's title, *Improving Upper Canada*, is meant as a nod to the colonial improvers who earnestly believed their promotion of agricultural improvement contributed to the economic and social progress of Upper Canada that they thought could be rapid if everyone engaged in the task. The title should not be mistaken as a statement that there exists data this study intends to analyse to prove that their promotion of improvements resulted in specific on-farm improvements in agricultural production and commerce that might be tabulated and reported. Much of the rhetoric of improvement espoused by improvers was overly ambitious for frontier farming in Upper Canada, as it mainly derived from British and American sources, or was of a type accessible only to those who possessed sufficient wealth to undertake it. Contributions to state formation proved to be the significant and identifiable outcome of the improvers' promotion of agricultural improvement in Upper Canada. I began this introduction with William Hutton because he represents the mid-nineteenth-century Upper Canadian agricultural improver, and his presentation of an essay at the provincial exhibition represented an outcome of decades of promotion by generations of Upper Canadian agricultural improvers. Yet, that 1851 moment also represents a disconnect between the award-winning ideals Hutton espoused and the daily toil undertaken by most settler farmers of the colony. It was a particular class of agricultural improvers who could afford to take a "Farmers' Holiday"[53] in autumn to travel to the provincial exhibition and, once having spent the day viewing what was on display, or collecting a prize from something they had entered for competition, took up the benefit of their paid membership in the AAUC to gain entry to an evening's meeting at which a Belleville farmer read his prize-winning essay, "On Agriculture and its Advantages as a Pursuit."

# 1 Transatlantic Improvers and the Niagara Agricultural Society, 1791–1807

In early spring 1792, the Agricultural Society at Quebec received a letter from individuals on the Niagara frontier announcing that, during the previous autumn, they had established a similar organization in their "part of the wilderness." The correspondents implied hope that their "Agricultural Society of the District of Nassau" would be considered a branch of the society at Quebec and maintain regular communication with it, a relationship similar to a branch society already established at Montreal.[1]

The letter had travelled some nine hundred kilometres from Niagara to Quebec during a significant moment of transformation for British North America. While its intention was to forge closer ties between a frontier settlement and the administrative centre of the Old Province of Quebec, Imperial legislation had reformed the colony in significant ways from the time of the Nassau District's agricultural society's autumn 1791 meeting to the writing of the late February 1792 letter announcing its establishment. Across the Atlantic Ocean, the British parliament's Constitutional Act took effect at year's end, dividing the expansive Old Province of Quebec into separate colonies of Upper Canada and Lower Canada, each with separate legislatures and civil administrations. Since the end of the American Revolution, settlers of the Niagara Peninsula (and elsewhere in the upper territory, composed of Indigenous populations plus a scattered collection of French and United Empire Loyalist settlements stretching eastward from Montreal to the Detroit River) sought leadership and governance from colonial administrators at distant Quebec. The name, "Agricultural Society of the District of Nassau," is representative of this fact. In 1788, British officials had divided the upper territory of the Province of Quebec into four districts in a rudimentary act of state formation. The Nassau District covered the huge expanse west from the Trent River along the north

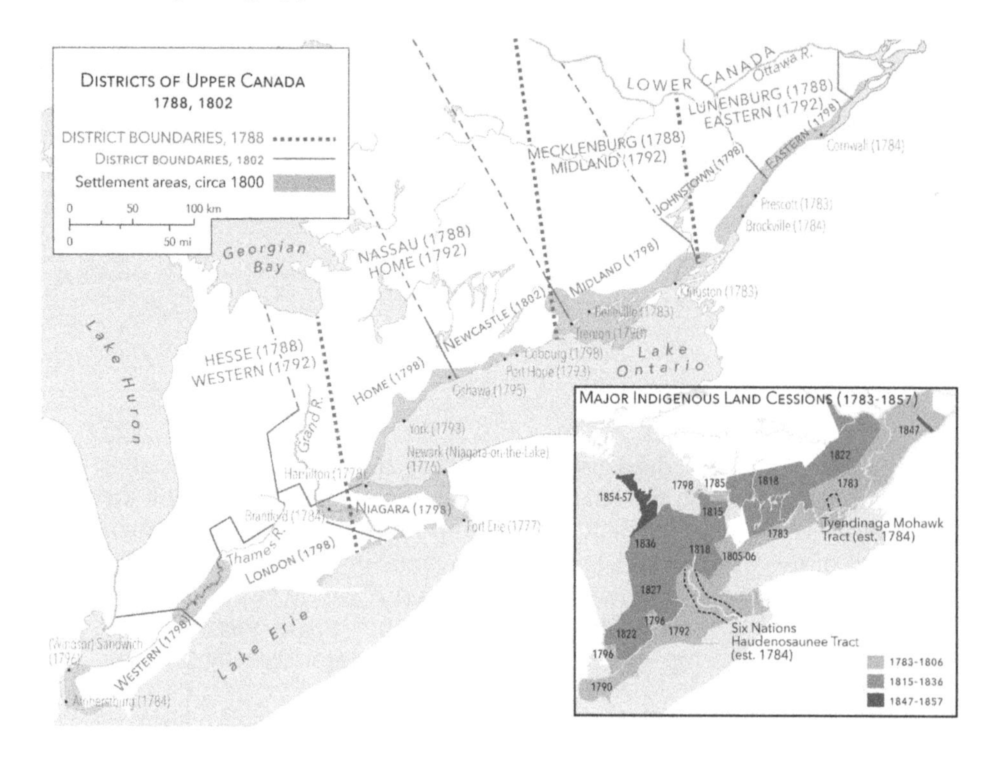


Map 1.1. Map of the districts of Upper Canada in 1788 (with revised 1792 names), also showing the province's districts as of 1802. The inset map identifies Indigenous land cessions that provided territory for colonists to settle Upper Canada. Numerous Indigenous communities continued living in these areas and the map identifies only the two major tracts established in 1784 (Inset map based on Map 30: Land Cessions 1783–1873 in Helen Hornbeck Tanner, ed., *Atlas of Great Lakes Indian History* [Norman: University of Oklahoma Press, 1987]) Map produced by Julie Witmer Custom Map Design.

shore of Lake Ontario, across the Niagara Peninsula, to a line running north from the tip of Long Point on Lake Erie.

The creation of Upper Canada reduced the urgency for the letter's authors to forge the bonds they sought with the gentlemen at Quebec. Soon, the gaze of the Niagara Peninsula elite would, from a civil administration point of view, reorient away from Quebec and inwards towards Upper Canada's new capital town of Niagara[2] to establish its members as Upper Canadian political and social leaders. In fact, when the letter was received at Quebec in the spring of 1792, the newly appointed lieutenant governor of Upper Canada, John Graves Simcoe, had sailed out from England and was waiting at Quebec for the St. Lawrence River to thaw so that he might travel upriver to assume his duties at the inland capital.

The formation of the Agricultural Society of the District of Nassau by leaders of the Niagara settlement offers a previously unopened window to view the formation of the Upper Canadian state immediately before and after Simcoe's arrival. It allows us to examine the agricultural society's attempts to infuse the ideology of improvement into the social, economic, and political leadership of the province at the colony's birth. By establishing this organization, the elite of Niagara were following a model of scientific and agricultural leadership that connected British North America with Britain and the United States. It fit tidily within their personal, business, and political connections, which were at once local, regional, American, and transatlantic. The settlement's leaders – primarily individuals possessing commercial, administrative, or military roles within the settlements on the British side of the Niagara River, across from Fort Niagara – viewed their economic and professional lifelines as reaching to Montreal and Quebec and then across the Atlantic Ocean to Britain. Many also maintained family and business connections with those in the United States.

The agricultural society established at Niagara stands as one of the earliest organizations in the province dedicated to colonial improvement. Following the creation of Upper Canada, the Niagara elite sought to transform their nascent agricultural society into *the* agricultural society for Upper Canada, led by them from the new colonial capital and infused with a province-wide mandate. Indeed, the individuals who founded this institution were overly ambitious and faced insurmountable obstacles to advancing the improving creed throughout an infant and wilderness province, but that failure itself is instructive. This chapter first considers the membership of the Agricultural Society of the Nassau District within the context of a coalescing Niagara oligarchy and the associational culture of the new settlements that existed prior to and developed during the years after Lieutenant Governor Simcoe's arrival.

Image 1.1. The Ontario Heritage Trust's plaque in Simcoe Park, Niagara-on-the-Lake that commemorates the Niagara Agricultural Society (Photo by author.)

The ideology of improvement played an important role facilitating a common goal among the disparate settlers of this frontier location and a lieutenant governor newly arrived from England. To properly understand this interaction, the chapter briefly examines Simcoe's life prior to his arrival to Upper Canada to emphasize the shared, transatlantic interest in agricultural improvement. It also highlights the networks formed by agricultural societies and other scientific societies throughout Britain, British North America, and the United States that sought to infuse the ideology of improvement into the agricultural practices of North American settlers. Having established this context, the chapter then explores the short history of the NAS, examining its leaders and their performative rituals of leadership, what the society hoped to achieve, and why such aspirations were overly ambitious for the first decades of the infant colony of Upper Canada.

Numerous historians have trumpeted the 27 October 1792 founding of the "Niagara Agricultural Society" at Niagara under the patronage of Lieutenant Governor Simcoe as a "first" for the Province of Ontario.[3] Indeed, the society was a "first," but its 26 February 1792 letter indicates that the lieutenant governor provided no input to the society's origin for he would not arrive at Niagara to assume office until late July. At that time, his priority was to appoint local leaders to political office and coordinate elections across the province to contest seats for the House of Assembly in advance of the colony's first session of legislature to be held from mid-September to mid-October. Thus, what has long been considered the founding of the "Niagara Agricultural Society" in October 1792 was likely its annual general meeting, for it had been in existence for about a year.[4] Presumably, Lieutenant Governor Simcoe accepted the honorary role as patron of the society at this meeting,[5] and with the selection of new executive members, the Agricultural Society of the Nassau District transformed itself into an organization with aspirations of a province-wide mandate to promote agricultural improvement throughout Upper Canada.

That the actual name of the agricultural society formed at Niagara remains unclear perhaps speaks to the central role the institution was to play in the function of colonial government and society. It was not meant to be merely a "local" society, denominated with a local name. It was to be *the* agricultural society, quite rightly located at the provincial capital and formed of the elite centred at that location. The society originated for the District of Nassau, but once Niagara became the capital of Upper Canada, any further descriptor was unnecessary. By way of comparison, a similar confusion surrounds the name of the agricultural society that Governor-in-Chief Lord Dorchester had formed for the Province of Quebec in 1789. Although sometimes referred to as the Quebec Agricultural Society, it was often only referred to as the "Agricultural Society" composed of the central society in Quebec and its branch in Montreal. A publication it produced in 1790 labelled the organization as "The Agricultural Society in Canada."[6] In Upper Canada, contemporary references to the institution at Niagara included "Agricultural Society," "Agricultural Society of this Province," or "Agricultural Society of Upper Canada." The society's president once called it the "Agricultural Society of Niagara" in a government petition, and the response he received was addressed to the "Agricultural Society of Newark." An Ontario Heritage Trust commemorative plaque in Simcoe Park, Niagara-on-the-Lake, uses the name "Niagara Agricultural Society" in its recognition of the institution as the province's first agricultural society. For the sake of clarity, I use this name to refer to the institution as it operated after October 1792.[7]

## The Agricultural Society of the Nassau District

Those who established the Agricultural Society of the District of Nassau in autumn 1791 tell us much about the nature and leadership of

Image 1.2. Robert Hamilton (1753–1809), n.d. Artist unknown (Toronto Public Library, JRR 1306 Cab.)

the Niagara settlements. John Butler was elected the society's president. Once an Indian agent who resided in the Mohawk Valley of New York, he departed for the upper territory of Quebec when the American Revolution began in 1775. Not long after his arrival to the British fort at Niagara, Butler secured control over much of the trade with Indigenous peoples, as deputy agent of Indian Affairs, as well as trade for the increasing number of Loyalists seeking refuge near the fort. He led his Butler's Rangers into numerous raids along the frontier during the revolution and, as the war concluded, began to coordinate the settlement of many of his rangers and others on the west side of the Niagara River, across from Fort Niagara. He also continued to perform influential roles within the Indian Department. When Britain subdivided the upper portion of the Province of Quebec into four large administrative districts in 1788, he was appointed a justice of the Court of Common Pleas for the District of Nassau, a member of the Land Board, and lieutenant colonel of the district militia. Along with his public roles, Butler also farmed and operated a mill. He became an elder statesman of the Niagara settlements and his connections to officials at Quebec were numerous and well placed, as were his social and political connections throughout the Niagara Peninsula. As founder of the settlement that would coalesce into the Town of Niagara and the neighbouring countryside, he was a logical choice to lead the agricultural society.[8]

Robert Hamilton and John Warren served as vice-presidents for the agricultural society. By 1791, Hamilton was well on his way to becoming the leading merchant along the Niagara River. He had arrived in the province from Scotland in 1779 to take part in the fur trade and, by 1780, had formed a partnership with Richard Cartwright, a loyalist from New York state who also had personal connections with Butler. The two merchants began provisioning Fort Niagara and the Indian Department. By the middle of the decade, Hamilton established himself at Queenston on the Niagara River, while Cartwright moved to Kingston at the eastern end of Lake Ontario. When they dissolved their partnership in 1790, the two held significant control of almost all goods travelling up and down the upper St. Lawrence–Great Lakes–Niagara River system. In 1791, Hamilton gained an even tighter grip on his end of the trade by securing the right to portage goods around Niagara Falls on the west side of the river in competition with the old route along the east bank, by then in US territory. Equally important was his contract to transship all military goods along this route. Such economic power and personal connections also brought him political appointments. One of the first justices of the peace to be appointed for the Niagara settlement, Hamilton had joined Butler in 1788 as a judge on the District Court of Common Pleas. He also served with Butler and three other individuals on the Land Board, which had the power to assign land to incoming settlers and relay its recommendations concerning other land claims to the Executive Council at Quebec for its adjudication.[9] Warren, who served as commissary at Fort Erie, was very much connected to Hamilton and Butler and other Niagara merchants, for his post was at the south end of the Niagara portage through which he ordered and received provisions for that fortification. He had also been appointed a justice of the peace in 1788, sat on the Land Board with Butler and Hamilton, and held other responsibilities, including district road commissioner.[10]

The secretary of the agricultural society was Francis Crooks, another Scottish merchant who had established a store in 1788 with his half-brother James at "The Bottoms," a collection of buildings outside the walls of Fort Niagara on the edge of the Niagara River. He also leased a building he owned for the "Yellow House" tavern.[11] The treasurer was William Dickson, Hamilton's cousin who had emigrated from

Scotland in 1785. Just twenty-two years of age, Dickson had made quite a fortune for himself in Hamilton's mercantile network by late 1791. He had built a large brick house at the capital, the first on the peninsula.[12]

Brothers Colin and John McNabb, along with Robert Kerr, acted as stewards of the society. Colin had settled on the Niagara Peninsula in 1787, having fought in loyalist regiments during the revolution. With the creation of the Nassau District, he had been appointed superintendent of inland navigation at Niagara, a position that involved him directly in the mercantile network, for he collected duties, registered vessels, and remained vigilant for smuggling. John was also a local office holder, including the position of road commissioner granted to him and several others by the Land Board in early 1790.[13] Kerr had settled in the Niagara area by 1789, having been recently appointed surgeon to the Indian Department. When required he employed his medical skills for many of the settlers of the peninsula as well as the military men stationed there.[14]

The twelve regular members of the agricultural society permit a glimpse at the networks of patronage that extended from those who held the executive offices of the society. For example, the following relatives of Butler, his fellow appointees to the Indian Department, and residents of the settlements he founded are recorded as members. Walter Butler Sheehan, Butler's nephew, had been recently appointed sheriff and had earlier secured the position of the Indian Department Storekeeper at Niagara through his uncle's influence.[15] Adam Vrooman had been a sergeant in Butler's Rangers and farmed in the settlement created by Butler, as did another member, William Johnston.[16] Ralfe Clench had also fought with Butler's Rangers and settled at Niagara in 1784 where, by 1791, he was appointed First Clerk of the Court of Common Pleas for the Nassau District.[17] Samuel Street was a merchant who, during the American Revolution, had relocated to Fort Niagara and provisioned both the British military and the Indian Department. By 1791, Street relied on Butler's patronage for provisioning the Indian Department, having been unable to secure military trade after the revolution like Hamilton and Cartwright. Street had also formed a trade partnership with Butler's son, Andrew. They constructed a sawmill in 1789 that they would sell to John Butler by 1792. Street had also been appointed a Justice of the Peace in 1788.[18] Thayendanegea (Joseph Brant), was a well-connected and respected Kanien'kehá:ka (Mohawk) leader of the Niagara area. He had risen to the status of a war chief just before the outbreak of the American Revolution and had worked with Indian Department officials since the late 1760s. Thayendanegea and his warriors had joined forces with Butler and his rangers for several battles in 1778 and 1779. After destructive attacks by American forces on Indigenous settlements of western New York, he and many other Kanien'kehá:ka refugees fled to the protection of Fort Niagara. After the revolution Thayendanegea was often absent from the Niagara Peninsula while he worked with others within the Six Nations Confederacy and other Indigenous peoples on the western frontier of the United States who were fighting to remain on their ancestral lands. Nevertheless, he was in regular communication with leading British officials and local officials, such as Butler of the Indian Department.[19] Another member of the society, James Muirhead, possessed early connections to Butler that are unclear, although this former army surgeon, whose medical skills were utilized by settlers throughout the Niagara settlements, would marry Butler's only daughter, Deborah, in 1795.[20]

Other regular members were connected through Robert Hamilton's mostly Scottish merchant network. These included: the previously mentioned John Warren, the commissary at Fort Erie; James Farquharson, the commissary at Fort Niagara; and Archibald Cunningham, Joseph Edwards, and George Forsyth, whose operations within Hamilton's mercantile network ranged from the Thousand Islands through to the Detroit River.[21]

While dividing the membership of the Agricultural Society of the District of Nassau in this fashion emphasizes the influence of Butler and Hamilton, it does not suggest that their two spheres of influence could be divided so tidily in practice. As the Niagara settlements developed prior to 1791, numerous personal connections were being made between families, resulting in some members of the society becoming connected just as closely to one as they were to the other. For example, Adam Vrooman, a former Butler's Ranger, teamed for Hamilton's new portaging contract along the west bank of the Niagara River.[22] In fact, the membership emphasizes that much about the Niagara economy was based on the military trade and the trade with Indigenous peoples, particularly the portaging of such goods around Niagara Falls. Philip Stedman was a member of the society, too. His company had just lost the lucrative contract of portaging government supplies to Hamilton and his partners, but it continued to portage goods along the east, American, bank of the Niagara River.[23]

Many of the members of the society were also farmers. As settlers continued to clear their frontier farms of trees and stumps, there was potash to send out through their merchant networks, and as the slowly cleared farms began to produce modest surpluses, there was also wheat and flour to trade.[24] Several members were on the verge of becoming major landholders, later speculators. Street, Dickson, and Hamilton would acquire tens of thousands of acres of land during their careers; the latter's total was more than 130,000 acres. Once an Upper Canadian administration was set in place, it would become easier for them to use their political connections to secure land grants and titles to large tracts of land.[25]

Membership in the Freemasons was another commonality shared among members of the Agricultural Society of the District of Nassau, in addition to their economic, administrative, and personal connections. A regimental lodge had been brought to Fort Niagara as early as 1775 by members of the 8th Regiment of Foot[26] in what was the "primary mechanism" responsible for establishing a network of lodges across the British Empire.[27] But civilians soon established lodges, too, starting with the St. John's Lodge of Friendship that John Butler and others formed near the fort in 1780. Butler had been made a Mason in 1766 at a lodge in Albany, New York, (a lodge in which Richard Cartwright was also a charter member). In 1787, as settlement progressed on the west side of the Niagara River, they petitioned the Grand Lodge of Quebec for a second lodge. This became St. John's Lodge No. 19. Butler was appointed Master, and meetings were held at Butler's Barracks. Charter member Ralfe Clench acted as secretary, a man recalled as "a pioneer in spreading freemasonry's influence throughout the peninsula."[28] Other members of the agricultural society were prominent Freemasons, too. Thayendanegea had been initiated in London during his travels there in 1775–6, and Robert Kerr had probably been initiated in Halifax during the revolution. George Forsyth had likely been initiated in Scotland before his departure for North America. He, along with Hamilton, who had probably been initiated into St. John's Lodge of Friendship prior to 1785, used their influence to relocate this lodge to Queenston in 1790.[29]

Freemasonry was "central to the building and cohesion of the empire," argues Jessica Harland-Jacobs, for it "was fundamentally imperial in its functions and fraternal in nature." Lodges not only fostered fraternalism and community, they provided critical forums "for imperial pioneers who devoted themselves to transforming remote locations into enclaves of British society." For loyalist refugees after the war, lodges helped overcome the trauma of dislocation and resettlement.[30] Freemasonry was an early voluntary association that, as Jeffery McNairn points out, provided "experiments in democratic sociability" and fostered new forums for interaction among colonists.[31] For these reasons, masonic lodges were a key element of the institutional development of Upper Canada.[32]

On the Niagara frontier, Freemasonry was also tied closely to practical aspects of state formation. In June 1791 the Land Board, with leading Masons Butler and Hamilton as members, authorized the construction of a public house in Niagara with a Masonic Lodge constructed next to it. The two-storey "Freemasons' Hall" permitted the Freemasons to host their meetings on the second floor, while the lower level was available for public use, such as social functions and religious services. This community centre would serve as the meeting place for the NAS, the one other institution of the Niagara settlement. Significantly, Freemasons' Hall also hosted Upper Canada's inaugural session of legislature, opened on 27 September 1792, with Simcoe addressing members from the lodge's Master's chair.[33]

## A Transatlantic Community of Improvers

Many members of this established, closely knit social, economic, and political elite welcomed John Graves Simcoe to Niagara as their new lieutenant governor on 26 July 1792 with a letter they presented to him identifying their economic, social, and political concerns about the Niagara community. It would be the first of many occasions in which the resident leaders of Upper Canada, who better understood the practical realities of the North American environment, their ties to the mercantile world of the St. Lawrence, as well as family and business connections in the United States, butted heads with this transplanted English administrator who was determined to shape Upper Canada into "the very transcript & image of the British People."[34]

Simcoe – the "intellectual magpie" as one biographer aptly characterized him – possessed a wide assortment of overly ambitious plans for his frontier colony, which he somehow expected to come to fruition in short order.[35] He is known mainly as the British soldier turned administrator whose primary interest in the frontier, borderlands colony was the organization of its defences against the ever-present threat of invasion from across the border and for his efforts to create a conservative British form of government, formed in stiff opposition to the republican ideology of the United States.[36]

But there is a danger in overstating the differences between Simcoe and the Niagara elite during his time in Upper Canada between 1792 and 1796. The lieutenant governor possessed a variety of tools to exert his leadership and to help endear himself to the Niagara elite immediately upon his arrival. First, he dispensed patronage. He appointed Butler, for example, as a lieutenant of Lincoln County shortly after the first meeting of the Upper Canadian legislature in 1792. Admittedly, he

Image 1.3. Wolford Lodge, Honiton, Devon. "The prospects from the house, and surrounding hills embrace a great extent and variety of richly-cultivated country." John Britton, *The Beauties of England and Wales*, vol. 4, (London, 1803), 298. (Private collection.)

did so to coopt the Niagara leadership into his visions for the future of the province. Second, like many of the Niagara colonists, Simcoe was a veteran of the American Revolution. Third, Simcoe was a Freemason, having been initiated in Exeter in December 1773. And Simcoe brought with him from England another veteran and freemason, William Jarvis, to serve as his provincial secretary. Prior to his departure, Jarvis had been appointed provincial Grand Master of the newly created Masonic Lodge of Upper Canada.[37] Lastly, the military officer and colonial administrator was also an agricultural improver, and his vision of an Upper Canada transformed from wilderness to a prosperous agricultural colony fit neatly into the efforts already underway at Niagara to lead improvement, namely, the Agricultural Society of the Nassau District. All would advance their fortunes by improving the agricultural output of Upper Canada's settlers, thereby increasing the export of agricultural produce and the import of manufactured goods. Their influence as elected and appointed officials of the colony's government would also grow with an expanded and prosperous Upper Canadian population. It was a shared transatlantic ideology of improvement that underpinned efforts to integrate the agricultural society at Niagara into the developing provincial government apparatus. In doing so, it was following the pattern of the Quebec Agricultural Society, patronized by Dorchester and populated by members of his governing council.

Simcoe, the gentleman-improver of the English Enlightenment, was as familiar with the art of agriculture as he was with the art of war, for he had spent years living as an English country gentleman. Invalided from the American Revolution and returned to England in 1781, he married Elizabeth Posthuma Gwillim, whose dowry provided him Wolford Lodge, a five-thousand-acre estate in Devon. The couple settled there in 1786 and, supported by his military income and Elizabeth's dowry, he hired John Scadding as his estate manager to set about improving his property. During this time, Simcoe purchased neighbouring properties to enlarge his holdings and he renovated and significantly expanded his manor house. In doing so, he joined many contemporaries within the English landed gentry who busied themselves with scientific improvements to their estates to advance the nation's centuries-old agricultural practices. Simcoe opened his home to guests on a regular basis and used his position as the gentleman of a manor seat to gain influence among the gentry of the surrounding area. For his efforts, he was elected as a member of the House of Commons in 1790.[38]

As a veteran of the American Revolution and as a gentleman improver and politician of Devon, Simcoe would have been fully aware of the impact to Britain that the loss of its colonies in North America had caused. From an agricultural point of view the revolution spurred efforts across Britain to improve farming practices with a goal to increase farm production, thereby supplanting the island's reliance on agricultural goods imported from its now-former American colonies. It also encouraged botanical exchanges with other parts of Britain's empire. Useful plants might be cultivated in other colonies, while new exotics discovered in new territories might be brought to England so that experiments could be conducted as to their suitability for food or manufacturing.[39]

In 1776, the Scottish social theorist Lord Kames wrote that it would be useless to despond over possible loss of the American colonies, "for if agriculture be carried on but to the perfection that our soil and climate readily admit, it will amply compensate the loss of these colonies." His treatise, *The Gentleman Farmer*, presented Kames's belief that the method to promote and implement agricultural reforms was to establish a "Board for Improving Agriculture ... eminent for patriotism and for skill in agriculture." The institution would be active in "promoting and propagating knowledge" of new agricultural practices. Significantly, Kames placed a great deal of emphasis on the "choice of proper members," concluding that this board could be open only to gentlemen "who serve for honour and not for profit."[40] Not long after Simcoe's arrival in Upper Canada, the British government would establish such a board.

Scotland's leading agricultural improver, Sir John Sinclair secured political support for a government-funded Board of Agriculture for Britain in 1793. During the House of Commons debate on his bill, Sinclair had echoed Lord Kames's suggestions, stating that the board should be composed of "respectable gentlemen, perfectly conversant in and acquainted with the subject," who would act "without reward or emolument." Such a board would facilitate "the establishing of free communication of the different improvements in agriculture, from one part of the country to another." Sinclair's plans came to fruition, thanks to the improving gentry who held seats in parliament and Henry Dundas, the Scottish-born Home Secretary in William Pitt the Younger's cabinet (and Simcoe's superior at Whitehall). The board's thirty-one members were all landed gentry or members of the aristocracy. As with earlier gentleman scientific societies of English Enlightenment the "half private, half public" Board of Agriculture and Internal Improvement, was funded by annual government grants but not answerable to the Treasury. Its constitution was similar to that of the Royal Society, and Sir Joseph Banks was offered membership to give the board respectability. Agricultural improvement was now an official concern of government, but the board's purpose was poorly defined. External critics claimed it to be too close to the government to be above politics, while internal critics thought it was not connected closely enough with the inner workings of the state to be useful. The board would continue to operate under varying degrees of government support until its demise in 1822.[41]

The Board of Agriculture's most significant, lasting success was the support it provided to secure passage of the General Enclosure Act of 1801, which simplified the process of land enclosure in Britain and hastened implementation of agricultural improvement. Another lasting contribution to British agriculture was the detailed statistical accounts it produced of agricultural practices within specific counties that documented the changing agricultural landscape and nature of its farming population. Each of these outcomes would have significant effects on Upper Canada. Enclosure would drive out many tenants to North America, where *they* were to be the improvers of the farms they brought into cultivation on their land grants, and as chapter 3 highlights, agricultural reports and statistical accounts would come to be a source of great consternation for an Upper Canadian elite attempting to maintain its fragile control on colonial leadership. But initially, and most fundamentally, this British model of leadership of agricultural improvement had a significant and sustained influence on Upper Canadian improvers who hoped a fledgling agricultural society at the colonial capital would develop a province-wide influence through some sort of direct connection to the colonial government, if not a direct place and role therein.[42]

Simcoe, the improving gentleman of Wolford Lodge and his estate manager, Scadding, represented the increasingly important partnership between the landowner, who provided the capital and innovations, and the estate manager and tenant farmers who performed the necessary labour to effect the transformation of agriculture across the British countryside.[43] It was the landed gentry who had the luxury of time, the land, and the wealth to devote to agricultural experiments. They could also afford to absorb the consequences of failed experiments with new crops or livestock breeding. A member of the Bath Society in 1780 best expressed the crucial paternalistic role that landed gentry were to play in the agricultural improvement of Britain: "Let Agriculture be studied by gentlemen of landed property, on philosophic principles; let it be taught to their tenants; and the happy consequences will soon be apparent throughout this island."[44]

Modelled after the existing tradition of gentlemanly scientific societies, agricultural societies became an important element in the landed gentry's cultural and political hegemony over the improvement of agriculture at home as well as in the colonies where reports of their pure-bred livestock, their crops, their experiments and observations, and the pleasing neatness of their estates were held up as the ideal for all to achieve. Not only were they interconnected in purpose and membership with the prestigious scientific societies of London, they also helped carry out the functions of the state, for members were often either connected to, or were themselves, government officials. Some even believed this network of agricultural societies could contribute to national security by way of improving landowners infusing their tenants with the pride of progress, thereby strengthening their loyalty and resolve to resist any potential invasion.[45] Between 1770 and 1790 at least thirty different agricultural societies were founded in England, Scotland and Wales, about forty new ones would be created in the 1790s alone, and nearly eighty more would be established during the following two decades.[46]

When Simcoe arrived at Quebec in late 1791, he returned to a North America that was busily making attempts to replicate and adapt British models of agricultural improvement to the New World. At Quebec, Lord Dorchester's agricultural society had been operating since February 1789, including its branches at Montreal and, recently, Niagara. Members included Dorchester's governing council and other elites from Quebec's political, judicial, and mercantile elite. The society was to be a semi-public extension of government, for members were to act as advisors to his governing council.[47] In his copy of the society's 1790 published papers, George Davison jotted down his synopsis of the society's purpose and the improving ideology as adapted for North America. A wealthy merchant, a member of Quebec's governing council, and a director of the agricultural society, Davison was one of its most active members, for he was also an agricultural improver on his 400-acre Lanton Farm west of Trois-Rivières, run by English farm managers.[48] In his January 1791 dedication to the London merchant, John Brickwood, Davison explained that the society had been established "for encouraging & disseminating A Taste of respect for Agriculture among the Gentry, A spirit of Industry of the best Practical Husbandry. Among the Famers; In the View of promoting the Domestic Wealth, the comercial [*sic*] prosperity & the National Value of that Extensive uncultur'd & improvable country." The society, he continued, would "attain its desirable Purposes through an Efficacious Co-operation of High Patronage, Spirited Example, Liberal Premiums, and an Authentic Publication of Facts."[49]

By late 1791, Dorchester's agricultural society was just one of several that had been established on the continent. In British North America, "The Society for Promoting Agriculture in Nova Scotia," had been established at Halifax in December 1789, under the patronage of Lieutenant Governor John Parr, and a similar institution at St. John, New Brunswick was founded in the following year by Dorchester's brother, Lieutenant Governor Thomas Carleton.[50] In the United States, prominent gentlemen had established agricultural societies in Pennsylvania and South Carolina (1785), Maine (1787), and New York (1791). A similar institution would soon be established in Massachusetts (1792).[51] In fact, President George Washington had become the model gentleman improver for the new republic. He was an avid farmer at his Mount Vernon estate, and as one historian reminds us, he had been "a colonial English country gentleman before he became the First Citizen of a new nation."[52]

Members of the Agricultural Society of the District of Nassau at Niagara understood that establishing an agricultural society in their "part of the wilderness" was their means to join the transatlantic community of improvers. By joining their society with the one at Quebec, they could draw on scientific knowledge from Britain as well as the United States through that agricultural society's publications and communications. Information from Britain was given primacy but reports of improvements conducted in the United States were important, too, for individuals there were actively adapting British discoveries, new plants, breeds of livestock, and technology to the challenges of the North American environment.[53] Upper Canada presented a set of harsh and immediate realities that restricted many activities the improving ideology might inform. Farming the frontier province was far removed from the task of farming the long-cultivated English countryside that Simcoe had recently departed. The colony contained only about ten thousand people scattered among a few small, isolated settlements. There was no coterie of British aristocrats, gentlemen scientists, and landed gentry who owned estates with estate managers and tenant farmers to conduct the work of experimentation. The colony possessed neither universities nor scientific organizations that might offer direction or assistance. There existed only the established mercantile, military, and administrative elite who might also farm but who had no leisure time nor capacity for financial risk to devote to agricultural experimentation. Their immediate concern upon Simcoe's arrival was to secure deeds to the lands they had been farming since their arrival during or after the American Revolution. As Anglo-Americans, many had had closer ties with families, friends, and commercial contacts in the United States than they did with Britain. Nevertheless, all were invested in the expansion of the colony, its governance and administration, and its agricultural production. To encourage such development, they could work with the lieutenant governor within their elected or appointed offices and employ the NAS to further extend their influence.

Simcoe's correspondence in the months following his summer 1790 appointment as lieutenant governor of Upper Canada demonstrates that he understood that improved agricultural techniques would play a significant role in exploiting fully the promise of the New World colony he was to govern. He was also well informed of efforts at leading improvement already underway in British North America. Reports to imperial authorities suggested Upper Canada possessed excellent soil and climate; therefore, agricultural production would become one of its greatest contributions.[54] Simcoe's January 1791 letter of introduction to Sir Joseph Banks, president of the Royal Society, outlined his goals for organizing Upper Canada's government and society according to the improving ideology. In it, he stated his opinion that the new colony "should in its very Foundations provide for every Assistance that can possibly be procured for the Arts and Sciences." He hoped that the imperial government, private benefactors, or perhaps even royalty, might financially support the founding of a public library in Upper Canada "composed of such books that might be useful to the Colony." Simcoe added: "In the Literary way I should be glad to lay the foundation stone of some Society that I trust might hereafter conduce to the extension of Science."[55] Banks was arguably the most influential scientific gentlemen within the British Empire,[56] and Simcoe hoped to avail himself of Banks's "ideas and patronage" so that he might procure for Upper Canada, "the great encouragement this Nation under His Majesty's auspices, affords to those Arts and Sciences which at once support and embellish our Country." Furthermore, he proposed a "Botanical Arrangement" that "might lead to the introduction of some Commodities in [Upper Canada] which Great Britain now procures from other Nations." He cited hemp and flax as two prime examples.[57] No response from Banks has been found.

As president of the Royal Society, Banks was at the very centre of planning for the exploitation of Britain's colonial resources. He directed scientific discoveries in newly explored territories and introduced new agricultural commodities to be grown in Britain's colonies around the globe. Such activities were conducted on behalf of the British state but implemented by his connections across a network of lesser associations throughout Britain and the colonies. Although the Royal Society was not formally part of the government, "it nonetheless performed some of the tasks of government thanks to the close ties that existed between Banks and the inner circles of government," notes Banks's biographer, John Gascoigne. Consequently, it was as part of the network of institutions that collectively made up the British state. Gascoigne explains that the distrust of bureaucracy and hostility to centralization due to its expense and potential for royal and ministerial patronage had long been a tradition in Britain. As a result, institutions that had only a loose connection with the central government were requested to perform many tasks of state. This informal system of government was successful because "most positions of power from local government upwards were held by members of the landed class bound together by ties of family, education and (at least in the late eighteenth century) a remarkable degree of consensus about basic political values."[58] The efforts by Banks and other learned gentlemen blurred the lines between the British state and private associations and scientific and agricultural societies of the gentry served as a part of this semi-public network of the British state.[59]

Along with his request to Banks, Simcoe had also secured funds from Lord Grenville, whom Dundas had just replaced as secretary of state for the Home Department, "to lay the Foundations of a Public Library." Simcoe intended "to procure the Encyclopedia & Books of that Description, Extracts from which might be published in the periodical Papers for the purposes of facilitating Commerce and Agriculture."[60]

**The NAS**

When Simcoe accepted the position as the NAS's patron in late 1792, he donated a set of Arthur Young's *Annals of Agriculture*, purchased with the funds he had acquired through Lord Grenville in 1791, as part of an £101 order of books for the colony.[61] The volumes were the works of Britain's premier agricultural statistician, spokesman for the landed gentry, and defender of agrarian interests. Simcoe, as patron of the NAS, also promised to subscribe ten guineas annually throughout his term as lieutenant governor, "to be disposed of in a premium for the benefit of agriculture in whatever manner the members think proper."[62] Members had agreed at its founding meeting in the fall of 1791 that each would pay a fee of 2s 6p at monthly meetings for the purchase of a library of agricultural works "beneficial to the settlement." As of early 1792, the society had already sent £20 to London for this purpose.[63] Thus, during its first fourteen years, the NAS acquired a sizeable collection of many of the major treatises on the scientific principles of agriculture published in Britain during the last half of the eighteenth-century to keep society members abreast of new innovations in agricultural practices and connected this frontier agricultural society to the transatlantic network of improvers. In particular, the society continued to purchase new volumes of Young's *Annals of Agriculture*, publications by the Royal Society of Arts, Manufactures, and Commerce (RSA), and those by agricultural societies at Bath and Dublin.[64] By 1805, the NAS possessed over fifty volumes, valued at some £32. We know this, because in that year the NAS donated its collection to the library at Niagara. Unfortunately, the collection was destroyed when, during the War of 1812, invading Americans set the town ablaze on 10 December 1813.[65]

As no minute books or other institutional records of the NAS survive, there remain only selected glimpses in newspapers and other sources as to the form and function of the institution. The society's letter, sent to Quebec in February 1792, listed its rituals of performative leadership. Members met once a month, with a dinner supplied by the produce of their own farms. Each member was allowed to bring "as a visitor, a decent intelligent farmer, from whom something may be gained worthy of being made public for the general good."[66] Once Niagara became the capital of Upper Canada and the lieutenant governor became patron of the NAS the need to perform the rituals of a gentlemanly society intensified.

As noted, Simcoe arrived at Niagara after a significant amount of consolidation of leadership among the local elite had already occurred. Yet, there was an important transition of power and influence occurring in 1792 and 1793. When the agricultural society was first formed in the fall of 1791, Butler had been the logical choice for president of the institution. Simcoe's appointment of Butler as lieutenant of Lincoln County recognized his status. Yet, the aging Butler was finding it increasingly difficult to use his reputation to draw income and dispense patronage as the Upper Canadian state took shape at the capital under Simcoe's guidance. Moreover, his health was failing, and he would die in 1796.[67] Robert Hamilton was elected president of the NAS by 1793, and upon Butler's passing, Simcoe offered Hamilton the county lieutenancy.

At the birth of the agricultural society in 1791, the thirty-eight-year-old Hamilton was just hitting his stride as a powerful merchant and office holder. By Simcoe's arrival, he was already regarded as the "most powerful, wealthy and respected" leader of the Niagara merchants and the settlers of the peninsula.[68] Despite his occupation, Hamilton enthusiastically presented himself as an agrarian gentleman. He managed an extensive farm at Queenston and resided there in one of the colony's finest homes, which "seemed, with more than a touch of symbolism, to dominate both the surrounding landscape and the river below."[69] The "fairly extensive farm" that surrounded his house "reinforced his public image of landed gentility," notes his biographer Bruce Wilson.[70] In contrast, Lieutenant Governor Simcoe lived at Navy Hall in Niagara, accommodations the French traveller the Duc de La Rochefoucault-Liancourt described as "a small miserable wooden house, which formerly was occupied by the commissaries."[71]

Joining Hamilton on the NAS executive as vice-presidents in 1793 were Surveyor General David W. Smith, Hamilton's cousin William Dickson, and the Reverend Robert Addison, the local Church of England clergyman, who was also the grand chaplain of St. John's Lodge of Friendship, No. 2.[72] Francis Crooks continued as secretary.[73] Of the twenty-seven members in 1793, twelve were merchants from the Niagara area of either Scottish or North American birth.[74] Society members, such as George Forsyth and Crooks, were principal merchants in the Niagara region. Each had a fine home at the capital, while in the surrounding rural areas, other members had homes of similar or greater stature. For example, there was Colin McNabb's home on the lakeshore and Daniel Servos's house near Four Mile Creek.[75]

A range of Hamilton's lesser associates and other merchants were also members of the agricultural society in the mid-1790s about whom little biographical information exists. They did not own grandiose homes, but their commercial connections or professions gained them membership in the institution and, in turn, provided them the opportunity for social mobility. At an agricultural society meeting, as in Masonic meetings, such individuals could interact socially with their superiors.[76]

Two reports of NAS meetings held in 1793, suggest that a "very respectable number of the Subscribers" attended the late April meeting, along with Simcoe "and suite." In late June, members dined at Freemasons' Hall. Several invited gentlemen also attended, "which with the Members of this laudable institution assembled, formed a very numerous party, – The utmost cheerfulness and conviviality prevailed on this occasion."[77] A British traveller to Upper Canada, named Hodgkinson, happened to be one of those gentlemen invited, and he recorded important glimpses into the activities and character of the NAS. He commented that "the table was abundantly supplied with the produce of [the members'] farms and plantations. Two stewards were in rotation for each meeting, who regulated for the day." Following the meal, "Every good purpose was answered by the opportunity [the meeting] afforded of chatting in parties ... on the state of crop, tillage, etc." But Hodgkinson qualified this remark, noting: "It is not supposed that in such an infant settlement, many essays would be produced on the theory of farming, or that much time would be taken up with deep deliberation."[78] Other records suggest that a "great silver snuff-box ornamented with the horn of plenty on its lid" was passed yearly to each elected president, and throughout the year, it "remained with the housekeeper who had to supply the next monthly dinner to the Agricultural Society."[79]

Hodgkinson was witnessing an invented tradition in development.[80] Members talked about their farms – and no doubt the weather – but did not engage in any scientific or theoretical discussions. He reported that "many of the merchants and others, unconnected with the country business" were members of the society. Yet he acknowledged the difficulty in placing distinct definitions on individuals in a colony such as Upper Canada: "There are other gentlemen through the settlement, whose early destination to commerce, took up that period, which is usually devoted to what is termed a regular education."[81] Merchants such as Robert Hamilton and his associates employed their membership in the NAS to self-identify as gentleman improvers. They may have been men of commerce, but they realized that a large part of their income was derived from supplying a growing agricultural population with imported goods, speculating in land, and brokering the grain trade.

Likewise, the "very useful building," that was Niagara's Freemasons' Hall was neither the chamber of a London learned society nor a grand estate of a British gentleman. From Hodgkinson's point of view, "the poor wretched straggling village" of Niagara attained neither the status of a proper capital nor even a decent frontier town. The "prospect of it to a stranger was less than gratifying," he continued. "It neither presents him with the regularity of ancient establishments, nor yet with the elegant simplicity of rural culture."[82] His is an interesting insight. This was about the highest form of ritual and spectacle of performative leadership that the frontier capital could produce, and the invited traveller recognized that the ceremony and content of the meeting was far from a match to those he knew from England. Nonetheless, he believed the monthly meetings of the NAS served what purpose they could on this Canadian frontier. The invented traditions made Hodgkinson feel welcome; it was like a travelling Freemason attending a colonial lodge meeting. These displays of authority were an important attempt to transform this disparate group of Niagara Peninsula elite into a self-identified set of agrarian gentlemen, and self-acknowledged leaders of Upper Canada's agricultural improvement connected to a transatlantic community of improvers.

Such rituals performed by the NAS fit a pattern defined and explained by other historians. Richard Bushman suggests that by the advent of the American Revolution the "great merchants and planters, the clergy and professionals, the officers of the courts and government" were expected to adhere to a well-established "genteel code." Maintaining the standards of this code of behaviour resulted in a "self-aware performance," which, in turn, led to the formation of "brilliant and harmonious societies where people came together to perform for one another." In such meetings, gentlemen members were "well aware of watching and being watched."[83] In her study, Tamara Plakins Thornton determines that membership in the contemporary Massachusetts Society for Promoting Agriculture was "a highly self-conscious act." Promoting agriculture was considered "a public service of the highest order that carried with it the moral prestige and national importance of agriculture itself."[84] The elite of Massachusetts "through a kind of cultural intuition" knew that being a gentleman farmer was appropriate to their "station and pretensions" and that interest in scientific agriculture was what defined a "proper gentleman," she argues.[85] Likewise, Donald Marti claims in his study of agrarian thought in New England and New York: "The endeavor for agricultural improvement enabled gentlemen to reach out to the ordinary farmers, displaying their learning and exacting deference. Men who could not be English lords could be American pedagogues."[86]

The NAS was not the only voice in Upper Canada promoting agricultural improvement. In late October 1796, "OBSERVER" addressed his lengthy public letter in the *Upper Canada Gazette* "To the farmers of Upper-Canada." It was a forceful assertion of the ideology of improvement that opened by telling farmers: "You are settled in a beautiful and fertile county, rich by nature, and designed by her to be respectable in population and wealth, so soon as you shall, on your part, do the little that is necessary to accomplish it." While some farmers had improved "50 acres, or more," he lamented that they cultivated such good land with "carelessness and sloth." "There is no fault in the seed or soil; it is in you," he declared. After excoriating them that only when farmers began to improve for themselves would Upper Canada improve in conditions and value, he instructed them: "Mend yourself first, then your neighbor. Strive to excell [*sic*] in improving in agriculture; the strife is honest and honorable. – Go immediately to the work of reformation, while it is *to-day*."[87]

Others attempted to engage the Upper Canadian farmer by adapting the improving ideology to the primitive forms of agriculture that many of the province's farmers practised while chopping their farms out of the North American forests. In March 1797, an unnamed correspondent explained how settlers of "this infant colony" might be "awakened from inactivity to lively enterprize [*sic*]." Prizes ought to be awarded for best sample of maple sugar, the greatest quantity of wheat raised on one acre of newly cleared land, and the greatest quantity of wheat grown on the same amount of old land, he argued. To educate others, he suggested that each award winner be made to describe in detail "the nature of the soil and process of the business in which he claims the premium." Agriculture could be "improved to the highest advantage" of Upper Canada through the spread of such information, he believed.[88] There is no evidence that the NAS acted on his suggestions.

Gideon Tiffany, King's Printer and editor of the province's official and only newspaper, the *Upper Canada Gazette*, printed advertisements and reports of the NAS meetings and, occasionally, items of agricultural interest selected for the Upper Canadian reader. In the spring of 1797, he published an "Agricultural Gazette" to fill space due to "the scantiness of political intelligence" from Britain and the United States. He introduced this feature with an editorial about the importance of circulating information related to improving agriculture. Tiffany, the American who had arrived in the Niagara settlement in 1794, recognized that "imported treatises" on agriculture from Europe were "applicable but in a very few instances" due to the differences in Upper Canada's soil and climate. Therefore, he promised to publish only "such approved experiments" that were applicable to the province and invited any correspondence that corrected "errors of the European methods." He concluded that small experiments each year with new grains or root vegetables were meaningful and useful and were "by no means beyond the power, or even conveniency [*sic*] of a farmer who cultivates a small or great number of acres." For example, Tiffany reprinted reports on English methods of fattening lambs, Ukrainian methods of obtaining fine lamb's wool, a method of growing potatoes, a method to prevent canker worms from destroying apple trees, and Scottish directions for growing flax that had been altered for conditions in Pennsylvania.[89]

But what did the NAS accomplish beyond providing a stage for the local elite to display their authority and patronage? The scant evidence suggests little. Reflecting on the efforts his society had made, President Hamilton commented in 1806 that its members had "assisted most materially in procuring to their country, the variety of Fruit, with which it now abounds."[90] In early spring 1794, he used his merchant connections to import fruit trees from a nursery on Long Island, New York. On this occasion, Hamilton noted to John Porteous, a merchant at Little Falls, New York, that the trees were for "a Society established here for the purpose of promoting Agriculture and Gardening in our New Country."[91] It appears that at least one member of the NAS might have taken advantage of such botanical imports. Ralfe Clench had the largest and finest orchard in Niagara, consisting of "114 trees producing six types of peach and five kinds of plum, as well as quinces, apricots, and nectarines." Unfortunately, American forces would destroy Clench's orchard during the War of 1812.[92]

One method that the NAS did *not* use to promote agricultural improvement was the hosting of an agricultural society "fair." Generations of historians, both antiquarian and academic, have claimed several newspaper announcements of fairs in the early Niagara newspapers to be activities sponsored by the NAS.[93] A first notice published in a November 1799 issue of the *Canada Constellation* announced: "The annual fair at Queenston, as established by proclamation, will be held on Saturday the ninth day of November in the present year. A park is provided, free of expence [*sic*] for the show of horses, cattle, sheep and hogs."[94] Two years later, the *Niagara Herald* announced the annual Queenston fair for the second Saturday of November, noting: "Preparations for the races and other amusements of the day are already making. A park for the show of cattle, hogs, &c. will be provided."[95] These annual events derived from the European model of a fair. As William Addison explains, from the Middle Ages in England, charters for fairs were granted on an annual basis to be held at a specific location on specific dates. Such fairs lasted several days, for example, a first for wholesale transactions, a second for the gentry to attend, a third for the public, and a fourth for a sheep market.[96] In turn-of-the-century Upper Canada, the fair was only a one-day event. Lieutenant Governor Simcoe had proclaimed the fair at Queenston into existence in 1792 because finding a buyer for agricultural produce in the absence of a local market (held weekly in the old English tradition) was difficult task for colonial farmers.[97] All classes attended the fair at once, and the livestock on "show" was likely being shown for sale, although quality breeding stock may have been brought for display to help increase and improve the livestock in the area. In any case, farmers were not entering their livestock into competition for prizes sponsored by the NAS, as some historians have implied. This is an important detail to understand because suggesting so misrepresents the character and activities of the NAS and distorts our understanding of the purpose, interests, and activities of this early Upper Canadian agricultural society. More fundamentally, implying that livestock competitions in Upper Canada had their beginnings during the Simcoe era suggests a greater presence of mixed farming and primacy of livestock husbandry being practised by frontier farmers in the 1790s than was the case. The colony's agricultural economy and mercantile trade was based on crops such as wheat, oats, and barley, and the few records available all suggest that the NAS was interested in cultivation and in introducing new plant species and varieties into the colony, not livestock husbandry.[98]

Late in 1795, Hamilton wrote to Simcoe concerning his plans "to modify the Agricultural Society as to form the basis of one which shall embrace the whole province."[99] Possibly, Hamilton's plans were influenced by the example set by the British Board of Agriculture, established two years earlier. Perhaps he had a second motivation. The provincial capital faced an uncertain future, with Simcoe's founding of the Town of York on Toronto Harbour in 1793 and Britain's impending handover of Fort Niagara to the United States under terms of the 1794 Jay's Treaty. Simcoe was determined to relocate the capital from Niagara to York, and if the NAS did not move to solidify itself as semi-public extension of the provincial government with a colony-wide mandate, it would effectively be limited to its current reality – little more than a local association of improvers and merchants on the Niagara Peninsula.

Simcoe responded to Hamilton's plan from York in early 1796 by stating that, while the matter was one that he had "much at heart," he felt that he should defer the issue until he could speak with Hamilton in person. It is unknown if the meeting occurred, and by February, Hamilton had to admit that the institution had been "for some time rather neglected by its members." Even Simcoe had not paid his annual membership. Hamilton suggested that his payment could wait until Simcoe's return to Niagara when, as the society's patron, he could offer some direction of how the money should be spent. But Simcoe departed the province for England on temporary leave in July 1796.[100] He would never return, accepting other imperial appointments instead. When back in Devon, he would, however, set about further improving and expanding his Wolford Lodge estate until his death in 1806.[101]

At Niagara, the agricultural society continued without its patron, but it did so in a rather somnolent state. Following the stipulations of Jay's Treaty, Britain did hand over Fort Niagara and other forts it held on the American side of the border to the United States in 1796, and in the following year, the Town of York was officially confirmed as the new capital of Upper Canada. But the NAS remained centred in Niagara, its provincial ambitions never realized. In the summer of 1797, it petitioned the provincial government for a "spott [*sic*] of Ground, on which as a Garden, or small Farm, they might, under their own management, make those experiments so essential in the Profession they are desirous to promote." It was granted a block of four acres within the town of Niagara. Here, the NAS planned to build a small building to house its considerable collection of books and to make room for other volumes it hoped to purchase. In his petition, President Hamilton expressed hope that the society's efforts would "have the best effect in disseminating knoledge [*sic*], & in promoting Industry, in Agriculture, which they Justly esteem[ed] the first Interest, as well as the chief Pride of Upper Canada." Hamilton also hoped that the NAS's initiative might be "followed by every district in the Province," although it took no action to encourage that outcome.[102] It is unclear whether the agricultural society actually built the depository for its library or if it ever cultivated this parcel of land, for the sources are silent as to any NAS activity between August 1797[103] and 1805 when the society donated its collection of books to the Niagara Library, a moment that has long been considered to be the agricultural society's final act.[104] It was not, and the next chapter contextualizes the final activities of the NAS in 1806 and 1807. But first, we must assess the meaning, character, and value of the NAS during its first fourteen years of existence.

The membership and meetings of the NAS provide important insights into the performative nature of the colonial leadership, for the organization provided one of the few stages on which the colonial gentlemen could perform their genteel rituals aimed at attaining respect and deference from the province's farmers. Operating at one corner of the British Empire, across the Niagara River from the United States, the NAS could neither be an exact replica of the British agricultural societies it tried to emulate nor the British Board of Agriculture it might have aspired to become. It had to adapt its membership and rituals to the realities of life in the rudimentary frontier settlements of Upper Canada and eastern North America. In the absence of any landed gentry to conduct agricultural experiments, the NAS provided an opportunity for the merchant, military, and administrative elite of Niagara to coalesce as members of an institution, invent themselves as agricultural leaders of a frontier colony, and present themselves at the provincial capital as the leaders of Upper Canada's agricultural improvement.

Moreover, the NAS provides valuable insights into early state formation in Upper Canada. Like contemporary British agricultural societies, it was private club, but its members were individuals who held public offices of various sorts. Such appointments were dispensed by the colony's lieutenant governor who also served as the agricultural society's patron. The NAS and its promotion of agricultural improvement drew together a disparate group of individuals under a mutually understood ideology, flexible enough to adapt to the aims of administrator, settler, and merchant. While improvement was an ideology to be practised locally by its members to demonstrate the benefits to farmer and colony, the agricultural society aimed to be provincial in scope. Founded at the colony's capital, its development was contemporaneous to that of the British Board of Agriculture. That the NAS did not expand across the province and that there are few recorded agricultural improvements actualized by the NAS underscores the overly ambitious goals of the NAS and the ideology it promoted. Most Upper Canadian farmers faced the labourious task of transforming forests into fields in a province that, in the 1790s, was a scattered collection of disconnected communities across an expansive territory. Such was the basic level of improvement underway in Upper Canada. Nevertheless, the NAS played the important, early role of rooting the improving ideology into the initial efforts to establish the Upper Canadian state at its original capital of Niagara. From the 1790s to 1830, establishing some means of government leadership of agricultural improvement continued to be a goal pursued by the provincial elite and increasingly championed by other commentators, both supporters and critics of the provincial government.

# 2 Imperial Defence, Agricultural Improvement, and the Upper Canada Agricultural and Commercial Society, 1801–1815

In early November 1806, Robert Hamilton wrote to Francis Gore, the province's new lieutenant governor, to ask if he would serve as patron of the NAS. Gore accepted the honour with enthusiasm, for he was "much flattered" and promised to offer "every assistance, and protection in [his] power."[1] At first blush, this request and acceptance appear innocuous. However, as this chapter explains, the fact that the *Upper Canada Gazette* published this exchange of letters indicates a deeper meaning at play. When the newspaper reached its subscribers, many among the provincial elite understood that Hamilton's offer and Gore's acceptance were a public rebuke of Justice Robert Thorpe and his Upper Canada Agricultural and Commercial Society (UCACS). By that date, Thorpe had been in Upper Canada for little more than a year, yet he had done considerable damage to the fragile networks that held together the provincial elite. He was a judicial appointment to the province's King's Bench, who, not long after his arrival to York, coordinated one of Upper Canada's first political opposition factions. Worse, this high-ranking judge blurred the lines between the political and judicial branches of government by announcing his candidacy for a seat in the House of Assembly. He fell quickly out of favour with all but a devoted few, for the insular elite of the capital and elsewhere in the province viewed Thorpe's open and aggressive criticism of the colonial executive as a betrayal from within. His suspension from office would end his time in the colony, but it would ignite a fierce debate about the leadership of agricultural improvement and how Upper Canada's agriculture might best be improved.

Robert Thorpe figures prominently in the histories of pre–War of 1812 Upper Canada,[2] yet none explore in depth the agricultural society he founded at York. By establishing an agricultural society that claimed to be province-wide in scope and authority, Thorpe adopted the model of semi-public gentlemanly scientific societies in Britain that assisted, informed, and implemented imperial agendas. The specific agenda to which Thorpe and his agricultural society would respond originated in 1800 when imperial authorities issued an emergency directive for British North American colonies to grow and export hemp to Britain to supply the Royal Navy, engaged in fighting France in the French Revolutionary Wars. A key aspect of the Upper Canadian government's response was its use of legislation and scarce public funds to encourage settlers to grow the crop, even if its initiatives and subsequent refinements generated little response across the province. When Thorpe arrived in York in September 1805, he seized on such failures as an opportunity to declare years of ongoing colonial government incompetence. Quickly, he attempted to assume leadership of Upper Canada's efforts to grow and export hemp by founding an agricultural society designed to be a semi-public institution by way of its membership and anticipated access to public funds. Cynically, he employed the language and actions of the transatlantic world of improvers for his own political and personal gain.

This chapter employs Robert Thorpe's establishment of his UCACS to continue the analysis of performative leadership of colonial agricultural improvement. Thorpe was no improver, however. Rather, he established his agricultural society as a stage on which to perform his leadership aimed at embarrassing the provincial administration and attracting the attention of imperial officials in London to advance his career. Once such cynicism of purpose was exposed, other provincial elites quickly abandoned their initial support of Thorpe and his agricultural society. Yet, the episode represents a continued effort by improvers to establish leadership of agricultural improvement as a role of the Upper Canadian government, assisted by a semi-public provincial agricultural society to implement its aims. Furthermore, the Thorpe imbroglio and its fallout is the link between the final days of the NAS and fresh debates about government efforts to lead agricultural improvement following the destruction and disruption inflicted by the War of 1812.

In 1800, Britain requested assistance from its North American colonies to supply it with a secure source of hemp, from which it manufactured rigging and cables for the Royal Navy. Long reliant on trade with Baltic ports for hemp and timber, Britain had become increasingly dependent on this supply during the ongoing French Revolutionary Wars, as new ships were being constructed and damaged ships repaired. Observers had raised warnings about the dangers of such reliance since the conclusion of the American Revolution, but at the end of the eighteenth century, Britain continued to import some thirty thousand tons of Russian hemp annually at a cost of more than £1,000,000.[3] Critics' fears were realized on 16 December 1800 when Russia left the coalition with Britain and joined Sweden to revive the 1780 League of Armed Neutrality. It blocked British ships from trade at Baltic ports, leaving Britain caught fighting a war without access to much of the timber and 90 per cent of the hemp necessary to maintain its navy.[4] Two days later, the Duke of Portland, Britain's Home Secretary, issued an emergency request to the lieutenant governors of the British North American colonies: "As the present Circumstances render it particularly desirable that the raising of Hemp should be encouraged in such parts of His Majesty's Dominions as are best adapted to its Cultivation, I am to desire you ... to take this object into your immediate Consideration." Lieutenant governors were to direct their legislatures to pass measures granting bounties to encourage farmers to grow hemp. In return, he promised, George III would reward successful hemp farmers with additional grants of land.[5]

In their haste to enlist Upper Canada's help, British officials gave little thought to how the young province might produce copious quantities of quality hemp at a competitive price in a short time. Upper Canadian farmers would have to produce a fibre equal or superior in strength and quality to the Russian product, and it would have to match the latter's price on the London market. Hemp was one of Russia's most lucrative exports, and centuries of trade had created an efficient, low-cost method of shipping top-quality hemp to foreign buyers. There, serf labour was a key factor in the reducing costs of the labour-intensive task of processing harvested hemp into fibres that could be spun into rigging and cables. They had also established a low-cost system to ship the hemp by barges to the Baltic ports each year.[6] At the turn of the century, Upper Canada simply could not compete in terms of labour costs or shipping capabilities. Its small population of about twenty thousand settlers remained scattered in pockets across the province, many on farms that were slowly emerging along the edges of the primaeval forest. Excess labour was scarce and therefore expensive. Mechanical means of processing were limited by the fact that the province's few mills were currently occupied with the immediate tasks of grinding grains and sawing lumber. And it would be many years before Upper Canada's age of canal building began, along with the construction of a reliable network of roads and bridges necessary to create a direct and integrated supply route from the empire's only inland colony to the London markets. Niagara Falls, the rapids of the St. Lawrence River above the Lower Canadian port of Montreal, and the repacking and loading hemp onto oceangoing vessels at this trans-shipment point all posed significant obstacles, added cost, and risked water damage to the raw hemp fibre. As the conflict in Europe transpired, the blockade would last but a few months and Britain would resume purchasing its naval supplies from Baltic ports.

None of these factors dampened the enthusiasm in York. Hemp promised a sizeable, lucrative, and potentially long-term contract that might pour substantial revenue into the cash-strapped colony. Upon hearing news of Britain's interest in hemp cultivation several gentlemen, including four of the capital's leading merchants, met at the Upper Canadian capital on 16 May 1801. All agreed to organize "a society for the encouragement of the culture of hemp in the Home District."[7] But their plans were soon interrupted by the arrival of Lieutenant Governor Peter Hunter from his winter stay in Quebec. He brought with him Portland's instructions, which he intended to present to the opening of the annual session of the Upper Canadian legislature. There was little point in establishing an agricultural society at York to encourage farmers of the Home District to grow hemp if the lieutenant governor hoped the provincial legislature would legislate and fund a province-wide program, backed by imperial authority. Plus, many of those who attended the May meeting in York could employ their roles in the House of Assembly, Legislative Council, and Executive Council to ensure such an outcome.

In his opening address, Lieutenant Governor Hunter informed members of the House of Assembly and the Legislative Council that the current war in Europe had made hemp an article "of the highest importance"; hence, he encouraged members "to secure the source of wealth which the success of the experiment will open to it." In its reply to Hunter's address, the House of Assembly hoped that a "solid foundation" could be laid for hemp cultivation without the use of public funds, whereas the Legislative Council, which contained more members and close associates of those who had attended the May meeting, offered full support for hemp cultivation in order "to render it at no great distance of time a staple of the Province, and thereby secure to Upper Canada an unquestionable source of wealth." As the House began to debate the matter, Hunter directed his Executive Council to prepare a report on the prospects of hemp cultivation so that he could submit an official response to Portland.[8]

Its report attempted to balance the promise of the imperial request with the realities of the state of agriculture in Upper Canada. British officials should not expect to see Upper Canadian hemp on the London market soon, it cautioned, for much of the land in the western part of the province, where hemp was most likely to grow, was "still in a State of Nature." Only a few small experiments in hemp cultivation had been conducted and the province possessed only limited quantities of hempseed and no implements to process the crop. Moreover, the Executive Council emphasized the considerable difficulty and expense of shipping hemp from the inland colony to Great Britain.[9]

Farmers required enticement. Not only would land devoted to hemp in any given year reduce acreage for grain crops, which provided food and income, but hemp's harvest and processing would also exacerbate the provincial labour shortage and take time and energy away from other farming duties, which included cutting down trees to clear new fields out of the forests. Whereas wheat could be cradled, winnowed, and stored in a granary, hemp had to be cut, gathered, and rotted in a pond for several days, after which the hemp fibres were separated from the stalk through the labour-intensive task of scutching and hackling. Next, the hemp fibres had to be bundled and dried in a waterproof barn, for moisture was the chief ruin of raw hemp. Only then could it be packed for shipment to market. Depending on where it was grown in Upper Canada, the shipping distance and cost could be substantial. For example, a few farmers along the Thames River in the western-most region of Upper Canada agreed to grow hemp. With shipment by water routes the only option available, their hemp would have to be shipped westward down the Thames River to Lake St Clair, then down the Detroit River and across Lake Erie, where it had to be portaged around Niagara Falls. From there, it crossed Lake Ontario to Kingston, where it was transferred to a bateau for shipping it through the rapids of the St. Lawrence River. At Montreal, it incurred warehousing and repacking costs before being loaded onto an oceangoing vessel for its final journey across the Atlantic Ocean to England.[10]

Spurred by the enthusiasm of the Legislative Council, the House of Assembly issued an address to Hunter, recommending that the lieutenant governor apply no more than £225 from the provincial treasury to purchase and distribute hempseed to the farmers of the province, as well as a sum of about £450 from which bounties for processed hemp could be paid.[11] Shortly after proroguing the legislature Hunter took action, appointing two executive councillors as hemp commissioners: the newly appointed inspector general, John McGill, who had served as the province's chief purchasing agent, and surveyor General David W. Smith. Their mandate was to purchase hempseed, distribute it at no cost to interested farmers, and pay out the government bounties to farmers for their crops as well as to the merchants contracted to ship the hemp to Montreal for export to England. As it was already midsummer, the commissioners could do little more than devise a plan to distribute seed for spring planting in 1802.

On 3 August 1801, McGill and Smith announced that hempseed, enough it was hoped to fulfil farmers' requests, would be distributed as soon as available. To do so, McGill and Smith engaged the military distribution networks operating in Upper Canada, directing the officer commanding Fort George to distribute seed to farmers from Niagara to Fort Erie early in the spring, and requesting that the commissary at Kingston send seed by bateau to the easternmost parts of the province. The commissioners divided the province into five "circles," namely, "The Home Circle (the Home District and the projected District of Newcastle); The Eastern Circle (the Eastern District and the District of Johnstown); The Midland Circle (the Midland District); The Niagara Circle (the Districts of Niagara and London); [and] The Western Circle (the Western District)." For the 1802 growing season, the commissioners promised to pay bounties to farmers for the first ten tons of hemp produced and dressed in each circle and delivered to the commissioners at York. Merchants who shipped the hemp to Montreal for export to London would receive a reward of just under £1 for every fifth of a ton.[12] Although the plan appeared comprehensive, the commissioners were overly optimistic that ten tons might be grown in each of the five circles during the first season.

The rather misplaced optimism in Upper Canada echoed a level of optimism among imperial officials and gentlemanly scientific societies of London that was rooted in an ignorance of colonial farming and infrastructure capabilities. The provincial government's use of public funds to encourage farmers to grow and process hemp anticipated not only the recognition for having contributed to the defence of the empire in a time of war but also the potential of long-term contracts with the Royal Navy to supply its sizeable annual hemp fibre requirements. With the proper encouragements, hemp might become a lucrative commodity for Upper Canada's farmers and merchants, thereby increasing the overall provincial economy. The colony could become the source of hemp within the British Empire that observers had long recommended be established. In the months and years ahead, continued optimism in London fueled hemp initiatives, thereby opening new connections between Upper Canadian leaders and gentlemanly societies of London that served as semi-public agents of government, drawing the colony further into the transatlantic world of improvers and improvement. Beyond the Colonial Office, the hemp initiative brought Upper Canada under the consideration of the Lords of Trade, the Admiralty, and the Navy Board. Each would weigh in with its opinions as to how best to procure hemp from the Canadas. So too, did London's RSA, a model of the clubbable world that played a semi-public bureaucratic role in the operations of the imperial state. The British Board of Agriculture also offered support. In response, the Upper Canadian government would package these additional supports from London with its own plans and funding in the hopes of better encouraging farmers to grow hemp.

London's RSA had, for decades, raised regular warnings about Britain's reliance on Russia for hemp. On 9 April 1801, it launched its own set of incentives to spur colonial farmers to grow hemp by announcing three prizes for the best samples of Canadian hemp grown in that year. Applicants were required to send twenty-eight pounds of hemp and two bushels of seed to the society's London offices, along with certificates bearing the seal of the lieutenant governor verifying the number of acres sown, and that the seed had been sown with a drill in rows eighteen inches apart. The farmer also had to produce a report on how his 1801 hemp crop had been cultivated and processed, its cost, and the type of soil. Submissions were to be delivered to the RSA headquarters in London prior to the last Tuesday in February 1802.[13] It was a supportive initiative but, in practical terms, the RSA's spring announcement offered little to Upper Canadian farmers in 1801 because it did not reach them, via handbills and publication in the *Upper Canada Gazette*, until late summer when the crops were already ripening in the fields. The RSA's announcement also made several gross assumptions that highlighted its members' ignorance of North American frontier realities and the challenges presented by transatlantic communications. First, it assumed that the Canadian farmer growing hemp would be literate enough to write a detailed report for the competition and, second, that he could actually harvest his hemp in September or October, undertake the labour-intensive task of processing it, acquire the proper documentation at York, and then ensure it was loaded onto a ship headed to England before the freeze-up of the St. Lawrence River. Regardless, the society would commit to offering yearly competitions, despite receiving few entries from Canadian farmers.

The provincial government continued its promotion of the crop through the publication of Lieutenant Governor Hunter's proclamation and the commissioners' announcement of bounties for hemp in each weekly issue of the *Upper Canada Gazette* throughout the summer and fall of 1801. Editor and King's Printer for Upper Canada, John Bennett also featured in his 29 August issue two articles: "Observations on the raising and dressing of Hemp," originally published by the American Philosophical Society in 1771, and a reprint of an article noting that Britain imported thirty or forty thousand tons of hemp annually from Russia. In mid-September, Bennett published "a compact little pamphlet" containing the notice of prizes offered by the RSA, the essay on raising and dressing hemp, and Hunter's proclamation.[14] At about the same time, Hunter received word from Lord Hobart, secretary of state for war and the colonies, that the Lords of the Privy Council had "observed with much satisfaction" the eagerness with which the government of Upper Canada had directed its attention to hemp and the "judicious measures" proposed for its encouragement.[15] By that time, however, some influential gentlemen on both sides of the Atlantic were beginning to question the likelihood of Upper Canada ever supplying hemp to the British market at a competitive price.

The president of the NAS, Robert Hamilton, wrote to the RSA in February 1802, outlining his concerns that few Upper Canadian farmers were likely to compete for the prizes. Not surprisingly, there had been no entries for the 1801 season and Hamilton worried that the society might cease its competition. The blame lay, he humbly pointed out, in the unreasonable regulations of the competition, including the one requiring a drill for seeding. In a colony "so lately cleared of large trees," he explained, stumps and roots precluded "the use of the drill plough, of which there is not yet one in the province." Hamilton suggested that prizes be offered for the greatest amount of hemp grown by one individual, no matter how sown. In response to Hamilton's concerns, the RSA made changes by expanding the number of competitions, decreasing the amount of land required to compete and allowing future competitions to cover two growing seasons to compensate for communication delays between the Canadas and London. Regardless, few farmers submitted entries.[16]

Opinions about the British North American colonies as a potential source of hemp were quite mixed among semi-official organizations in London in 1803. As the RSA revised its competitions to encourage more Canadian farmers to grow hemp and compete, Sir Joseph Banks, still president of the Royal Society, wrote to Lord Glenbervie, vice-president of the Board of Trade, expressing his concerns that Upper Canada was likely to produce low-quality hemp at far too great a price. In part, Banks's conclusion was based on the report submitted by Upper Canada's Executive Council.[17] Britain's Board of Agriculture offered its support of hemp cultivation in the British North American colonies. The board's president, Lord Sheffield, wrote to Hunter in June 1803, forwarding him news of a 50-guinea prize the board was offering to the individual who submitted to the board the "most satisfactory report" on the state of the cultivation of hemp in those provinces, giving attention to soils, the growth and processing of hemp, the cost of labour, and a statement of obstacles and the best means of removing them to promote the crop. It is not clear how Hunter advertised the board's competition to Upper Canadians after he received it in November of that year, and it appears that no such prize was awarded, although the board would offer the prize again in 1804 and 1808.[18]

In Upper Canada, officials continued to debate how the colonial government might more effectively encourage hemp cultivation and its export. In 1804, the provincial legislature amended the original hemp program in significant ways. Gone were the rewards for farmers who grew hemp. Instead, Lieutenant Governor Hunter appointed four hemp commissioners based on their "fidelity, diligence and ability": James (Jacques) Baby of Sandwich, Robert Hamilton of Queenston, Richard Cartwright of Kingston, and William Allan of York. All were prominent merchants who were to purchase hemp from farmers for export to London. The first three were legislative councillors, and Baby was an executive councillor. The government committed £900 for this new program, an amount that by comparison, exceeded what the legislature designated for the improvement of roads. But the hemp commissioners were to replenish the fund with proceeds from the sale of Upper Canadian hemp delivered to England; thus, the government expected reimbursement for its initial outlay of public funds.[19] Regardless, most farmers remained disinterested in taking on the risk of growing hemp and the commissioners had little work to accomplish. The government's 1805 increase in its price per ton of hemp also accomplished little result.[20]

Meanwhile in London, the Lords of Trade engaged Charles Taylor, secretary of the RSA, in September 1805 to prepare a concise account of the cultivation and manufacture of hemp "suitable to the soil and climate of Canada." In November, Taylor's essay, hempseed, and engravings of hemp machinery to guide local manufacturers arrived in Quebec. Copies were forwarded to the lieutenant governor at York.[21] As these materials arrived, the Colonial Office issued new instructions to the lieutenant governors of Upper and Lower Canada, based on advice received from the Lords of Trade, who were "strongly impressed with the Importance of obtaining, as far as it shall be possible, a supply of Hemp from the British Dominions."[22] By the spring of 1806, six hand machines and six hackles for processing hemp arrived at Quebec for distribution as models for local manufacturers to copy. Good hempseed remained a problem, however, for the shipment that had arrived at Quebec in November was "completely spoiled and good for nothing."[23]

### Robert Thorpe and the UCACS

Robert Thorpe arrived in York in September 1805 amid these transatlantic efforts to encourage hemp cultivation in Upper Canada. He was

Image 2.1. The Honourable Robert Thorpe, (ca. 1764–1836), n.d. Artist unknown (McCord Museum, Montreal, M22349.)

to assume his recent appointment as puisné justice of the Court of the King's Bench, one of two top judges in Upper Canada, second only in status to the chief justice. Unbeknownst to Upper Canadian officials, Thorpe was picking up where he had left off in his last position as the chief justice of Prince Edward Island. From his arrival there in 1801 to his departure three years later, Thorpe had proven himself a quarrelsome individual prone to telling anyone who would listen to his plans for widespread reform of the colony.[24] Soon, he began to write a steady stream of letters to officials in the Colonial Office and elsewhere expressing his disgust at the dreadful condition of the provincial economy and what he perceived to be the lack of direction offered by a self-interested colonial administration. After only four months in Upper Canada, Thorpe would complain to Edward Cooke, undersecretary of state for war and the colonies, that "[n]othing has been done for the Colony, no roads, bad water communication, no Post, no Religion, no Morals, no Education, no Trade, no Industry attended to."[25] To a certain degree, this was true.

Thorpe arrived during a difficult and transitional period of leadership in Upper Canada. Lieutenant Governor Hunter had died just weeks earlier, replaced by the most senior member of the Executive Council, Alexander Grant, who was little more than an elderly caretaker. The new judge exploited the power vacuum to assert his influence as much as possible in the colony. He developed a friendship with fellow Irishman, Peter Russell, and it quickly secured him an entrance into the tight, nearly closed society of York. In fact, the former provincial administrator likely skewed Thorpe's first impressions of the province, for not only had Russell been passed over as Hunter's temporary replacement, but he was also being passed over for the appointment he dearly longed for, the lieutenant governorship of the province.[26] Thus, Thorpe employed the influence of his senior judicial role to voice his concerns and those of others, and he managed to do so with little check until the new lieutenant governor arrived. To Cooke, he bragged again in January 1806 that he could employ his influence to secure within a year any legislative measure that Cooke desired to see passed due to the "wild" state of the colonial legislature.[27] Thorpe sent this letter just as Bennett published Taylor's observations as a supplement to his *Upper Canada Gazette*, plus one hundred pamphlets to be distributed by the hemp commissioners. Bennett sent additional copies, which included the engravings of hemp-dressing machinery, to clerks of the peace in each district of Upper Canada for public reference.[28]

Seizing on the new Lords of Trade initiative, Thorpe founded his UCACS at York on 22 February 1806, declaring it would "impress an early attention to Hemp" throughout Upper Canada. Five hundred copies of the society's initial proceedings were to be printed and distributed throughout the province, and members were requested to exert their influence by engaging "their neighbours and acquaintance among the Farmers, to cultivate annually a portion of Ground (however small) with Hemp." The society itself would be centred in York, with its annual meeting scheduled to coincide with the legislative session and its quarterly meetings to be held at the sitting of the Quarter Sessions. Both occasions were the most regular opportunities for provincial officials to meet at York, having travelled to the capital for the conduct of their legislative and judicial duties. A corresponding committee would receive and disseminate reports from members across the province as to how they had grown their hemp, in what quantities it had been harvested, and what expense was incurred in preparing it for market.[29]

Thorpe proudly forwarded the resolutions of his society to Lord Castlereagh, the colonial secretary. Drawing upon only second-hand knowledge of the Hunter administration, Thorpe misrepresented to

Castlereagh the Upper Canadian government's efforts to promote hemp cultivation under the late lieutenant governor's leadership, declaring it had neglected the subject "like every other thing that could tend to serve the Province, or render it valuable to Great Britain." His UCACS would remedy that neglect, he claimed. Thorpe forwarded Cooke the UCACS resolutions, too, and in his enclosed letter requested him to make "the Societies in England" aware of his organization and its efforts so that they might assist the UCACS and direct its attention "to anything serviceable to Great Britain."[30] Later, Thorpe forwarded a printed copy of the UCACS's pamphlet to Castlereagh, noting the great need for a variety of agricultural implements, particularly the machinery needed to process hemp fibre. Cattle and seeds of every kind, both grass and grain, were all wanting in the province, too. While this was most certainly true, Thorpe criticized the general population of Upper Canada, claiming that the chief cause of the decline was the "total inability of the Inhabitants to procure any change."[31]

Never known as a gentleman improver, Thorpe's intent and purpose were becoming clear. His was a cynical effort to advance the improving ideology for his own personal gain. He had hoped to supply the Colonial Office with his skewed versions of colonial activity at York to secure patronage and prestige from his high-ranking correspondents in London. In writing to Castlereagh in May 1806, Thorpe reported that members of his society had agreed that the provincial government's grant for the encouragement of hemp was insufficient and requested a "small sum to be allowed the Society for the purpose of procuring from England such things as may be necessary to promote the advancement of Hemp and other agricultural improvement in the Province." He also requested that Castlereagh direct Upper Canadian officials offer, in addition to the present bounty on hemp, "a small quantity of waste land, free of expence, to excite emulation and compensate exertion."[32] Thorpe, by way of chairing his UCACS, also hoped to become an Upper Canadian landowner.

Thorpe did not want just any plot of land, for Thayendanegea, now a member of the UCACS, had offered the society fifty acres from the Kanien'kehá:ka reserve along the Grand River for the purpose of planting hemp. On two occasions, Thayendanegea travelled to York to attend a society meeting to invest Thorpe's organization with the land. Each time, Thorpe refused to acknowledge Thayendanegea's intentions, thereby rebuffing his offer. Thorpe had in mind another seven-mile square parcel of land in the extreme western corner of the province between Sandwich and Amherstburg, also belonging to a local Indigenous population. Thorpe wanted the provincial government to purchase this land for the cultivation of hemp, because of its location near a ropewalk that two enterprising individuals were operating next to the provincial marine's naval yard at Fort Amherstburg. It was making cordage for a local market, both Upper Canadian and American, from the hemp fibre provided by a few farmers in the south-west corner of the province who had responded to the government's call to grow the crop. After a meeting he had with Thorpe in October 1806, Francis Gore, Upper Canada's new lieutenant governor, noted that the judge had been unclear as to how the parcel of land would benefit the province, and surmised that the land was for Thorpe's own personal gain. A wary and wise Gore offered Thorpe no support nor land.[33]

Setting aside Thorpe's personal ambitions for a moment, we need to explore the structure of his UCACS to understand the stage he and others had established for his performative leadership of improvement. The UCACS was structured like the "circles" that the hemp commissioners had created in 1801. Seven members were appointed to establish a branch society for each of the province's districts outside the Home District. In the case of the Niagara and the Western Districts, the individuals chosen were the government-appointed hemp commissioners, Robert Hamilton and James Baby.[34] Significantly, it appears that Hamilton joined his NAS with the new organization at York under the name of the "Niagara Branch of the Upper Canada Agricultural Society."[35] This decision, plus those by numerous other individuals to join the organization emphasize the fact that Thorpe's formation of the UCACS at a February 1806 meeting occurred within a small window of opportunity.

Initially, Thorpe possessed the status and official connections to attract to his cause a solid core of the province's elite. The attendance of more than fifty "[g]entlemen from different parts of the Province" at the inaugural UCACS meeting was due in large measure to it being scheduled for the second-to-last Saturday of the legislative session at the capital. This occasion, rather than Thorpe's invitation, drew together at the capital individuals representing the disparate groups of local elites from across Upper Canada. For many potential agricultural society members, the meeting provided their first opportunity to meet Thorpe. Those who joined the UCACS on that day included the chief justice, the province's other puisné justice, the attorney general, the solicitor general, the receiver general, the provincial secretary, the surveyor general, the deputy surveyor general and the Survey Office's first clerk, the sheriff of the Home District, the master in Chancery, the clerk of the Executive Council, the clerk of the Crown in Chancery, the King's Printer, the chaplains of the Legislative Council and the House of

Assembly, the sergeant-at-arms of the Assembly, four of the province's six executive councillors, three of its eight legislative councillors, thirteen of the nineteen members of the House of the Assembly, and several justices of the peace. Also joining the society were two hemp commissioners, several of the province's leading merchants, and a proprietor of the ropewalk at Amherstburg.[36]

Viewed with the benefit of hindsight, the executive members of the UCACS portended the factions that Thorpe's personality and actions would soon bring out into the open. Thorpe was chair of his society, and the directors who joined him in the Corresponding Committee included Peter Russell, Solicitor General D'Arcy Boulton, Attorney General Thomas Scott, the Reverend George Okill Stuart, and William Dummer Powell, Thorpe's peer on the King's Bench. Another member, William Weekes, appears quite out of place, for he was certainly not of the same social and political status as its other members. Recently elected the member of the House of Assembly (MHA) for Durham, Simcoe, and East York, Weekes (also Irish) was perhaps the most influential voice in forming Thorpe's opinion of the political state of the province.[37] Thus, his place within the executive of the new society makes some sense when placed in the context of how highly Thorpe regarded him, as well as two other members appointed to the UCACS executive: Treasurer Charles B. Wyatt, Upper Canada's surveyor general, and Secretary John Small, clerk of the Executive Council and registrar of the Court of Chancery. Wyatt had held his office since 1804, but financial difficulties, his loud clamouring for an increase in fees, and his complaints about the men employed in his office were causing some in York to question his abilities and gentlemanly character.[38] Small, too, had been an outsider to the social life in York despite his government appointments. In 1800, he had killed the previous Attorney General, John White, in a duel that arose from White's comments about the morals of Small's wife, Elizabeth. The gossip gained a certain amount of credibility, isolating them from the rest of the York elite.[39] Thus, 22 February 1806 represents the zenith of Thorpe's ability to draw the provincial elite to his cause. At any date thereafter the composition of the executive of the UCACS ceases to make sense because Thorpe and his cabal of oppositionists brought the growing rifts within the ranks of the provincial elite into open conflict.

In the five months following his arrival, Thorpe had not only criticized the government he represented, but he had also entered the small and fragile web of social relationships in the capital much like a bull into a china shop. This was not difficult to do, and Thorpe excelled at the task. York contained an elite society so insular and defensive that its social functions were, as Katherine McKenna states in her examination of the Powell family, "not mere diversions but battlegrounds on which fights over social position were won and lost."[40] After the death of William Weekes in an October 1806 duel, Thorpe ran in a by-election for Weekes's now-vacant House of Assembly seat on an oppositionist platform, assisted by two UCACS members, Wyatt and Joseph Willcocks, the Home District sheriff. It was the first organized political opposition faction in the province's history. Most provincial elites felt a deep sense of betrayal and distanced themselves from interaction with Thorpe if they had not already done so. None responded more effectively than Robert Hamilton.

Initially, Hamilton had considered himself an active member of Thorpe's UCACS, both as a hemp commissioner and by reforming his NAS as a branch of the society at the capital. He also had his branch members distribute copies of a circular published by the UCACS throughout the Niagara Peninsula.[41] No doubt Hamilton saw potential for profit by portaging future hemp crops around Niagara Falls and shipping them through his commercial partners to London. However, when Thorpe announced that he would contest for a seat in the House of Assembly, Hamilton went on the attack to distance himself and his society from the judge's politics and his UCACS. In his attack, he had willing accomplices in Lieutenant Governor Gore and the editor of the *Upper Canada Gazette*, John Bennett.

This moment returns us to Hamilton's November 1806 request that Gore become patron of the NAS. In his letter, Hamilton boasted that his institution had "subsisted with the utmost harmony for upwards of Twenty years." And of its membership, Hamilton proclaimed: "If they have not made a great deal of noise, they flatter themselves that they have done some little good, and they have enjoyed much comfort." From the scant evidence available, the society had been in existence for only fifteen years, with inactivity characterizing its most recent decade. An admission of this fact was Hamilton's signature to the letter that noted he was "President *pro tem.*" Regardless, Gore enthusiastically offered his patronage and the assistance his role might provide.[42]

By highlighting and exaggerating the age of his society, Hamilton was presenting his NAS – not the Niagara Branch of the UCACS – to the newly arrived lieutenant governor as the premier agricultural society of the province. Hamilton could outperform Thorpe. None other than the province's first lieutenant governor had been his agricultural society's original patron. Hamilton could offer Gore honorary membership in a society that had a long history, previous patronage, and a current membership of respectable, loyal individuals connected to the St. Lawrence trade for the hemp initiatives. By November 1806, this was something that Thorpe was unable to offer as a senior judge campaigning on an oppositionist platform for a seat in the House of Assembly. That Bennett published this correspondence in the *Upper Canada Gazette* as a public attack on Thorpe is suggested by the timing of its publication and the fact that the judge had yet to pay Bennett for printing the hundreds of copies of the UCACS's pamphlet (and he never did).

Perhaps reading the November 1806 exchange between Hamilton and Gore was what prompted Thorpe to write an early December letter to an official in London that in Upper Canada, the "scotch Pedlars [*sic*] ... have so long irritated & oppressed the people; there is a chain of them linked from Halifax to Quebec, Montreal, Kingston, York, Niagara & so on to Detroit – this Shopkeeper Aristocracy has stunted the prosperity of the Province & goaded the people until they have turned from the greatest loyalty to the utmost disaffection."[43] It may have been his expressed frustration over the fact that any success of his hemp scheme would have to rely on Hamilton and the province's other merchants to ship it to England as dictated by provincial legislation and the lieutenant governor's appointments as hemp commissioners. Around the same time, Thorpe also claimed to the same correspondent that Gore was "privately trying to overturn" his agricultural society, which he had "labored to promote" and was doing an "infinity of good."[44]

What Thorpe failed to understand, or chose to ignore, when organizing his UCACS was that conservative provincial officials were not about to take part in any organization that sought to circumvent legislative authority. They expected a provincial agricultural society promoted by a high-ranking judge to serve as an extension of government; it would foster agricultural improvement while reinforcing the hierarchical order of an agrarian society. If the UCACS was to be successful, it would do so as a colonial version of the RSA or Board of Agriculture and serve as a semi-public institution in support of the government, employing its official and merchant membership to advance the government's initiatives. Few would support an organization with a leader who made no secret of using his leadership of the UCACS to further his personal and political ambitions. His oppositionist politics and tactics made him an "apostate" to the provincial elite because he had betrayed the system "which had favoured and trusted him with a highly respectable position."[45] Thorpe won his seat in the by-election, and although it was distasteful for a member of the judiciary to take a seat in the House of Assembly, an attempt to have Thorpe unseated as an MHA proved unsuccessful. Hence, in July 1807, the Executive Council concluded that Thorpe's conduct had a "uniform tendency to degrade, embarrass,

& vilify his Majesty's Servants & Government." Lieutenant Governor Gore put an end to the judge's political and judicial career in Upper Canada by suspending him from office.[46]

In contrast, Hamilton and the NAS reestablished themselves as the province's pre-eminent society. That same July saw Gore and his wife attending a dinner at Niagara for some two hundred guests to celebrate the king's birthday. Hamilton led off the after-dinner dancing with Annabella Gore as his partner. A few days later, Hamilton hosted an NAS dinner at his Queenston home with the lieutenant governor, as patron, in attendance. Although reports of the dinner suggest that it was "prepared for a large company," it was one of the last recorded activities of the institution.[47]

By the time of Thorpe's departure from the province in late 1807, his UCACS was in shambles and the remaining members were left to sort out the damage he had inflicted. In September, "A Member" inserted a notice in the *Upper Canada Gazette* wondering who was acting as treasurer of the UCACS and what had happened to the society's funds. The newspaper's new editor and King's Printer, John Cameron, added his own notice, stating that he, too, was curious about the finances, especially "A Member's" reference to the fact that his predecessor, Bennett, had yet to be paid for printing copies of the society's pamphlet. The UCACS held a meeting during the last week of 1807 to investigate the question of the missing funds and, as the *Gazette* reported mockingly, it was discovered that they had "found their way into the hands of the most *able* and *upright* President, who in the multiplicity of *momentous* concerns in which he was engaged, forgot to remind the society where its cash was deposited."[48] Cameron announced his withdrawal from the society. He could no longer associate himself with the UCACS in light of the "piracy" of the society by Thorpe for an "insidious purpose," though he remained proud of the model of a hemp-processing machine he had manufactured.[49] Later, Cameron would publish an angry, rambling account in his *Upper Canada Almanac for 1810* that highlighted the depth of bitterness that Thorpe's conduct had left among the York elite. Cameron recalled that "on our second and third meetings, the impolitic President not occultly, hoodwinked the society's title so far that, on the part of himself and his *assigns*, it dwindled into an electioneering club."[50] The final meetings of the UCACS were held in January and February 1808, the latter being its annual general meeting at which the UCACS was dissolved because members had "neglected to comply with one of the leading & principal resolutions." Few members were paying their membership dues, and few, if any, were planting any hemp.[51]

Robert Thorpe departed for England where he attempted to appeal his suspension from office without success. An associate of his, John Mills Jackson, would draw on Thorpe's complaints to write his scathing account of the political scene in Upper Canada, published in London in 1809. Although Jackson's pamphlet caused little stir in London, it was so critical of Lieutenant Governor Gore's conduct that it further infuriated the provincial government who had thought it was rid of Thorpe. In response to the attack on his character, Gore drafted a reply and obtained leave in 1810 to travel to London to personally defend his conduct. He would not return to Upper Canada until 1816, for the War of 1812 and the need for military leadership of the colony intervened.

Jackson had arrived in Upper Canada just a few months after Thorpe and had become a close associate of the judge, along with Wyatt, Weekes, and Willcocks. Within the year, he had returned to England informing officials of his grievances pertaining to the government of Upper Canada. In part, his pamphlet characterized the efforts of the UCACS in much the same manner as Thorpe had: the provincial government had received the Colonial Office's plan for promoting the growth of hemp in the Canadas, but it had accomplished little. Thorpe's society, he suggested, had done the noble thing of trying to do what the government had not. Several appendices and footnotes testified to individuals who were faithfully growing hemp for export. Furthermore, Jackson claimed, the UCACS had been labelled "a Jacobin club, though not a word of politics was uttered, or anything transacted but what exclusively related to hemp and agriculture in general."[52] Not surprisingly, Jackson's inflammatory pamphlet generated several earnest responses from the Upper Canadian elite to address and counter his claims.

One such response to Jackson's pamphlet presents us with the final recorded activity of the NAS. Although its long-time president, Robert Hamilton, had died in 1809, the following year, the society purchased and distributed for free one hundred copies of the anonymously published *Letters from an American Loyalist in Upper Canada to his friend in England* written in response to Jackson's pamphlet. The NAS hoped to counter not only the political opinions expressed by Jackson's pamphlet but also those of another Thorpe associate, Joseph Willcocks, who was printing his criticisms in the columns of his newly established *Upper Canada Guardian* at Niagara.[53] The anonymous author of the letters was Richard Cartwright, the gentleman merchant of Kingston and one of the province's hemp commissioners. Through his letters, he offered the final assessment on both the UCACS and the NAS. A Loyalist with a keen sense that the Upper Canadian identity was unique to North America and not merely a copy of British society and institutions, Cartwright had become a staunch supporter of the provincial government, especially that led by Lieutenant Governor Gore. He had been deeply offended by Robert Thorpe's actions and Jackson's pamphlet.

Cartwright corrected Jackson's "self-delusion" concerning the potential for Upper Canada to grow and export hemp, drawing detailed information from his position as the hemp commissioner at Kingston. In contrast to the wild claims of Jackson, he countered, colonial farmers had produced little hemp and that which had been shipped across the Atlantic failed to be sold at a rate that could overcome the costs of packaging and transportation. Both branches of the provincial parliament had "done every thing that was practicable," he claimed, yet experience showed all that they must lower expectations that hemp might become a staple of the province. In effect, he underscored what Sir Joseph Banks had concluded in 1802.

Cartwright also addressed Jackson's complaints about the treatment received by Thorpe's UCACS. In fact, he levelled pointed criticisms at any attempt to establish an agricultural society anywhere in the province. It was no surprise that Thorpe's society failed, he argued, and the reasons had nothing to do with any action of government, perceived or real, for it was not at all obvious to him what benefits the province's agriculture might have derived from the agricultural society. Cartwright argued that

> the country is yet too young for such Societies. That class of men who have time and money to devote to such public spirited institutions, is not yet sufficiently numerous; and there could be little scope for the improvements which such a Society might suggest, in a country where the best cultivated grounds are hardly yet cleared of their timber; where labour is more requisite than skill; and where the cultivator having no rents to pay, is not urged by necessity to change his accustomed modes of tillage, for others held out to him as more productive.

Dismissing the performative leadership on which Hamilton's presidency of the NAS was founded, Cartwright claimed rightly that while the agricultural society still operated "for convivial purposes," it had "always been compleatly [*sic*] inefficient as to the professed object of its institution."

Cartwright held back from the brink of an outright dismissal of this form of agricultural leadership, however. In a moment characterized by hyperbole and bitter recrimination, Cartwright presented a level-headed opinion as to the prospects of an agricultural society leading the sorts of improvements its members and imperial authorities hoped it might encourage. The merchant whose wealth derived from his expansive business enterprises in Upper Canada believed the province possessed promise of "great improvements," but population and capital had to increase before the efforts of government or agricultural societies could achieve their intended success. After all, he reminded readers, it had been just twenty-six years earlier that the entire province had been a wilderness. Observers should be amazed at the progress of the province, he argued, rather than complaining, like Jackson, that Upper Canada did not possess all the advantages of a longer-settled, more populous country.[54]

By the time Cartwright wrote his assessment, war with the United States was looming. It would descend on Upper Canada in the summer of 1812 and continue for more than two years. The war drained the province of farm labour, and it left the provincial agricultural economy and many farms in ruin. In a cruel irony to the pre-war hemp initiatives, retreating British forces destroyed the ropewalk at Amherstburg in September 1813, plus several schooners docked at Chatham loaded with hemp ready for export.[55] Rare successes of the province's hemp initiatives, spawned by a European war, were destroyed in the North American battles.

As the following chapter examines, the war's aftermath spawned desperate calls for immediate government leadership of agricultural improvements. New critics levelled fresh complaints about government inaction on this important front. Robert Gourlay, who arrived in Upper Canada in 1817, employed Jackson's critique to substantiate his post-war attack on the provincial government and its supporters. Gourlay ridiculed the pre-war provincial hemp legislation as a waste of public money, although he admitted that "[t]his absurdity we must not wholly rest on the shoulders of the simple Canadians ... it was a patriotic measure; and blindness may be allowable in matters so elevated and pure." Instead, imperial authorities "who should have known better" were to blame. "A little inquiry, and a little consideration, and a little pen and ink calculation, might have convinced our ministers that their hope of getting cheap hemp from Canada was vain."[56] But blaming only colonial legislators and British ministers was an overly simplistic assessment.

The Upper Canadian government's response to Britain's emergency request to grow and export hemp set a precedent of funding a specific crop in the hopes it would become a staple for export. Although the plans were ultimately a failure, such actions would inform a broader post-war debate about the government's role in leading agricultural improvements to expand the provincial agricultural economy. Moreover, the support it received from the RSA that it incorporated into its government initiatives connected Upper Canada to the transatlantic world of agricultural improvers in which semi-public societies were engaged to inform and implement the aims of government. As a result, there had been ready support for Thorpe's UCACS among provincial elites to further the aims of the Upper Canadian government's hemp initiatives through his province-wide organization elevated to a semi-public status by way of members' colonial appointments or elected offices and the society's expected access to public funds. But the society was never more than a stage for Thorpe's opposition politics and career ambitions. This ensured its end shortly after its birth, which was long before imperial and provincial authorities gave up on their attempts to encourage hemp cultivation in Upper Canada. All the while Britain remained engaged in a European war, then a North American war, and continued to spend millions of pounds to procure its supply of hemp from Baltic ports.

# 3 Robert Gourlay, the Upper Canada Agricultural Society, and Independents, 1815–1830

On the first day of December 1818, the Reverend John Strachan, rector of the Anglican church at York, wrote a letter to his mentor, the Reverend James Brown, a professor at the University of Glasgow. In it, Strachan beamed with pride that many former students, whom he had taught at his Cornwall Grammar School in Upper Canada prior to his move to the provincial capital six years earlier, were "now the leading Characters in many parts of the Province." He was particularly proud that several had responded with what he believed to be an appropriate, forceful opposition to the actions of Robert Gourlay, the Scottish gentleman who had recently caused a considerable political stir in the province. Each had denounced what Strachan termed as Gourlay's "seditious publications" and had taken admirable measures to prevent Gourlay from stirring up more discontent among Upper Canadians. Strachan claimed Gourlay had "done harm by exciting uneasiness irritations & exciting unreasonable hopes" in a "quiet Colony" like Upper Canada "where there is little or no spirit of inquiry & very little knowledge." He complained that the provincial executive, of which he was a member, had been far too lax in its approach to Gourlay's "mischief," although he believed the new lieutenant governor, Sir Peregrine Maitland, was beginning to reestablish proper order. The reverend concluded happily that things were now "Falling Back to their peaceful state and as we have in truth no grievances, the people are beginning to discover it is so."[1]

Such was Strachan's cleverly myopic view of the province, issued from his newly built "Palace," York's grandest brick home. He failed to mention several important things to his mentor, including that one of his former students had publicly assaulted Gourlay in Cornwall; that another had horsewhipped Gourlay in the streets of Kingston; that the deputy postmaster of that same town may have interfered with the delivery of mail to Gourlay; that Strachan, himself, had employed all of his political, social, and religious authority to ensure trumped up charges of seditious libel were initiated against Gourlay; that after the accused's acquittal on those charges, Strachan had ensured the newly arrived lieutenant governor shared his negative opinion of Gourlay; that Strachan was behind Maitland's instructions to the provincial legislature for the passage of a recent "gagging act" to make illegal Gourlay's style of political agitation; that he continued to work feverishly to secure the arrest and conviction of Gourlay to rid himself and the province of the Scottish "gadfly";[2] and finally – and most relevant to this study – he failed to mention that he had recently helped organize a provincial agricultural society as a direct response to the criticisms Gourlay had levelled against provincial authorities for doing little to effectively lead the province's agricultural improvement.

Strachan's suggestion that Upper Canada was a quiet place where no one had any grievances was as insulting as it was laughable. The destruction of properties and families from the War of 1812 left many settled areas of the province struggling to restore shattered lives with a depleted supply of labour to do so. After the war, Upper Canada spiralled rapidly into the depths of an economic depression. Military provisioning during the war had both inflated the colonial economy and stripped much of the province of its crops and livestock. Agricultural prices plummeted once British soldiers withdrew, and the colonial militia was sent home. From the farm to the market and then to the wharf, multiple aspects of the province's agricultural economy required urgent improvements.

After 1815, not only was reconstruction of war-torn areas a priority but the necessity of additional infrastructure – roads, bridges, canals – also remained at the forefront of political debate, which was fierce because the provincial government was starved of revenue and the colony wallowed in debt. The inland colony relied on Lower Canadian officials to collect duties at St. Lawrence River ports and transfer to Upper Canada its share. As of 1817, a short-term arrangement had been made with Lower Canada for it to transfer a portion of revenue to the upper province, but shortly thereafter, the Lower Canadian assembly failed to honour that agreement. After protests to imperial authorities, who considered and then withdrew plans to unite Upper and Lower Canada into a single colony, the upper province eventually received remedy in 1822 via imperial legislation, which set an annual transfer agreement that remained in effect until 1841.[3] But this was a later, long-term solution to just one aspect of Upper Canada's economic problems. As Strachan wrote his letter to Brown in December

1818, more local solutions promising more immediate results were desperately needed, too.

At that moment, relations among Upper Canada's political leaders were strained to the breaking point, the result of a failing economy, plus intense debates about compensation for wartime losses and the uncertain future of American settlers in the province. During the yearly session of the Upper Canadian legislature in 1817, a House of Assembly committee investigating the state of the province issued resolutions critical of the government and recommended the solution of encouraging American immigration to the province. Before the House could vote on the entire list of resolutions, Lieutenant Governor Gore, recently returned from London, intervened and prorogued the session out of fears that criticisms of his administration and suggestions of closer relations with the United States would only increase the province's vulnerability. His actions were a clear indication to members of the House of Assembly and all political observers throughout the province that an insecure provincial executive was not about to tolerate direct criticism of its leadership. In June, Gore made a hasty departure for England, tired of colonial politics and seeking a less combative imperial appointment in London. At almost the same time, Robert Gourlay was escaping his own difficult situation in England by departing for Upper Canada.

This chapter examines the ideology of improvement within the context of economic development debates during the years immediately following the War of 1812, viewing it particularly as a central facet of, and response to, the criticisms that Robert Gourlay levelled at the state of the province and its political leadership. It then examines the aftermath of this important political episode by analysing several important, yet failed, attempts by the provincial elite to assert its leadership of agricultural improvement across the province in direct response to Gourlay's caustic criticisms. (Chapter 4 analyses additional responses by the colonial legislature and others.) The Scottish agricultural improver understood the value of the word *improvement*, both as an ideology and in terms of its practical implementation on the farm. Likewise, the vicious response by the York oligarchy and its supporters to Gourlay inquiring of Upper Canadians about what they thought needed improvement in their part of the province indicates that they, too, knew the value of that word and the ideology that it represented. Gourlay's imprisonment and banishment from Upper Canada are well-known responses to his criticisms of the colonial government and general state of the province,[4] but the organization of the Upper Canada Agricultural Society (UCAS). in 1818 was an equally significant and calculated response by the Upper Canadian elites to exert their hegemony. It was a post-war revival of the belief that a province-wide agricultural society centred in the capital of York would provide an effective solution to the province's agricultural woes. But that goal proved impossible. In a colony that was still a collection of local settlements, local elites elsewhere in the province asserted their independence by actively rebuffing the elites at York. Some established their own local agricultural societies. Such acts of independence from York only increased during the 1820s as new immigrants to Upper Canada with social status and education saw no obligation to defer to the oligarchy at the provincial capital, thus exacerbating the difficulties to form a truly provincial agricultural society there. But to understand the provincial elite's motivation to establish a new provincial agricultural society, we must first examine Robert Gourlay's reasons for departing Britain for Upper Canada.

## Robert Gourlay, Agricultural Improver

What had sparked the thirty-nine-year-old Gourlay's interest in visiting Upper Canada in the spring of 1817? His wife, Jean, was the niece of the late Robert Hamilton, and she had inherited 866 acres of land in Dereham Township, Oxford County, in the London District. It was her portion of 2,600 acres that had been given by her uncle to his widowed sister and her two children. Seven years earlier, William Dickson, the executor of Hamilton's estate and Upper Canadian legislative councillor, had visited Robert and Jean in England to convince them to immigrate. At that point, Gourlay declined, although he did purchase another five hundred acres adjacent to the land his wife had inherited. Then, in 1814, another Upper Canadian cousin, Legislative Councillor Thomas Clark, visited Jean and encouraged them to immigrate. Again, the Gourlays declined the invitation.[5] Three years later, however, Gourlay would view the opportunity to immigrate to Upper Canada as a useful escape from the financial and political mess he had created for himself in Scotland and England.

A brief introduction to Gourlay's pre-Upper Canadian years is in order, for as in the previous chapter, the context of a well-known political crisis in Upper Canada needs to be understood with a broader framework than tempestuous colonial politics alone. Robert Gourlay was a university-educated, innovative agricultural improver with considerable practical farming experience. When he arrived in Upper Canada there were few, if any, men in the province who could match his knowledge of the latest theories and methods of scientific agriculture. Born to a prominent Scottish Whig landowner, an eager Gourlay studied husbandry at age nineteen by touring the estates of his father's friends during the summer of 1797. That fall, he enrolled at the University of Edinburgh's recently established program in scientific agriculture. Following his first year of classes, Gourlay undertook a walking tour of Scotland and Ireland to gain a better understanding of the general state of agricultural practices. Two years later, he toured England with the same purpose, noting that, in some areas, he saw farming methods little changed since the Middle Ages. Two years of wet weather had also devastated the crops in many areas, giving Gourlay a glimpse of the devastation that famine could cause. These were visions of rural life he would not forget and, following that trip, the amelioration of the lives of Britain's landless poor would be the heart of Gourlay's political thought.[6]

With his father's connections to support him, Gourlay gained an introduction to Scotland's most prestigious improver, Sir John Sinclair, president of the British Board of Agriculture. In 1799, the board commissioned Gourlay to perform a survey of Lincolnshire and the County of Rutland in the East Midlands. It was an undertaking certain to draw prestige for the young gentleman from Fife, but Gourlay's enthusiasm for the task and his concern for the landless poor brought him notoriety instead. Although the board agreed to publish his statistical findings, it refused to print Gourlay's liberal report as to how the plight of the poor might be remedied. President Sinclair, who had been outvoted on this matter, later secured publication of Gourlay's report in Secretary Arthur Young's *Annals of Agriculture*. And later, Thomas Malthus would cite Gourlay's report in his influential consideration of poverty and population. But, for Gourlay, his employment with the Board of Agriculture was finished.[7]

Continuing his travels, Gourlay visited the south of England. There, word of his presence reached the Duke of Somerset, who wanted his farms to be run by Scottish managers, for they had a reputation of being among the best sort of agricultural improvers. Instead, Gourlay returned north to his father's Craigrothie estate in Fife. During Robert's absence, his father had purchased three more farms, and gave Gourlay the opportunity to manage one named "Pratis." Here, he set out to practice on his farm the latest scientific agriculture that he had spent years learning at university and observing during his travels. In 1803, the Duke of Somerset again invited him to rent one of his farms in Wiltshire, but Gourlay declined. His life as a prominent young Scottish gentleman was set, and in August 1807 he married Jean Henderson, whose dowry would eventually lead him to Upper Canada.

This peaceable life was brief, for in 1808, Gourlay made known his political views about the plight of Britain's poor tenant farmers, making public pronouncements in favour of tenant farmers having a greater voice in the process of land leasing. Although the local Fife gentry turned against him, Gourlay pushed on with his views, publishing a pamphlet that expressed his views and called out the reluctance of his peers to accept his arguments. The result was a threat of legal action, and the young farmer fled Fife for Wiltshire, this time accepting the Duke of Somerset's repeated requests to manage one of his farms.

At the duke's 750-acre Deptford Farm, Gourlay planned to accomplish two things. First, he would transform its production as rapidly as possible with his new farming methods. Second, he would continue to observe English poverty, for there was no shortage of it in Wiltshire. Gourlay soon faced significant obstacles to accomplishing his first goal. The complex nature of land leasing meant that Gourlay had little room to manoeuvre in his efforts to transform the farm operation while still generating sufficient income from farm production to pay rent. Moreover, the duke, seemingly more interested in personal wealth than in the advancement of agriculture, viewed Gourlay's chosen course of husbandry with scepticism and pressed for the money he was due. By 1812, the terms of his lease turned into a legal contest lodged against the duke that was won by Gourlay. But with no requirement for the duke to settle immediately, Gourlay found himself battling bankruptcy and fighting to keep himself above the level of the rural poor tenants that he had wished to study.[8]

Writing in 1813, Gourlay noted: "I cannot stay my hand from improvement; it is a weakness inherent in my nature." Soon after, he joined two newly formed agricultural societies, the Bath and West of England Society and the Wiltshire Agricultural Society. Always the critic, he found the former society to be less progressive than he wished. To rectify the situation, Gourlay announced three contests of his own. At the same time, he published an anonymous call for tenant farmers to organize against the great landlords by electing delegates to meet him at a convention in London. Gourlay did not maintain his anonymity for very long, and the result was his expulsion from the Bath and Wiltshire Agricultural Society. Shortly thereafter, Gourlay found himself disinherited, for his father viewed him as a serious liability to the strength of the Gourlay name.[9] Push factors to immigrate to North America now equalled, if not exceeded, the pull factors of his wife's land in Upper Canada. But his escape could not be a clean break. He may have left his financial problems – and Jean – behind, but it was not possible to leave behind his personal trait that determined he must right wrongs and speak out for the underprivileged against the privilege of his own class regardless of the expected damage to his reputation and career.

That trait had not served him well in Britain, and it would undermine his attempts to advance agricultural improvement in Upper Canada, particularly considering the timing of his arrival to Upper Canada and the people who welcomed him into the province.

Gourlay had planned for a trip to Niagara via New York and then an autumn return to England from Quebec. But he missed a New York–bound ship when he reached Liverpool, so he boarded another departing for Quebec. Upon his arrival in North America, Gourlay's inquisitiveness led him on a detour off the direct route to Niagara. He walked from Cornwall to Brockville and then up and back along the bush road stretching north to the newly established settlement of Perth. In each of these places, as well as Kingston from where he took a steamboat to Niagara, Gourlay began questioning locals about their situations in anticipation of creating a statistical account of Upper Canada to inform potential immigrants. When Gourlay finally arrived at the residence of his wife's cousin Thomas Clark in July 1817, he spent six weeks recovering from the mosquito bites he had suffered travelling the Upper Canadian frontier. During this time, he learned about the state of Upper Canadian politics as told to him by the Niagara elite. Clark, Dickson, and the late Robert Hamilton's two sons, Robert Jr. and George, provided him with their politically charged opinions fueled by the recent protests within the legislature, which had been led by the Niagara MHA Robert Nichol and had caused Lieutenant Governor Gore to prorogue the session.

The old Niagara Peninsula settlements of Robert Hamilton's era had suffered great ruin and distress during the war. Numerous military campaigns had destroyed towns, lives, and farms, including the late Hamilton's Queenston estate. The local leadership was now formed of a generation of men – including Hamilton's sons – who had risked their lives to defend the territory. It seemed to them that, despite all the loyal rhetoric being spewed from the provincial executive at the capital, York remained unsympathetic to the plight of the peninsula. What Niagara and similarly ruined locations throughout the western portion of the province required was financial compensation for the losses sustained in the defence against the American invaders. Moreover, Gore had forbidden commissioners to administer the oath of allegiance after the war as a response to fears of an increase in the already sizeable population of Americans settling in the province. William Dickson had openly defied Gore's instructions regarding this and had been summarily dismissed as a justice of the peace for his actions. The Niagara elite and others saw the effect of this measure to be the end of land sales to Americans, the very people many believed possessed the settler skills necessary to improve the colony's agricultural economy. Land owned by the Niagara elite remained unsold, and when its members visited Gourlay as he recuperated, they no doubt expressed anger at this circumstance of not being able to generate income when they desperately needed more money. Gourlay, too, was desperate for income. He hoped to profit from his wife's Niagara lands and, thus, was easily swayed to a prejudiced opinion of the departed lieutenant governor as well as those Upper Canadian authorities who sought to preserve the ban on American immigration.[10]

For the political elite at York, and their followers scattered throughout the province, the War of 1812 had produced a deeply conservative outlook for the province's political and social future. A "kind of Messianic Toryism inflexible in its insistence upon unbending adherence to orthodox values" is how S.F. Wise describes it. After the war, their conception of loyalty was allegiance to the British Crown, plus an "adherence to the social, political, religious, and cultural values essential, in Tory eyes, to the preservation of the province."[11] Robert L. Fraser points out that the provincial elites after the war "increasingly pinned their hopes not on a nascent social structure but on the Constitutional Act and its provision for an appointed legislative council, and on the rule of law, the security of private property, the magistracy, and the legal profession." To them, he argues, these institutions "provided the best, and seemingly the only defence of order in Upper Canada." They continued to attempt a recreation of a stratified British society, despite the fact that Upper Canada possessed neither an aristocracy nor a "settled province-wide, as opposed to local or regional, social structure of any sort." The result, Fraser claims, was a political structure that was "unworkable, and hence, unstable."[12]

It was into this politically unstable situation that Gourlay charged headlong. When recuperated, he fused his role as a frustrated Upper Canadian landowner with his ideas for promoting a system of immigration for the mutual benefit of the English poor and the improvement of Upper Canada. Towards this goal, Gourlay returned to skills he had performed as a board of agriculture surveyor. He drafted a list of thirty-one questions for Upper Canadians to answer about their current situation. He travelled to the capital, met with several of York's political elite and received permission from Samuel Smith, the provincial administrator who was filling the gap between Gore's departure and the arrival of a new lieutenant governor, for the publication of his survey questionnaire along with an address to the inhabitants of Upper Canada in the pages of the *Upper Canada Gazette*. He also had seven hundred copies of his address and questionnaire printed and mailed to officials of each township across the province for distribution in each locality.[13]

These questions and the answers that settlers provided have been central to any political analysis of the Gourlay episode in Upper Canadian history but overlooked is the fact that Gourlay's questionnaire differed little from those used by the British Board of Agriculture, reports of which, undoubtedly some of the Upper Canada elite were aware and perhaps had read. But it was Gourlay's thirty-first and final question, plus Gourlay's process of collecting information, that raised the hackles of the provincial executive. "What in your opinion, retards the improvement of your township in particular, or the province in general; and what would most contribute to the same?"[14] were loaded questions. First, its premise identified to the insecure provincial executive that it had not been performing its self-appointed task of leading and implementing improvements. Second, it contained the democratic suggestion that any respondent across the colony possessed the capability to understand what was best for the improvement of their township and the province as a whole. Third, Gourlay had requested township meetings be held to complete his questionnaire and that they return the information to him for compilation into the report he would publish. This deviated widely from the British Board of Agriculture's process of having an appointed surveyor collect data in person and having a government-approved board issue a report of the findings – or censor that report when a surveyor such as Gourlay might challenge it to do so. To a defensive colonial elite, township meetings also smacked of republicanism and threatened a provincial executive already smarting from Nichol's recent agitation on the improvement issue within the House of Assembly. Thus, Gourlay's actions only amplified reasons why the executive should feel threatened, especially when he issued a second address to the inhabitants of Upper Canada that condemned Gore for having prevented American immigration to the province and called for a legislative inquiry into the state of the province with a commission to be sent to England with the results. Here, as in Scotland and England, Gourlay the scientific surveyor and agricultural improver became Gourlay the gadfly.[15]

The new legislative session opened at York in early February 1818, just as Gourlay published the first instalment of his second address. Although Gore had long departed the province, the acrimony from the first session continued to boil. The House of Assembly again voted to form a committee on the state of the province, exacerbating tensions between the elected chamber and the provincial executive. In the face of sustained hostility, Smith followed Gore's lead from the previous session and prorogued the legislature on 1 April. All the while, Gourlay continued to voice his criticisms, which were now even directed at his wife's cousins, Dickson and Clark, for having tried to distance themselves from his open attacks on the York elite and the latter's recommendation that Gourlay leave the province. At York, Strachan and other men of influence now believed Gourlay to be a "dangerous incendiary," particularly because his actions encouraged expressions of disloyal opposition from members of the House of Assembly. They believed it behoved them to eliminate the threat that Gourlay and his actions posed to their leadership.[16]

Gourlay's third address, published in early April 1818, levelled stinging criticisms at Smith for having prorogued the legislature. Not only had his actions denied an inquiry into the state of the province, Gourlay argued, but repeated prorogations were also a threat to the constitution. In contrast, the provincial elite believed Gourlay's calls for town, township, and district meetings represented that very same danger because they were aimed at holding a convention at York at which delegates would approve a petition to the Prince Regent pleading for constitutional change in Upper Canada. Following an April meeting in Niagara at which the petition was drafted, Gourlay had a thousand copies printed for distribution and began to tour the province. It was at this point when the Reverend Strachan, his former students, and others took action. Gourlay was harassed in Prescott, assaulted in Cornwall, and horsewhipped in Kingston. In each of these latter towns, charges of seditious libel were laid against him. Acquitted on both counts, Gourlay forged ahead with his reform platform, hosting a "Convention to the Friends of Enquiry" at York during the first week of July 1818.[17]

Arriving in the middle of these crises in August 1818, Upper Canada's new lieutenant governor, Sir Peregrine Maitland, was presented with the convention's petition. He simply refused to accept it, arguing that the British constitution provided no place for such meetings. After all, the House of Assembly was the forum for people's duly elected representatives. For its part, the Legislative Council advised Maitland that such conventions endangered the constitution. Newly arrived from England, Maitland viewed Gourlay's actions in terms of the demands for parliamentary reforms that he had recently witnessed in Britain. He did have a point. Gourlay admired Britain's leading reform agitator, William Cobbett, and had adopted his inflammatory rhetoric about the landed class's abuse of the British constitution at the expense of the lower orders.[18] Thus, when Maitland recommended legislation to ban such "seditious" meetings and dismissed or denied civil and military appointments to any supporter of Gourlay, he had the provincial executive's full support. Not surprisingly, Strachan would write glowingly of Maitland by year's end.[19]

In Maitland's eyes, the solution was clear. The Upper Canadian government needed to implement a "gagging act" such as the imperial parliament had passed the previous year to quell that nation's reform agitators.[20] To the provincial executive, Gourlay's continued activities required firm action on their part to demonstrate to colonists and imperial authorities that Upper Canada was no breeding ground for sedition. Gourlay's time in Upper Canada neared an end.

In his speech at the opening of parliament in October 1818, Lieutenant Governor Maitland encouraged the legislature to ban all future meetings such as Gourlay's convention. In accordance with this wish, the House of Assembly passed its own version of the British Gagging Acts by way of an act "to prevent CERTAIN MEETINGS within this Province." Persons electing or elected to assemblies "purporting to represent the people" that were "deliberating on matters of public concern" would be "guilty of a high misdemeanor."[21] Effective 27 November 1818, the act aimed to prevent Gourlay or anyone else from holding any future public meetings in Upper Canada. Yet, Maitland and his Executive Council remained concerned about Gourlay's continued presence in Upper Canada. Thus, the leaders of the province's judiciary, Chief Justice William Dummer Powell, Attorney General D'Arcy Boulton, and Justice William Campbell agreed on one additional measure. On 19 December 1818, they employed the terms of the Sedition Act of 1804 to arrest Gourlay as an alien of the province. For refusing to leave Upper Canada as commanded, Gourlay was committed to the Niagara jail in early 1819 where he would remain until 20 August when he was banished from Upper Canada and released in New York state. By then, his physical and mental health had deteriorated significantly.[22]

## The UCAS

At the same time as Lieutenant Governor Maitland and his judiciary were plotting their anti-Gourlay legal strategy, the *Upper Canada Gazette* of 10 December 1818 published an announcement that the "propriety of establishing an Agricultural Society in the Home District" had been "lately discussed at a meeting of some gentlemen" of York. Other gentlemen who wished to be involved were requested to meet the following week, "for the purpose of taking into consideration the most effectual and practical means to encourage, promote and improve the Agriculture of the Province."[23] Had the just-implemented provincial gagging act not banned such meetings? Was such a gathering not designed for "deliberating on matters of public concern"? Would responding to the advertisement not be a "high misdemeanour," because those in attendance would be forming an extra-governmental body "purporting to represent the people"? Clearly, the "several Magistrates and other gentlemen" who met on 14 December 1818 "to now and from hence forward unite and associate" in the UCAS[24] had no fear of government retribution under the terms of the gagging act, for several were the very same "gentlemen" who were devising continued legal action to be taken against Robert Gourlay.[25] Clearly, none of its organizers believed the agricultural society would not be "an extra-governmental body"; it would be a semi-public extension of the government itself.

Despite his imprisonment, we can imagine that, at some point, Gourlay might have chuckled at the irony that his political agitation had caused the formation of an Upper Canadian agricultural society. His experience as a member of the Bath and West of England Agricultural Society had caused him to view such institutions as "worse than useless." Gourlay would later claim: "Agricultural Societies might have done good in [promoting dexterity and skill among farmers] but their objects have never been sufficiently defined or substantial; and, respecting too little the grand principles which govern all men, they have invariably disgusted the practical farmer, attempting to lead them by slender virtues, – by empirical pretensions and coxcombical exhibitions."[26] As it turned out, his last phrase might have served as a motto for the UCAS.

The call for an agricultural society was a significant part of the Gourlay episode, for it was a response intertwined with the legal actions of the province's judiciary. Gourlay's criticisms had struck a nerve, and while the provincial executive and judiciary had silenced Gourlay and any similar critics, it now behoved them to promote agricultural improvement throughout the province. Although, *prima facie*, the formation of the UCAS contravened the colony's Gagging Act, this agricultural society was to be the means to extend the provincial elite's paternalistic authority to its clients in every district of the province and from those locations to the back concessions of every township by promoting branch societies in each district of the province. The agricultural society may not have been established by legislation or funded with public money, but it was an equally important reaction by the provincial executive to Gourlay and the perceived threats to its leadership that his stinging criticisms posed. Its founding and subsequent meetings "purporting to represent the people" were tacitly sanctioned, considering the overlapping membership between the UCAS's executive and the provincial executive, not to mention its patron.

William Campbell, one of the province's two puisné justices, was elected president of the UCAS. His equal on the bench, D'Arcy Boulton Sr., was chosen as vice-president, along with Jacques Duperron Baby, the province's inspector general and member of the Executive and Legislative Councils. In consequence of his recent promotion to the bench, the elder Boulton had recently secured the position of acting solicitor general for his son, Henry John, who was elected the agricultural society's treasurer. The UCAS selected as its secretary, the King's Printer and editor of the *Upper Canada Gazette*, Robert C. Horne. Lieutenant Governor Sir Peregrine Maitland agreed to serve as the society's patron.

The directors' list was also heavily populated by the York elite and their connections beyond the capital. The Reverend Strachan, the executive councillor, joined the province's chief justice, William Dummer Powell, along with Thomas Scott whom Powell had recently replaced in this top post. Other directors included Peter Robinson from Newmarket. He was the MHA for the riding of York East and Simcoe at the north end of the Home District. A businessman with many interests in milling, land development, and fur trading, he operated a store in York with his brother-in-law D'Arcy Boulton, Jr.[27] Lieutenant Colonel Joseph Wells was a recent arrival to York society. After the war ended in Europe, he was appointed inspecting field officer of militia in Upper Canada, but when he arrived in the colony with his wife and young children, he discovered the position had been abolished. He was placed on half-pay and given a grant of 1,200 acres of land. By 1819, Wells's military rank and record of service had opened many doors into the close-knit society of the Upper Canadian capital.[28] The province's receiver general, George Crookshank, was also a director. His estate house that fronted his property to the west of York was a well-known landmark for anyone who sailed into Toronto Harbour. Levius Peters Sherwood rounded out the list of directors. The son of a prominent Loyalist family from Brockville at the eastern end of the province, Sherwood had been at York successfully defending two Métis clients in a case stemming from the 1816 violence at the Red River colony. He had also proven himself to be an active opponent of Gourlay's agitation during the previous summer.[29] At first, he would seem like an odd individual to elect as an officer for an agricultural society at York. But there was a purpose: those who wished the UCAS to be provincial in scope and authority hoped this rising star of the Johnstown District would employ his influence in that part of the province on behalf of the UCAS.

The founders of the UCAS announced the institution would provide an effective and practical means to "encourage, promote and improve the Agriculture of the Province." Members promised to adopt "all such means within our power, as will tend to create competition and emulation among the Farmers of this Province" and to enable them "to excel in the various branches of Agriculture and Rural Economy." To accomplish this goal, the board of directors would, at times, host competitions at public exhibitions and offer winners "[p]remiums, prize medals, or other pecuniary or honorary marks of distinction."[30]

Strong on political and social authority, the UCAS executive had limited authority in agricultural improvement, at least to the level of Gourlay's knowledge and experience. Several were improving their park lot estates on the outskirts of York, such as Powell at Caer Howell, Crookshank at his harbour-front estate, and James Baby with his two-hundred-acre lot near York, plus their sizeable land holdings throughout the province. D'Arcy Boulton had been given a two-hundred-acre land grant and accumulated other lands, including grants for each of his children. In 1817, his son D'Arcy Jr. had built one of the finest Georgian brick homes in the province on his one-hundred-acre estate, The Grange. Sherwood and Robinson had many business interests in their districts that were connected to farmers, but their ambitions were directed to political or judicial appointments. Wells would become owner of the two-hundred-acre Davenport estate but not for another three years.[31]

Given the nature of the UCAS's founding, it is not surprising that its constitution set substantial membership dues to ensure it would remain a gentlemanly organization. It was, after all, meant to provide a distinct contrast in leadership to the democratic township meetings and convention that Gourlay had proposed for addressing Upper Canada's shortcomings. Basic membership, by either paying £1 per year or making a one-time donation of £5 or more, permitted one vote at all general meetings and permitted a member to stand for election to any office of the society. A subscription of £2 per year or a donation of £10 entitled a member to two votes, a system that continued in proportion for every additional subscription of £1 or donation of £5. Such membership dues, well beyond the financial capability of most farmers in an economically downtrodden colony like Upper Canada, underscored the inherent paternalism of the UCAS.

To reach out to gentlemen in the "out Districts" of the province, the UCAS promised that, if a district raised £20 or more in annual subscriptions, a "sufficient number" of directors would be elected from that district with powers to form a branch society. In recognition of the branch societies it expected to incubate, the UCAS set its annual general meeting for the second Saturday of the yearly session of the provincial legislature, when many local leaders from ridings across the province would be at York.[32]

This "patriotic institution" printed a circular to be distributed across the province that claimed: "Amongst the various objects that have occupied the attention of Sir Peregrine Maitland, for the benefit of this Province, perhaps none may eventually be of more importance, by contributing to its general improvement and prosperity, than the establishment of a Provincial Agricultural Society."[33] The latter part had some truth. During late 1818, Maitland instigated some improvements to Upper Canada landholding: he had addressed the problem of absentee landownership in the province; he had reformed the land board to clear out a backlog of applications for land; he had worked to limit the liberal granting of Crown lands, and by 1819, he had secured passage of a bill to tax uncultivated land as a measure to stem land speculation and open wild lands to settlement and agriculture.[34]

Despite the authority that the UCAS sought to assert from the capital, its influence across the province proved to be minimal. At the society's first general meeting in February 1819, members agreed that some of the original rules were "inexpedient, and inadequate to the carrying into effect the views & purposes of the institution." As a result, members agreed on a set of supplementary rules, which were more democratic and aimed at soliciting new members to secure province-wide support. It appears to have been a turning point for the UCAS: the immediate threat that Gourlay posed had passed, and now the society needed to transform itself from a group of defensive elite reactionaries into a legitimate provincial organization supported by a broader range of Upper Canadians interested in improving the province's agriculture.

The UCAS dispensed with the schedule of membership dues and votes, and it halved the yearly membership rate to 10s or a one-time donation of £5 for one vote at all general meetings. But if a member wished to stand for election to any of the offices of the society, he was required to pay £1 5s per year or make a one-time donation of £10. This second level permitted an expansion of the board of directors, for donors of this latter amount would be automatically provided with a seat on the board. The society also clarified the process by which district branches could be formed, as well as their relation to the executive at York. Following the collection of £20 in any district, individuals were authorized to elect two district vice-presidents, a district clerk, and a district treasurer. Serving in yearly terms, they and at least three directors would form a district board of directors with the power to hold annual meetings timed with their district's general session of the peace, the best opportunity to draw individuals to the district town. This district board would also have full authority to use their funds and schedule district competitions and exhibitions as they saw fit, in accordance with the general aims of the UCAS. Up to 20 per cent of the funds they raised, however, would be contributed to the general funds of the provincial organization.[35]

The UCAS executive requested those who received one of the three hundred copies of the society's circular to exert their "personal influence" among their neighbours to solicit memberships and report regularly to Secretary Horne at York. They believed optimistically that "every respectable person throughout the Province, (farmers particularly), will be happy to enroll their names in so useful and patriotic an institution." Yet, at the same time, the UCAS made it crystal clear that the organization at York took precedent, stating: "That as this Society is a Provincial Institution, under the especial sanction & patronage of His Excellency the Lieut. Governor, all General Meetings of the Society, shall be held in the Home District, and not elsewhere." President Campbell and Vice-Presidents Baby and Boulton further defended this position by noting that "although the general direction and management of the Society, must necessarily be established at the seat of Government, the benefits resulting from it will equally and impartially extend over the whole Province."[36]

The UCAS fit a wider pattern of post-war efforts by the provincial elite to consolidate their influence across the province. As S.F. Wise demonstrates, the tory elite hoped to unify the province by "formulating provincial goals of a distinctive kind and by bequeathing their special sense of mission to the Canadian political culture." The leadership, centred in York, would create "a genuinely provincial political system, based upon the alliance of the central bureaucracy with regional power groups." For example, the plan for branch societies of the UCAS fits the pattern of leadership established by the Loyal and Patriotic Society of Upper Canada, which had been created in 1812 by several of the same gentlemen to offer relief to those in distress due to the war. A general board of directors of this society of patriotic gentlemen had appointed committees "of their own members, residing in the different districts." The district boards were responsible for collecting the actual information and for submitting their reports to the general board so "that unity may be preserved in the Society throughout the Province."[37] Likewise, the UCAS would allow the York elite to direct agricultural improvements throughout the province without ever having to leave the capital. But the UCAS had set itself an impossibly tall order, for post-war Upper Canada remained, as Wise notes, "a welter of parochialisms, of disparate groups cut off from one another by differences of origin, religion and language, and by poor communication."[38]

It appears that the Newcastle District Agricultural Society became the only branch of the UCAS, founded by individuals who attended a late March 1819 meeting at the courthouse in Hamilton Township. While more than £40 was collected in memberships and donations, it is unclear how long the society survived or what activities it sponsored. Zacheus Burnham chaired the meeting, and at a subsequent meeting in April, he would be elected as one of the society's two vice-presidents, as well as its treasurer. Burnham was an American who had immigrated to Upper Canada in the late 1790s. Settling in Hamilton Township, he became one of the largest landowners in the area and set about improving his "Amherst House" farm near the emerging town of Cobourg. His farm was much larger in size and production than most in Upper Canada, and Burnham hired a farm manager and other labourers to complete his seasonal tasks. Significantly, this gentlemen farmer had ties to York, particularly John Strachan, for Burnham's son was a pupil of the reverend's Home District Grammar School. Likewise, the agricultural society's secretary, George Strange Boulton, was not only the son of D'Arcy Boulton Sr., but he had also attended Strachan's former school at Cornwall. A lawyer, George had recently moved to Port Hope to begin his practice. These two veterans of the War of 1812 were joined on the executive by a third, Colonel Richard Bullock, late of the 41st Regiment of Foot, who was elected the society's second vice-president. Following the war, Bullock had settled at "Springfield Park," Murray Township, near Carrying Place.[39]

Although the initiative of the York elite was copied by local elites elsewhere in the province, these local oligarchies remained steadfastly independent of the UCAS. When an Agricultural Society of the District of Johnstown was formed in Brockville in May 1819, it showed little deference to the elite at York and did not join as a branch of the UCAS, despite Levius Peters Sherwood having been chosen a director of the UCAS for the district. Charles Jones chaired the inaugural meeting, and those present would elect him as the society's president. A Loyalist and member of one of the founding families of Brockville, Jones owned a considerable amount of property within the town and the surrounding township of Elizabethtown, and he owned a flour mill in Brockville. Jones was a firm believer in Upper Canada's agricultural future and a friend of the York elite, such as the Reverend Strachan, whose shared tory political views provided Jones with several local government appointments.[40] Nevertheless, it seems there was no compulsion to connect – or, perhaps, subject – the Agricultural Society of the District of Johnstown to the efforts of the elite at York. The society hosted cattle shows at Brockville in October 1819, October 1820, and January 1821, but it is unclear if the society survived beyond that date.[41]

Likewise, the gentlemen of Kingston who founded the Agricultural Society of the Midland District (ASMD) in February 1819 expressly refused to be subordinate to a York-based organization. Although they acknowledged the actions taken by the gentlemen in York, the "Merchants and Gentlemen" who founded the Midland District's first agricultural society chose to do so independently. Keeping "in view the vast extent of the Province," the new society informed the public that it did not "seem feasible" that the ASMD "should have any closer connection" with its counterpart at York "than that which an occasional correspondence may form."[42] This is not surprising. Although York was the provincial capital, Kingston remained the most populous and largest commercial centre of Upper Canada after the War of 1812. The conflict had generated a Kingston-centred boom, then bust, for military, commercial, and administrative operations. Approximately 250 new houses were built between 1812 and 1820, doubling the size of the town. However, the British troops garrisoned at Kingston, which had numbered 4,000 in 1814, numbered less than 1,000 in 1817 and, by 1824, only 478. This reduction affected the town economically and socially. Consequentially, by 1819, the mercantile and professional gentlemen of the town looked for ways to lead Kingston and the Midland District, of which it was the principal centre, out of economic distress.[43] The ASMD would facilitate this oligarchy's command of postwar renewal and improvement.

Strachan (a director of the UCAS) lamented the ASMD's "show of Independence" in a letter to his former student, John Macaulay, editor of the *Kingston Chronicle*. He believed that independence had been "purchased too dearly by sacrificing the best advantages" that he perceived to be "infinite" to Upper Canada. Strachan stressed that there had been no attempt on the part of the York society to profit from such an arrangement and explained to Macaulay that he had foreseen problems in a scheme that required Kingston elite to unite with those at York in an institution founded by the latter. Strachan claimed that he attempted to dissuade his fellow members from broaching the subject of union; however, "the benefits appeared so great & the impossibility of objection so obvious" to other members of the UCAS, and he was outvoted on the issue.[44]

While Gourlay was clearly the spark that ignited interest in founding the UCAS and other agricultural societies in the province, the ideology of improvement had spawned other similar associations elsewhere in

British North America. Agricultural societies had been founded in the districts of Montreal and Quebec during 1817, following crop failures throughout the province in the previous year. In response, the Lower Canadian assembly passed "[a]n Act for the encouragement of Agriculture in this Province" during its early 1818 session that offered £2,000 in public funds to the agricultural societies for use as prizes offered for best specimens of crops, machinery, or improved systems of farming. Legislators also recommended that an exhibition of improved breeds of horses, cattle, sheep, and swine be organized and advertised within each district, with prizes awarded. A subsequent act in 1821 offered another grant of money for similar purposes and for the purchase of books, seed grain, and models of improved implements, to the agricultural societies now established in Montréal, Trois-Rivières, Québec, and Gaspé. Gentlemen at York and elsewhere in Upper Canada were made aware of these developments by the *Upper Canada Gazette*'s publication of prize competitions offered by the "Quebec Agricultural Society" and, later, the prizewinners in January and February 1819.[45]

Likewise, the *Gazette* also informed its Upper Canadian readers of the agricultural society established at Halifax, with the governor of Nova Scotia – and Scottish improver – George Ramsay, Ninth Earl of Dalhousie, elected its first president, and the province's Chief Justice Sampson Salter Blowers elected vice-president. The earl, who possessed "an insatiable interest in agricultural improvement," donated £100 of the £350 offered by the 120 gentlemen who became members – a very competitive target for Upper Canadian improvers to attain.[46]

Upper Canadians were also aware of another agricultural improver from Nova Scotia, one of the most vocal about the task at hand for British North American settlers. John Young, writing under the pseudonym "Agricola," published his first letter from Agricola in the *Acadian Recorder* in July 1818. He championed the founding of local agricultural societies and, having once been an employee of Sir John Sinclair in Scotland when he was president of the British Board of Agriculture, it should not be surprising that Agricola also called for the establishment of a central board of agriculture. Although it is difficult to judge the degree to which Agricola's writings and arguments influenced the York elite in establishing a provincial agricultural society, his writings were available in Upper Canada. John Macaulay, editor of the *Kingston Chronicle* published selected letters of Agricola in his newspaper and acted as an agent for subscription to Agricola's publication of his letters in book form.[47]

Recent gentlemen emigrants from England to Upper Canada also understood the new form of improvement being championed by the gentlemen at York. For example, John M. Flindall wrote to the *Upper Canada Gazette* in April 1819 in response to the circular sent out by the UCAS announcing its formation. His letter made clear his understanding of the latest improved methods of agriculture and expressed his anticipation of the benefits that would result from their application to the Upper Canadian environment by the "truly patriotic endeavours" of those who were actively founding the agricultural societies. Prior to his emigration following the War of 1812, he had been a London bookseller, author, and associate of Charles Taylor of the RSA. He may also have been a member of that society. When his business collapsed in the economic crisis that accompanied peace in 1815, he relied on Taylor to provide him with the testimonial that helped him receive his grant of one hundred acres in Murray Township near Carrying Place in the Newcastle District. Soon after his arrival later in the year, Flindall began conducting agricultural experiments on his farm.[48] Although supportive of the cause of improvement, Flindall represented the general threat to the elites who had formed the UCAS. He, like others of this class of learned, well-connected, influential gentlemen who settled in the colony after the war, might be interested in improvement but saw no reason to offer deference to an oligarchy at York.

## The Cattle Show

While the root cause of the UCAS's creation languished in the Niagara jail, the agricultural society hosted its inaugural cattle show at York's market on 17 June 1819.[49] It was likely the first cattle show held in Upper Canada.[50] The UCAS offered first-, second-, and third-place prizes for competitions in categories for bulls, cows, and rams between three and five years of age brought for exhibition at the market. The Upper Canadian legislature was in session and the society's patron, Lieutenant Governor Maitland attended the show as did most members of the assembly and the councils. Thus, the cattle show provided the UCAS an important opportunity to display its pretensions to province-wide agricultural leadership and encourage leaders of the province's localities to return home and establish a branch society of the UCAS. As "a first effort," noted Secretary Horne, the exhibition "appeared to give much satisfaction" to the many York residents and "farmers from the country" who attended. Farmers "appeared to be deeply impressed with the importance and general utility of the association and expressed their determination to exhibit some fine cattle at the next show." Horne concluded glowingly, that "aided by the patriotism of the country," the UCAS "cannot fail to produce the most beneficial effects."[51]

Before examining the overconfidence contained in Horne's glowing report, we need to understand that, by hosting a cattle show, the York elite were emulating the agricultural improvers among the gentry of Britain. There, cattle breeding had become the very symbol of improvement in England. Shorthorn beef breeds such as the Durham had become particularly prized. In particular, the Smithfield Club's cattle show had become the premier event of its kind in Britain. The exhibition traced its origins to Francis Russell, Fifth Duke of Bedford, an original member of the British Board of Agriculture who devoted his life to managing his estates, which were already run in a highly efficient manner. In 1797, he began to host annual exhibitions of sheep shearing at his four-thousand-acre model farm at Woburn. Here, he had previously initiated a "pioneering experiment" in cattle and sheep breeding as well as methods of growing crops. His exhibition became an annual event and grew to include a variety of examples of improvements, such as ploughing competitions, for which he awarded prizes.[52] A year later, Bedford became the founding president of the Smithfield Club, which included fellow Board of Agriculture members, Sir Joseph Banks and Arthur Young. The society viewed itself as a national society dedicated to the improvement of the standard of British livestock breeds, and its name derived from the exhibitions of improved livestock it hosted in conjunction with the annual Smithfield Christmas cattle market of London. Yearly competitions hosted by the club offered substantial prize money for different classes of cattle, sheep, and pigs. By 1806, the competition between gentlemen had grown, as had the number of interested spectators, causing the exhibition to become its own separate event at a more commodious location in London.[53]

As Harriet Ritvo argues in *The Animal Estate*, cattle breeding had become the very symbol of agricultural improvement, for larger animals provided more beef for the rapidly expanding population of Britain. Yet, for all their claims to patriotism, she notes, the actions of the elite improvers of Britain were self-serving. They competed for prizes among themselves, had exaggerated forms of their prize-winning livestock sketched or painted for the walls of their estates or for the pages of agricultural periodicals, and spent their personal wealth on procuring breeding stock and raising it on their private estates. How they improved the beef inventory of Britain was unclear, for they mostly lamented that regular farmers did not join their societies, nor did they purchase or rent their cattle for breeding purposes. In fact, Ritvo singles out the Smithfield Club, noting how its events "celebrated and reaffirmed the position of the wealthy and powerful magnates who headed it, by parading the symbols of their magnificence in the form of extraordinarily large beasts. Their rhetoric of service reminded ordinary farmers that the men who could afford to raise prize animals were their natural leader, at the same time that the opulence of the display underlined the exclusiveness of high stock breeding."[54] The entry of prize-winning livestock established one's status as an agricultural improver, and the exhibition as a whole was based upon the educative value of "emulation." Farmers who viewed prize-winning livestock, it was believed, would return to their farms and work on emulating similar qualities within the livestock on their own farms by purchasing new breeding stock or employing the better-quality breeding stock of their neighbours.

At York, an advertisement published several weeks before the UCAS cattle show had encouraged farmers to become members so that they might enter their livestock for competition. But few farmers responded, and the competition for cattle was rather weak. In his report, Secretary Horne offered a rather poor excuse for this outcome, suggesting that it resulted from the "people of the town and country, each supposing that the other would bring forward a large number of these animals." In the end, the cattle show of June 1819 seemed to undermine the UCAS's claim to provincial leadership and even to any satisfactory degree of local authority.[55]

As a result, the UCAS undertook a third round of amendments to its constitution which, taken as a whole, suggest a contradictory approach to increasing support. While the society halved the price of annual membership again (this time to 5s) to encourage membership, it also increased the regulations for exhibition livestock at future cattle shows. Those who intended to enter competitions at future cattle shows were required to notify the secretary of their intent at least three weeks prior to the event.[56]

The revised constitution did little to alter the agricultural society's influence across the Home District or the province. Its second cattle show, held at York in May 1820, was its last recorded event. Horne claimed that the "Show of Horses and Bulls did great credit to the District" and noted the show was attended by a "very large number of Spectators, many whom came from a considerable distance" (this time, the legislature was not in session). But he lamented that "a larger number of practical farmers" had not joined the society "which was formed for their immediate benefit." Seemingly incapable of connecting the dots between the demands of the UCAS constitution and lack of farmer support, Horne also noted that several exhibitors had been turned away because they had not made known their intentions to exhibit livestock prior to the three-week deadline. In doing so, he struck on the true problem of the UCAS with no apparent self-awareness: "Hitherto, the Society has been almost entirely supported by the inhabitants of this town, who are only indirectly interested, from motives of public spirit."[57] Gourlay's banishment from the province eliminated the immediate purpose for which the UCAS had been formed and the society's strength rested solely on the paternalism the York elite and its closest associates to spread throughout the province. But it proved to be incapable of extending its influence throughout the farmers of the Home District, let alone across the province, where the local elite had outright refused to join as a junior district branch to the UCAS in York. The independence of the Kingston oligarchy's ASMD provides important insights into this substantial obstacle. Moreover, that society's model of cattle shows tells us much about Upper Canada's dual heritage, British and American, as viewed through the lens of the transatlantic world of agricultural improvement.

**The ASMD**

Thomas Markland initiated the call for an agricultural society. As one of Kingston's two principal merchants and a major landowner who was well connected to other prominent families of the town, the sixty-two-year-old Loyalist was a leading figure of Kingston society. He served as the Midland District treasurer and held several other administrative and judicial appointments. Although he possessed no direct ties to York, Markland had been "a firm supporter of the executive during the debates centering on Robert Gourlay." His son George, who was rapidly rising through the ranks of the provincial public service, had been a former student of Strachan's at Cornwall, and kept close contacts with the capital through his generation of provincial elite.[58]

Markland chaired the 8 February 1819 meeting at Kingston to discuss the proposed agricultural society, at which time a committee was formed to draft a constitution for presentation to a subsequent meeting scheduled for several days later in the village of Bath, just west of Kingston on the Lake Ontario shore.[59] The gentlemen who assisted Markland in framing the society's constitution reflected the commercial and naval character of Kingston more than they did the agricultural community farming in the town's hinterland. The Reverend Rowland Grove Curtois was chaplain of the forces at Kingston; the Reverend John Wilson was headmaster of the Midland District School at Kingston; Benjamin Whitney was a Kingston merchant; Alexander Pringle was a co-proprietor of the *Kingston Chronicle*; Anthony Marshall was a Kingston surgeon; and John M. Balfour was a retired army lieutenant who lived in the town.[60] None of these individuals were farmers; Markland's vast lands were held primarily for speculation. Nevertheless, the economic revival of postwar Kingston relied on the agricultural and merchant economy, and each gentleman could bring to an agricultural society their social, political, and economic influence that they believed would attract other agricultural improvers from across the expansive district.

Although the second meeting was intentionally held at Bath because it was in the heart of agricultural settlement to the west of Kingston, those in attendance elected the Kingston gentleman Allan McLean, the MHA for Frontenac and Speaker of the House of Assembly, as president of the ASMD. He was a Loyalist, like Thomas Markland, and was also one of Kingston's main property owners. He held various public offices and was involved in many other philanthropic projects with Markland and other Kingston gentlemen. Markland, himself, was chosen to be one of three vice-presidents, along with Thomas Alexander Fisher, a Kingston magistrate in the Court of Quarter Sessions, and James Cotter, the MHA for Prince Edward County. Markland's son George was elected Treasurer and Secretary of the society.

The ASMD constitution was ambitious. It called for a business committee to be composed of two representatives from each of the twelve townships in the Midland District, plus ten representatives from the more settled Kingston Township next to Kingston. Of the latter ten, most gentlemen belonged to the town, not the township. They included Whitney and Balfour, two of the gentlemen who had drafted the constitution, plus George Markland. Also chosen from Kingston Township were William Mitchell, a Kingston gentleman involved with Markland in establishing educational and banking institutions for Kingston; Lawrence Herchmer, a Kingston gentleman and justice of the peace (who would die later that year); Anthony McGuin, another justice of the peace, schooner owner, and road master; John Kirby, a merchant, road master and militia officer, who was involved with Markland on philanthropic community projects; Smith Bartlett, a Kingston merchant and brewer; and Micajah Purdy, a farmer in Kingston Township, road master, and individual of some prominence, having been a candidate for the provincial assembly in the 1812 elections. Among the list of locally prominent committee members from the other townships of the Midland District were three House of Assembly members: James McNabb, representing the riding of Hastings and Ameliasburg Township; James Cotter, representing the riding of Prince Edward; and Isaac Fraser, representing the riding of Lennox and Addington. They were joined by Allan Macpherson the man who was becoming known as the "Laird of Napanee," a village located in the County of Lennox and Addington. These locally prominent gentlemen were to distribute their portion of the two hundred copies of the agricultural society's constitution to solicit memberships to the society.[61]

Like Horne's publicization of the UCAS's activities in his *Upper Canada Gazette*, the ASMD received enthusiastic support from John Macaulay in the columns of his *Kingston Chronicle*.[62] Macaulay viewed the ASMD as an important aspect of Kingston's commercial greatness, and it fit within his vision of the Midland District's development. He championed the efforts of the organization, often berating the farmers of the district for not showing more interest in the society's efforts. His was an influential voice, for he was the scion of one of Kingston's earliest Loyalist settlers and one of its principal merchants. He had benefitted from the education and personal connections made while a student at Strachan's school in Cornwall, and his editorials made no secret of his support of the Upper Canadian elite and the maintenance of a hierarchical society. So, too, had his public opposition to Robert Gourlay, which may have included using his appointment as deputy postmaster to permit the removal of a mailed response to Gourlay's questionnaire from the Kingston post office before Gourlay could collect it.[63]

In the 2 July 1819 edition of his *Chronicle*, Macaulay wondered why the farmers of the Midland District had not offered their support to the new agricultural society, declaring they should be blushing at the "apathy and want of public spirit." Had the farmers of the district no interest in improving their methods of cultivation or their livestock? Were they not dismayed by the poor opinion of their wheat held in every market? Were they not upset that American farmers supplied Kingston with fresh meat? The "*badness of the times*, and *scarcity of money*" could not be reasons for failing to support the ASMD, he argued, for they were excuses that could be employed in any part of the province. England provided the best proof for why farmers needed to join the ASMD: "Societies have long been established … on the most extensive scale, whereby the Agriculture of that country has been carried on to a height of perfection, unequalled at any period by any other nation." Moreover, agricultural societies had been formed in the United States, too, he noted, "spreading rapidly over this vast continent."[64] "A Correspondent" soon answered Macaulay's questions, suggesting that the problem rested with the society's executive. Several of the ASMD township officers had yet to circulate their subscription lists to enlist new members. Macaulay, too, had since heard a "few slight objections" to the society's rules, but he begged the farmers not to allow the new institution "to languish at its commencement, and finally to sink into oblivion."[65]

A "District Show" held on 18 October 1819 at Adolphustown demonstrated an attempt by the ASMD to involve those from beyond the immediate Kingston area. Adolphustown was central to the Loyalist settlements in the western townships of the Midland District.[66] Macaulay reported that "a number of the most respectable Farmers" from the local townships had attended and suggested that others in the district at further distances from Adolphustown would have also attended had the show been on any other day of the week than Monday. He understood that several individuals "very properly declined leaving their homes to drive their cattle on the Sunday." In his next issue, Macaulay recommended the ASMD host cattle shows in each county of the district so that farmers could "drive their cattle to the place appointed without either inconvenience or expence [*sic*].[67] But the society was not capable of facilitating such changes.

By the end of the society's first year, only four of the thirty committee members had actively sought out memberships in their respective counties, providing the treasurer a total of £49 10s.[68] Despite a month's notice for the society's first annual meeting at Kingston's courthouse in February 1820, it had to be adjourned for want of members.[69] When the annual meeting was finally held in April, Secretary George Markland acknowledged to those assembled that the society's first year had been disappointing. The lack of membership received and the expenditure of prize money after the first cattle show forced the ASMD to cancel a planned second event for early 1820. To make the organization "more popular among the farmers of the District," members agreed to reduce the annual membership fee, and future annual general meetings would coincide with the April Quarter Sessions at Kingston in the hopes that it would supply better attendance. The ASMD lowered its ambitions to just one district cattle show per year, hosted anywhere in the Midland District the business committee deemed appropriate, though any county that raised £25 in subscriptions would be permitted to hold its own show.[70]

In its second year, the ASMD expanded the society's business committee to fifty members, or ten per county (Frontenac, Addington, Lennox, Prince Edward, and Hastings), and provided for one vice-president to be elected from each county. As a result, a different sort of gentleman stepped to the forefront of the society in 1820. Past president Allan McLean withdrew from office in the society altogether, replaced by former vice-president Alexander Fisher. But the major change in membership was made at the vice-president level. Benjamin Whitney was elected to represent Frontenac County, while Thomas Dorland (Lennox), Ebenezer Washburn (Prince Edward), John W. Meyers (Hastings), and Thomas Fenny (Addington) joined Whitney as vice-presidents. Both Washburn and Meyers had attended local meetings to draft a response to Gourlay's questionnaire (Washburn was chair of his meeting), and both had sons who had attended Gourlay's convention. In addition, both Washburn and Dorland, as MHAs for the ridings of Prince Edward and Lennox and Addington respectively, had been associated with Thorpe's opposition campaign prior to the war.

The election of the new vice-presidents emphasized that the Kingston elite's authority did not extend very far out in the Midland District west of Kingston. In that town, Gourlay may have been horsewhipped and prevented from receiving his mail, but the Scotsman's criticisms and queries struck a chord in the townships and settlements in the district. These locally prominent individuals did not possess the same worldview as the mercantile gentlemen of Kingston, and they had even less connection to the provincial executive at York. The sixty-four-year-old Ebenezer Washburn resided at Hallowell Bridge (today's Picton), the main shipping and mercantile centre of Prince Edward County, and was one of the largest landowners in Hallowell Township. He was a local justice of the peace, who had represented the people of his county in parliament from 1800 to 1808. Sixty-one-year-old Thomas Dorland had represented the people of Lennox and Addington in parliament from 1804 to 1812 and, due to his political leanings, had fallen out of favour with the provincial executive prior to the war. Serving as a captain of the Lennox Militia during the war certainly brought him back into favour, and he soon came to be known as one of the foremost justices of the peace in the Midland District. John Meyers's milling complex on the Moira River in Thurlow Township had become the nucleus for the community of Belleville. Owning more than three thousand acres of land and having used clay from his own farms to build one of the first brick houses in Upper Canada, Meyers was certainly a prominent gentleman of the district. his hospitality was famous, and his war record from the American Revolution was legendary. The seventy-four-year-old Meyers would help found the Hastings County Agricultural Society in May 1820; however, his contribution would be cut short by his death in late 1821.[71]

Evidence suggests that the ASMD survived through 1820, hosting a cattle show at Napanee in October. It claimed the event a success based on both the quality of competition and attendance, although admitted that no individuals from either Prince Edward or Hastings County had attended.[72] The ASMD's May 1821 general meeting (not

April, as had been the schedule set in 1820) had to be adjourned for a lack of attendance. When it was held later that month, members elected a new set of officers. Alexander Fisher retained the office of president to which he had been elected in 1820, George Markland returned to his post as secretary after a year's absence,[73] while Christopher Hagerman was chosen treasurer. The ASMD planned to satisfy farmers of the western counties of the district by planning a county cattle show in Belleville in early October, with the district cattle show to follow later that month in Bath. Despite these ambitious plans, the cattle show held at Belleville in October is the last record of any ASMD activity, and it is not clear if the district society was even in existence by that time.[74]

Instead, by a process that is unclear, the ASMD seems to have devolved into three county agricultural societies to continue the work of promoting improvement. Soon after the May 1821 ASMD annual general meeting, another meeting at Waterloo (present-day Cataraqui), chaired by Allan McLean, passed resolutions in favour of raising more funds for the ASMD by way of soliciting additional memberships throughout Frontenac County and for securing a cattle show for their county. The meeting reconvened in late June. Chaired by Markland, it adopted a constitution for a "Frontenac Agricultural Society." Markland was elected president, and the society's constitution, published in the report of the 30 June meeting, made no mention of the district society whatsoever.[75] Many of the same Kingston-area individuals elected to the ASMD executive from the county society's executive. Was the ASMD terminated in favour of county societies? Or did the new county societies erode the core support for the purported district institution? The scant records provide few answers. As noted, a Hastings County Agricultural Society was founded at Belleville on 9 May 1820, and an Addington County Agricultural Society was formed at Ernestown on 9 June 1821.[76] At its founding meeting, the Frontenac Agricultural Society scheduled its annual cattle show for Waterloo in October, but there exists no evidence of such a show hosted in 1821. A cattle show of 9 October 1822 featured President George Markland expressing his deep interest and attachment to Frontenac County during a lengthy address to those attending the event. However, the legislative councillor had been appointed an honorary member of the Executive Council, and shortly thereafter, he departed Upper Canada to spend several years in England.[77] A 12 October 1825 cattle show stands as the Frontenac Agricultural Society's last recorded event.[78]

**A Different Model of Cattle Show**

The agricultural societies spawned in reaction to Gourlay were short-lived, but they provide insight into the improving ideology as applied to the Upper Canadian world. All hosted "cattle shows" at which the competitions for cattle on show were given top billing. The ASMD declared that it did so because the war had "created so great a demand that scarcely any cattle were left to continue the breed." In fact, Macaulay argued for rewards to be provided to those who imported improved, better breeds of livestock into the district and that the ASMD require any prize-winning livestock be retained in the district for at least six months for breeding purposes.[79] As noted earlier, the UCAS modelled its livestock-only exhibitions on the cattle shows hosted by the Smithfield Club in England while, at the eastern end of the province, the cattle shows of the ASMD and the Agricultural Society of the District of Johnstown appear to have drawn on an American model. It was yet another factor that emphasized why the dream of a united provincial agricultural society centred in York was an impossible one to achieve.

That those in Kingston and Brockville would have looked to the United States for a model of agricultural exhibition is not surprising. Neighbouring New York state was a special example for the province, and its actions were watched with particular interest by Upper Canadians. As Jane Errington points out, the recent war had not seriously interrupted the exchange of ideas between the elite of Kingston and their counterparts in New York and New England. Great Britain was "the ideal to be followed," but the United States became Upper Canada's "immediate and constant point of reference. It was a yardstick which Upper Canadians frequently used to measure their own success."[80]

The most obvious model for the cattle shows hosted by the ASMD and the Agricultural Society of the District of Johnstown was the "Berkshire" agricultural exhibition, introduced by Elkanah Watson of Massachusetts during the previous decade. Watson sought to establish agricultural societies on a more democratic basis than the British model to make the cattle show "a distinctively American institution with an educational purpose." In October 1810, he hosted the "Berkshire Cattle Show" near his farm in Pittsfield, Berkshire County, Massachusetts and, in the years that followed, the Berkshire County Agricultural Society added an exhibition of domestic manufactures to its cattle show. Watson returned to his original home in Albany, New York, in 1816 and encouraged the establishment of county agricultural societies along his model. He was instrumental in securing the legislative establishment of the New York

State Board of Agriculture in April 1819 (an institution discussed in chapter 4).[81]

The ASMD's first cattle show offered first and second prizes for a bull raised in the province and owned in the district, a cow, a ram, an ewe, and a boar and single prizes for yearling steer or heifer, breeding sow, and three-year-old heifer. However, like the Berkshire model, the ASMD cattle show also offered prizes for the best samples of wheat, barley, and white peas; three prizes in a ploughing match; and a prize for the "best improved Plough, suited to the agriculture of the country."[82] Macaulay believed the list could be expanded even further, adding a wider range of competitions in domestic manufactures, as well as prizes for fruit and best-cultivated orchards.[83] The Agricultural Society of the District of Johnstown's first cattle show at Brockville, also held in October 1819, did just that. It provided a range of competitions for livestock, including best bull, cow, ox, stud horse, ram, ewe, boar, sow, and fattened hog. It also offered prizes for the best sample of home-manufactured woollen cloth, linen cloth, and six pairs of woollen socks. And it paid attention to crops, too, by offering prizes for the best field of winter wheat grown on "old land," field of Indian corn, field of barley, field of potatoes, cleaned flaxseed for feed, and cultivated farm.[84]

None of the Upper Canadian agricultural societies discussed thus far survived beyond 1825. Yet, significantly, these agricultural societies introduced the cattle show into Upper Canada, both the Smithfield and Berkshire models. When the provincial government established a system of district agricultural societies in 1830, the legislation to do so would be also drawn from a blend of British and American influences as to an agricultural society's purpose, the composition of its executive offices, and its place and role within the governance of a state. Spring and autumn cattle shows in Upper Canada, from 1830 onward, would also adopt the Berkshire model. Before analysing the origins of the 1830 agricultural societies legislation in the next chapter, a final section here provides more evidence from the late 1820s as to why the UCAS, led by the insular and defensive elite of York, had little chance of attaining or sustaining itself as a province-wide institution.

## The Agricultural Society of Upper Canada and New Local Societies

Newly appointed King's Printer Robert Stanton argued in an October 1826 issue of the *Upper Canada Gazette* that the "establishment and encouragement of agricultural societies throughout the Province" would be "the most effectual method" of diffusing information about "the most modern improvement in husbandry." He hoped that some "leading Farmers" would organize an agricultural society for the Home District and that those in other districts would show similar initiative. He pleaded with his readers to consider the example provided by Britain. "In the Mother Country," he argued in his first editorial, "the means which such Societies afford, not only of increasing Agricultural knowledge, but giving an impulse to the exertions of the Farmer, is fully demonstrated and acknowledged, and … they receive a general support from all classes of society."[85] Stanton was somewhat out of touch.

The health of agricultural societies in England and Scotland was a mixed record. The British Board of Agriculture had ceased to exist in 1822 and, despite its call for improved cattle breeds to feed the nation, the Smithfield Club struggled on in a very limited existence until revived in 1826. Elsewhere, the Highland and Agricultural Society of Scotland remained in full operation, hosting annual exhibitions, as did a myriad of local agricultural societies throughout the Scottish and English countryside.[86] All British North American examples were at a fragile state of existence. Lieutenant Governor Sir Howard Douglas had recently established a provincial agricultural society in New Brunswick, whereas the Nova Scotia government ceased funding of its central board of agriculture when its charter required renewal in 1826. John Young had been elected as a Nova Scotia legislator in 1825; thus, Agricola's earnest voice of improvement fell silent. Most county societies in Nova Scotia, spawned by the availability of public funds, also soon ceased operation. In 1825, the Lower Canadian legislature offered another one-time grant to that province's district agricultural societies, though the amounts were roughly two-thirds of what had been offered in 1821.[87] In the United States, the New York state government had recently cut funding to the network of agricultural societies it had fostered, and without financial support, most could not continue. By contrast, the Massachusetts Society for Promoting Agriculture, which could rely on the patronage of wealthy merchant members, was in a phase of revitalization during the fifteen years following the War of 1812.[88]

No one acted on Stanton's recommendations to establish a new agricultural society. But a year later, the editor made the same recommendation. During the recent assizes at York, he claimed that he had met "with several persons from the Country" and that they concurred that an agricultural society for the Home District should be organized. Stanton suggested that it be done at a meeting during the coming Quarter Sessions, "when a number of people from different parts of the Country will necessarily be in attendance at Court."[89] This time, however, Stanton's motives may have been less than altruistic, for his interest might have been piqued by the knowledge that, two months earlier, John Galt had founded the "Agricultural Society of Upper Canada" at newly established town of Guelph in the neighbouring district of Gore.

Galt is best known in Upper Canadian history for his involvement in the Canada Company's million-and-one-third-acre scheme to sell Upper Canada's Crown land reserves, primarily concentrated within the Huron Tract, but he had been a gentleman member of the improving world of England. As a member of the RSA, Galt had published his "Statistical Account of Upper Canada" in an 1807 issue of London's prestigious *Philosophical Magazine*. It was a report of information supplied to him from Upper Canada by his cousin William Gilkison who had immigrated in the 1790s. Upon Galt's arrival to North America in 1826, he travelled for a short time with the Earl and Countess Dalhousie and accompanied Dalhousie to a dinner of the agricultural society at Quebec. The governor-in-chief of British North America was an avid agricultural improver.[90]

On 12 August 1827, Galt established the Agricultural Society of Upper Canada at Guelph in the District of Gore, a town that Galt and the Canada Company had founded just months earlier as its headquarters for settling the Huron Tract. By the summer of 1827, the York oligarchy and their supporters were increasingly suspicious of what Galt's biographers call his "vision and purpose" and "welter of activity." They interpreted Galt's naming of his new society the "The Agricultural Society of Upper Canada" as a direct challenge, especially considering its membership, its political orientation, and the connection to the Gourlay incident a decade earlier.

Galt had encouraged forty-three gentlemen to become members, and using personal connections unavailable to most York gentlemen, he had three British agriculturists elected as honorary members of his agricultural society. The first was Thomas Coke, the most widely recognized agriculturist of his age and host of the famous annual Holkham sheep shearing (177[8]–1821). The second was Sir John Sinclair, renowned Scottish improver and former president of the British Board of Agriculture, and third was John C. Curwen, a well-known agricultural improver and member of the RSA who had founded the Workington Agricultural Society and a gentleman who had a major influence on the improvement of agriculture both in the county of Cumberland and on the Isle of Man.[91]

Galt's last choice for an honorary member is illustrative of his problems with the tory elite of York. Curwen may have been a prominent agricultural improver, but he was a Whig who, at times, was known to support radical causes in British politics. Similarly, Galt's political leanings troubled the tory elite of York. In Upper Canada, he associated with political reformers, and those who were socially out of favour with the York elite. One of the principal reasons for the elite's concern was Galt's unintentional connection with William Lyon Mackenzie, the reform-minded editor of the *Colonial Advocate* and sharp critic of the tory elite of York. When Mackenzie had learned of the Canada Company's creation, he had sent a complete file of the *Advocate* to Galt while on a visit to the province. Perhaps unaware that Mackenzie's editorials were already becoming a significant thorn in the side of the provincial establishment, Galt wrote a cordial letter to a fellow Scotsman. This resulted in a very cold response from Lieutenant Governor Maitland and others at York when Galt arrived in the province in 1826. To his horror, Galt learned that his letter had been used as evidence in the trial resulting from the destruction of Mackenzie's printing press and shop in June 1826 by sons of York's elite families. Although Galt tried to clear his name, the damage had already been done.[92]

The list of officers for Galt's new agricultural society would raise eyebrows at York, too. Many were gentlemen who had challenged the government elite at York with competing visions of Upper Canada's improvement. William Dickson, George Hamilton, and Tekarihogen (John Brant) were chosen to form a committee to manage the affairs of the society, while the society's treasurer was Thomas Smith, an accountant and cashier to the Canada Company.[93]

The Honourable William Dickson, now aged fifty-eight, may have been one of the magistrates who had ordered Robert Gourlay to leave the province, but the incident did not leave him unblemished in the eyes of some of the York elite. Dickson was the cousin of Gourlay's wife, and he had influenced Gourlay's views about the province and supported his efforts to publish a statistical account of the province. Once a member of the NAS and a cousin of its late president, Robert Hamilton, Dickson's wealth and prestige stemmed from his status in Hamilton's mercantile network, and his world was that of the Niagara Peninsula and his lands along the Grand River centred in the village of Galt, which he developed with Absalom Shade, another member of Galt's new agricultural society.[94]

George Hamilton had also been a supporter of Gourlay. A moderate reform member of the House of Assembly, Hamilton was at the zenith of his influence and fortunes when he joined Galt's agricultural society. His village of Hamilton, situated on Hamilton Harbour, was a rapidly developing commercial centre, offering stiff competition to the town of York. The Agricultural Society of Upper Canada's Secretary, Tekarihogen, was the Kanien'kehá:ka leader of the Six Nations of the Grand

River and resident of his late father Thayendanegea's home on Burlington Bay. Tekarihogen and Galt had worked together in 1825 to protest to the Colonial Office about the Six Nation's difficulties in settling land claims with the Upper Canadian government. Galt's sympathies for Indigenous peoples, his biographers note, also "hindered his acceptance into the narrow and partisan society of York."[95]

Despite its pretension to province-wide support, Galt's agricultural society proved to be ephemeral. Upon the arrival of Sir John Colborne to Upper Canada in 1828 as Lieutenant Governor Maitland's replacement, Galt requested his and the government's patronage of the Agricultural Society of Upper Canada. Seemingly, Galt and the other members had come to realize what previous provincial agricultural societies had already discovered: "without the countenance of Government, we can expect to attain little," Galt pleaded, "and yet with that, how much may we not hope to effect; where Patronage is so rarely claimed, and where it is so much wanted." His petition received no response. Moreover, the agricultural society's treasurer, Thomas Smith, had turned out to be more than just a Canada Company employee. He had been sent from England to investigate Galt's conduct and, in 1828, had returned to London with some of the company's papers. Galt himself returned to England in early 1829 only to be greeted by the embarrassment of being recalled from his position with the Canada Company. Further humiliation awaited him; he was soon placed in debtor's prison for being unable to meet his creditor's demands.[96]

Two new agricultural societies founded in the spring of 1828 were the products of retired British officers who had settled in Upper Canada following the end of the Napoleonic Wars. Connected to influential individuals in England and drawing a military pension, these gentlemen possessed the land grants and finances to set themselves about improving farms in a manner befitting a local squire. In March 1828, a meeting of gentlemen at a tavern outside the town of Perth founded the County of Lanark Farming and Agricultural Society for the seven townships of the county. It was chaired by a local miller, William Tully. The secretary was Henry F. Lelièvre, son of Captain Francois Tito Lelièvre, a former French naval officer who had transferred his allegiance and his frigate to Britain in 1798 during the French Revolutionary Wars. Another war brought him to Upper Canada, and he was involved in the defence of York against the American invaders in April 1813. When peace was declared, he was granted lands in Upper Canada. A testament to the military community at Perth, the March meeting ended with toasts to the king, the British army, the british navy, and Lieutenant Governor Maitland whom the new society attempted to secure as its patron. There is no evidence that Maitland provided such support, and he departed the province later that year. There is also little evidence that this society existed for very long or undertook any initiatives to improve the agricultural practices of the area.[97]

Two months later, a meeting of gentlemen at Cramahe organized the County of Northumberland Agricultural Society for the chief purpose of "spreading the theoretical and practical knowledge of farming among [the farmers of the county]." This was something that the society's president, Benjamin Whitney, claimed in his inaugural address was "much wanted" to close the wide disparity between most farms of the county and the few that were well cultivated by prosperous farmers.[98] A former executive with the ASMD, "Squire Whitney" had recently moved from Kingston to Cramahe under a cloud of accusation. When the Bank of Upper Canada had been established at Kingston in 1819, Whitney was elected its president. By August 1822, Whitney's conduct in that office was under investigation, although it was likely spawned by a personal spat between Whitney and the bank's cashier, Smith Bartlett (a founder of the ASMD). All charges were found to be groundless. Nevertheless, Bartlett resigned in September 1822, just days before the struggling bank collapsed.[99]

The vice-president of the agricultural society, Captain Francis Brockell Spilsbury, was a half-pay naval officer who had settled near the village of Colborne to create his estate, "Osmondthorpe Hall." Here, his biographer notes: "Like many of his class who had small incomes and pretension to become gentry, Spilsbury indulged an interest in agricultural improvement," although such actions "bore little practical relation to the economic realities of the colony."[100] While both Whitney and Spilsbury would die within a few short years, other members continued to advance the interests of the County of Northumberland Agricultural Society. It, too, made a cattle show the major focus of its efforts. The society's first such event, held in the town square of Colborne in October 1829, took the character of the Berkshire model, with its lengthy list of prize categories that ranged from livestock to butter, and from a ploughing match to an essay on the culture of wheat.[101] As the 1820s ended, it was the only agricultural society operating in Upper Canada, and it made no pretensions of being provincial in scope or authority.

In 1818, a clique of threatened Upper Canadian elite may have succeeded in banishing Gourlay from the colony, but they could not disregard the Scottish improver's call to address the lack of agricultural improvement in the struggling province. The general response to his actions proved Strachan's December 1818 assessment to be entirely misleading. Settlers of the "quiet Colony" had many grievances and raised them when someone who sought change inquired. Established by members of the provincial executive, the UCAS was both a response to the perceived threat Gourlay's criticism posed and an attempt to reassert control of the province's agricultural improvement by means of a cohesive, province-wide agricultural leadership. These efforts failed in short order once Gourlay had been banished from the province. The reasons for this suggested that Richard Cartwright's 1810 observations still had credence. Upper Canada of the 1820s continued to possess too few gentlemen with the leisure, finances, and willingness to properly support agricultural societies, no matter the social stature or political authority individual members of a society's executive might possess. The UCAS's attempt to assert itself as a province-wide organization, gathering local leaders under its administrative umbrella planted firmly at the capital, only served to highlight the lack of interest in being subservient to the tory elite of York. The UCAS and its contemporaries did, however, introduce to Upper Canada the cattle show as the chief function of an agricultural society. Yet, comparisons of the UCAS, the ASMD, and the Agricultural Society of the District of Johnstown highlight varied levels of influence by models of improvers elsewhere in Britain or the United States. As the decade closed, new immigrants with no ties to the colonial elite made the dream of a province-wide agricultural society even more unattainable. Despite the institutional manifestations of agricultural improvement in Upper Canada being mostly short-lived, provincial legislators throughout the 1820s and across the political spectrum agreed that agricultural improvement should be a function of government. How and why Upper Canadian legislators, both tory and reform, agreed on legislation in 1830 to fund district agricultural societies, and the renewal of public support during that decade and the 1840s is the subject of the following chapter.

# 4 Agricultural Societies as State Formation, 1821–1851

Early in the legislative session of 1830, Charles Fothergill rose in the House of Assembly to give notice of his intentions to introduce a bill "in aid of the Agricultural Interests" of Upper Canada and "to relieve our Farmers from the difficulties arising from the scarcity of money." Three days later, members began debate on "the best means to promote the institution and prosperity of Agricultural Societies, in this Province."[1] It's a legislative moment that has been mostly discounted. Fothergill's biographer, Paul Romney, terms the assemblyman's legislative efforts in this matter his "chief legislative legacy" and his "swan-song" from public office, suggesting that the agricultural societies' legislation that Fothergill sponsored is an act "nowadays thought to have done little to encourage agricultural progress in the province." Upper Canada's agricultural societies, Romney concludes, "stood as so many monuments to [Fothergill's] inability to understand the rank-and-file farmers he lived among in Durham County and claimed to represent," for they "did little or nothing to foster the technical innovations that their promoter had preached."[2] True, Fothergill's legislation provided little immediate relief to the agricultural or financial concerns of the province and he died in 1840, years before later amendments to his legislation began to gain the widespread, permanent support that he and other supporters hoped to secure in 1830. But Fothergill's legislation did establish the foundation of the central method by which the provincial government would support agricultural improvement through to Confederation and beyond.

His legislation provided an Upper Canadian solution to a failure in the leadership of agricultural improvement, and its passage highlights an important transition in the colonial government's approach to supporting that improvement. The success of Fothergill's bill in 1830 also provides important insights into an Upper Canadian legislature otherwise described as increasingly divisive and dysfunctional during the 1815–30 period due to a deepening rift between the elected House of Assembly and the appointed Legislative and Executive Councils, plus a growing opposition group of reformers that the Upper Canadian electorate voted to a majority of assembly seats in 1828, a first in the province's history. For these reasons, Fothergill's efforts need to be reconsidered in context of the transition his bill represented from older failed models of supposed provincial agricultural leadership; how the legislation bridged two models of an agricultural society, from England and from New York state; and its use of public funds to support the work of such institutions in the promotion and implementation of agricultural improvement. Critically, his legislation did not create a single province-wide agricultural society or board. Instead, it provided the opportunity for multiple agricultural societies to be founded and funded across the province. Later amendments to his legislation during the 1830s and 1840s demonstrate a mostly continuous commitment to agricultural societies as publicly funded organizations devoted to the promotion of agricultural improvement.

To expose and analyse this important aspect of Upper Canadian state formation, this chapter places Fothergill's 1830 bill in a broader legislative context, beginning in the early 1820s, arcing through his legislative moment in 1830, and onward to amendments made prior to mid-century, by which time the provincial government would establish a board of agriculture, drawing leadership of agricultural improvement ever more completely into the responsibilities and functions of the colonial state.

In 1821, the House of Assembly appointed a select committee "to take into Consideration the Internal Resources of the Province of Upper Canada in its agriculture and exports and the practicability and means of enlarging them, also to consider the expediency of granting encouragement to domestic manufactures." The committee's main recommendation was to investigate the feasibility of constructing canals to remove the barriers to shipping exports created by Niagara Falls and the treacherous rapids of the St. Lawrence River east of Kingston. In his analysis of the report, Robert L. Fraser argues the colonial gentry realized the important role of commerce in reviving the provincial economy, and canal construction represented their chief political and financial leadership in response to the post-war economic turmoil. By 1824, a private Welland Canal Company had been created, and after lengthy considerations, the provincial government began to invest large portions of its scarce finances into the project in 1826. A rudimentary canal carried its first ships in 1829, with multiple expansions and improvements to follow. After the completion of the Welland Canal, the provincial government assisted the construction of additional canals on the St. Lawrence River. Yet, the committee made other recommendations that provide important insight into the range of legislative supports considered to effectively promote agricultural improvement while supporting the construction of infrastructure.[3]

Robert Nichol, MHA for Norfolk, chaired the committee. He was a merchant, a former member of the NAS, and as noted in the previous chapter, he had become a vociferous opponent of the administration during the post-war provincial parliament (1816–20). In the new legislature of early 1821, Nichol continued his role as critic, including serving as chair of the committee on internal resources. Proposed by reformer William Warren Baldwin, it was the first such committee struck by the provincial assembly, and its report established a framework for economic recovery during a period of severe economic depression that had begun in 1819.[4] A central focus of debate during the last weeks of the early 1821 parliamentary session was the committee's chief recommendations to study the feasibility, routes, and cost of canals.[5] But canals required commodities to ship through them for export and the committee's report also recommended "the propriety holding out some Legislative encouragement" that might induce the province's farmers to turn their attention to the "important object" of cultivating hemp, flax seed, hops, and tobacco – all items "of ready sale in the British Islands."[6]

When Lieutenant Governor Maitland recalled the legislature in November 1821, to address "the general depression" of Upper Canada's agriculture and trade, his throne speech drew upon subjects contained in Nichol's report of the previous session: "I cannot but feel desirous it might be found possible to afford some effectual encouragement to the experiment of cultivating hemp and other production as would find a more constant and profitable demand in the parent State." He emphasized the critical need for the two chambers to cooperate "to create confidence in our measures and give weight to our representations."[7] Accordingly, the assembly appointed a seven-member special committee "to take into consideration the best means of encouraging the Growth of Hemp, Flax and Tobacco, and such other Agricultural Products of this Province as will find the most ready sale in the Mother Country and the British Colonies."[8] Chaired by Christopher Hagerman, MHA for Kingston, and an executive member of both the ASMD and the Fountenac County Agricultural Society (FCAS), the committee reported with a bill designed to promote the encouragement of hemp alone.[9] It did so despite the fact that, in 1818, the legislature had repealed an act that offered bounties to hemp cultivators. No claims had been made during the two-year existence of that legislation and the £1,000 once appropriated for that purpose had been desperately required for more pressing public expenditures.[10] Regardless, in January 1822, royal assent was given to "[a]n Act granting to His Majesty a sum of money for the purpose of purchasing and erecting machinery within this Province to prepare Hemp for exportation" that offered £300 to construct a public hemp mill that would process fibres for export, as well as £50 per year for three years to maintain the mill. Shortly thereafter, imperial authorities issued instructions for naval vessels stationed on the Great Lakes and the St. Lawrence River to be rigged with hemp grown in the Canadas, inviting the export of surplus supplies to Britain for use at dockyards there. Although the instructions spurred the Upper Canadian government to offer fresh encouragements to farmers to grow hemp and although an individual would use the public money to construct a hemp mill near Cobourg in 1830, once again, Upper Canadian farmers produced insignificant amounts of hemp. Moreover, shortly after the mill's construction, the Admiralty would close its naval facilities in Upper and Lower Canada as Britain shifted focus to land-based defences for British North America.[11]

Despite the failures of these initiatives, the Upper Canadian assembly's 1821 debates on the improvement of the province's economy raised a familiar theme: assemblymen of all political persuasions agreed that agricultural societies were valuable agents in fostering agricultural improvement across the province. Charles Jones, MHA for Leeds (and formerly president of the Agricultural Society of the District of Johnstown), believed agricultural societies were "the most proper" manner of leadership to attain the confidence of farmers. James Wilson, MHA for Prince Edward, concurred. He believed that a small sum of £50 "in the hands of an Agricultural Society, and well applied," could do more than £1,000 "improperly laid out." Although Henry Ruttan, MHA for Northumberland, believed that encouraging hemp "appeared rather speculative" and that the province should focus on developing its own domestic manufactures, he agreed with the member from Leeds about the value of agricultural societies to promote improvement and as the best method to encourage the development of domestic manufactures.[12] But this was all talk and no substantive action. Legislators appeared to be incapable of squaring their desire to employ agricultural societies as agents of government-led agricultural improvement initiatives with the fact that any agricultural society founded in Upper Canada had proved only that there existed insufficient private philanthropy to sustain such organizations at either a provincial or a local level.

Contemporary accounts suggest how far Upper Canada's farming practices were from achieving the idealized improved state of agriculture. Both visitors and residents found much to criticize and little to praise. Such reports, of course, need to be considered with a critical eye, for the authors tended to be promoters of the mixed-farming ideal that promoted livestock and crops, with the former producing the manure to provide the nutrients to grow increased harvests of the latter. Douglas McCalla notes that postwar Upper Canada was a colony that grew crops, particularly wheat, while Kenneth Kelly points out that mixed farming was an ideal promoted by British agricultural writers that could only take hold in Upper Canada among those who had the money to finance it. Keeping a field for pasture and erecting buildings to shelter livestock – and to collect manure – did not return profits in the short term. If a settler had livestock, the woods were where the animals grazed.[13] Thus, it is not surprising that authors who expected to see farms practising mixed farming as they travelled the province were disappointed. Nevertheless, their comments highlight the mission they saw ahead to improve the agricultural practices of Upper Canada.

In 1821, traveller John Howison criticized Upper Canadian farmers for having "no system in their agricultural operations." In his view, the "great barrier to improvement" was "an obstinate contentment and unmoveable fatuity, which would resist every attempt that was made to improve them." Other, less acerbic observers supported Howison's criticisms. Botanist John Goldie had observed Upper Canadian famers haying in 1819 and commented that he "did not think that a more effectual method could be adopted to rot the hay completely." From the Midland District, Kingston newspaperman John Macaulay would observe in 1822 that "too much capital and labour ha[d] been employed in the first settled townships in the clearing of land, and by far too little in the cultivation and improvement of the cleared land." He suggested it was "not unusual to see farms from eighty to a hundred and fifty acres, where one fourth, if not one third, of the land [was] lying waste, exhausted by a continued succession of crops without manure, and which, of course, after the expense of clearing, ceases to yield any profit to the proprietor."[14]

In 1824, several others would join the chorus. The Reverend William Bell of Perth asserted that the province's agriculture was "still in a very backward state, even in the old settlements." Edward Allen Talbot concurred, stating he had "never observed a single acre of land … that was so cultivated as to produce more than two-thirds of the grain, which, under more judicious management, it would certainly have been found to yield." And the Kings' Printer, Charles Fothergill, noted that in the

Newcastle District, "many examples are found wherein wheat has been raised on the same ground for 16 or 18 years successively without the application of manure."[15] In that same year, a new critic would emerge in the province, and he would launch a blistering and sustained attack on the Upper Canadian administration for its lack of leadership of improvements, agriculture and otherwise.

"On you alone, Farmers, does Canada rely," declared William Lyon Mackenzie in the first issue of his *Colonial Advocate and Journal of Agriculture, Manufactures and Commerce*, published at Queenston in May 1824. He planned for his newspaper to promote political reform but promised to also devote "considerable space" to the "useful art" of agriculture, by publishing items of "practical utility" for farmers. He told readers that he would carefully watch the progress of the art of agriculture in other countries and that, "in particular," he would "give due attention to the improved modes of farming from the United States. From these Americans," he argued, "we have much to learn, ere we can rival them in the practice of Agriculture." If led properly, Upper Canada could be every bit as developed as the United States and could easily rival its neighbour in agricultural production, he claimed.[16] Although the last portion of the newspaper's title would disappear by October, Mackenzie's "Journal of Agriculture, Manufactures and Commerce" remained a regular column in the *Colonial Advocate* until 1826, when Mackenzie considered changing the newspaper to an agricultural journal or selling it. He did neither. Instead, facing low circulation numbers and financial failure, Mackenzie moved his printing office to the capital at York. Subsequently, political matters took up most of his newspaper.[17]

Mackenzie's lengthy introductory issue of the *Colonial Advocate*, published 18 May 1824, is instructive because he devoted sixteen pages to an "Editor's Address to the Public" that outlined his opinions about Upper Canadian farmers, the state of agricultural practices, and the agricultural leadership in the province. Mackenzie, a Scotsman who had arrived in Upper Canada in 1820, was unquestionably well versed in with the improving ideology. Prior to his departure for North America at age twenty-five, he had been a member of a scientific society, and claimed to have read some 950 books.[18]

Mackenzie centred much of his commentary on a quote from Fothergill's recent speech to the electors of Durham in his campaign for a seat in the House of Assembly. Fothergill had claimed that "this fine country has so long languished in a state of comparative stupor and inactivity, whilst our more enterprising neighbours are laughing us to scorn." Mackenzie agreed wholeheartedly, criticizing both the province's agricultural practices and, in particular, the lack of agricultural leadership offered by the lieutenant governor. Major General Sir Peregrine Maitland was not an improving gentleman, he claimed, but "a right valiant and most excellent military chieftain," who had done nothing in recent memory of "a public nature worth recording." If Maitland was the patron of agriculture, Mackenzie chided, "it must be as one of the members of the agricultural society of London, for here he is as little known and as little spoken of, as the obscure Irish or Scotch emigrant, that may have landed a week ago." In sharp contrast, Mackenzie applauded the governor-in-chief of British North America, Lord Dalhousie, as a British improving gentlemen deserving of farmers' admiration. Mackenzie had seen his farm near Quebec and called Dalhousie "the best friend to Canadian agriculture, that England has ever sent to us."[19]

To be clear, Mackenzie underrepresented Maitland's actual efforts to reform landholding in the province and the lieutenant governor's requests to the assembly for measures to improve agriculture as a foil to amplify his praise of New York State Governor, De Witt Clinton. As of 1821, thirty-five of that state's forty-six counties had organized agricultural societies, and by 1824, Mackenzie could celebrate the connection of those county agricultural societies under a state Board of Agriculture established in 1819, all fully funded by the New York state legislature under Governor Clinton's guidance.[20] Upper Canada's neighbour had industry, domestic manufacturing, canals, and agricultural associations. But as for Maitland, Mackenzie queried: "What road has he made? What Canal has been begun in his time? Of what agricultural society is he patron, president or benefactor? What do the domestic manufactures of the Province owe him?"[21]

From his lengthy attacks, Mackenzie drew several important conclusions that, considering the ongoing, heated Upper Canadian political debates surrounding the Alien Question, must have irritated Upper Canada's tory elite as they read the first issue of this new periodical. He asserted that "a very great proportion indeed of our most and useful and effectuel artizans [*sic*] and agriculturists are to be found among the emigrants from the country that gave birth to a Franklin, a Washington, a Hamilton and a Clinton." Thus, he concluded, the Upper Canadian executive did the province no favours by blocking "industrious farmers from the States to settle in our country."[22]

Mackenzie drew deep from the rhetoric of the "agrarian myth," which Richard Hofstadter claims was near its zenith of acceptance in the United States at that time. The ideology of agricultural improvement in the United States, with origins in the English Enlightenment and the learned societies of London, had become heavily coloured with the republican ideology of the early nineteenth century in which the yeoman farmer was identified as the backbone of nation. By the 1820s, the agrarian myth was a creed that characterized the God-fearing, independent yeoman farmer as the foundation of republican society. He produced the primary source of America's wealth, he had been the first to defend his country, and he provided the religious and moral virtue upon which American civilization was based. A literary, not popular, notion, the agrarian myth had originated with the upper-class improvers and was given a wider voice by an emerging agricultural press. The moral fibre of the nation, it argued, rested with the yeoman farmer. On his self-sufficient farm, he communed daily with nature in an honest toil. His role was pre-eminent in the republic, for not only did he feed the nation, but the strength of commerce and industry also relied on, and profited from, his efforts. Farmers needed to be praised to keep them on the farm, because many farmers in the north-eastern states were leaving their farms for industrial or commercial opportunities in the cities of the eastern seaboard. Agricultural pursuits were upheld as the most morally pure, and raw commercial activity for financial gain was considered the least moral. For the upper classes who were involved in commercial and industrial endeavours, the agrarian myth provided a sort of soul cleansing that compelled them to promote improved agriculture either on their rural estates or as a member of an agricultural society.[23]

New York state, itself a model of the agrarian myth in motion, offered powerful and practical examples of agricultural improvement to Upper Canada. By the 1820s, it possessed a network of government-supported agricultural societies connected to a state agricultural association and an articulate agricultural press. The columns of several agricultural newspapers were filled with the rich rhetoric of the yeoman farmer and his position as the foundation of the American republic. Upper Canadians subscribed to these journals (a likely means by which the Berkshire model of a cattle show had found support in Upper Canada), and editors of Upper Canada's few newspapers had copied snippets of American editorials praising the yeoman into their columns of for some time.

Mackenzie "earnestly desired" to see "efficient societies for the improvement of arts and manufactures" established throughout British North America. The arts, science, manufactures, and commerce had advanced greatly of late in both Europe and America; thus, Upper Canada should lose no time or let "party spirit and narrow sectarian motives" interfere with the general diffusion of knowledge. He hoped that Upper Canada's political leaders might soon open their eyes to the agricultural and commercial advances of New York state so that "we may ere long behold Agricultural associations in Upper Canada; Agricultural improvements properly rewarded; Agricultural professorships, and a practical, yet scientific race of Farmers."[24] Although rooted in the American world, the ideas Mackenzie offered for reforming Upper Canada's agriculture were by no means radical.[25] Yes, he looked to the American republic for a benchmark of improvement that Upper Canada ought to attain, but his first editorial centred on the words of a conservative reform candidate in the province, Charles Fothergill, himself a scientific amateur and gentleman of the English Enlightenment.

Fothergill is an important figure to the history of agricultural improvement in 1820s' Upper Canada, for not only did he publish editorials and reprint columns about agricultural improvement in the *Upper Canada Gazette* years before an agricultural press developed in the province, but he also campaigned on the issue when he stood for election. As Romney suggests, Fothergill's "conservative reform" views "and his image of gentility and respectability [were] useful to the emergent reform movement at a time when many people still equated 'party' activity with disloyalty." A "savant with a sense of duty," as Romney describes him, Fothergill had been a gentleman of the English Enlightenment, a naturalist, and an author prior to his departure from England in 1817. (He was also an actor, and a profligate who had squandered his money on racehorse breeding.) Upon his arrival to Upper Canada, he settled near Smith's Creek (today's Port Hope) on a sizeable land grant he had received from Lieutenant Governor Gore for the purposes of founding a settlement of English gentlemen. As with many of his projects, the settlement failed to materialize. He applied for the position of King's Printer, and in 1822, he moved to York to undertake his first public appointment. But his opposition to the provincial executive resulted his removal from office in early 1826. From 1825 to 1831, Fothergill served as the MHA for Durham, and it had been his initial election campaign motto "AGRICULTURE and INTERNAL IMPROVEMENT, without the aid of those who EAT more than they EARN" that attracted the attention of Mackenzie in 1824. As an MHA during the late 1820s, Fothergill became recognized as a leading member of the opposition members of the legislature who were becoming known as "reformers."[26]

The provincial election of 1828 returned to the House of Assembly the first majority of candidates who were in opposition to the provincial executive. This moment is generally viewed as the start of a deepening and bitter political feud between tory and reform factions, with little being accomplished before the session came to an early end in 1830 upon receipt of news of George IV's death.[27] Indeed, the House of

Image 4.1. Charles Fothergill (1782–1840), 1834, Grove Sheldon Gilbert. (Courtesy of the Royal Ontario Museum 960.12.)

Assembly spent much of its time debating fifty-three bills that were ultimately rejected by the Legislative Council. Yet, it was in 1830 that provincial legislators, both elected and appointed, both tory and reform, recognized the utility of government-legislated agricultural societies as publicly funded agents of improvement for the colony. By agreeing on this matter, the colonial legislature transformed failed attempts at founding a province-wide agricultural society into an Upper Canadian method of making agricultural improvement a state responsibility.

Fothergill initiated a House of Assembly debate on "the best means to promote the institution and prosperity of Agricultural Societies," by way of a series of resolutions he presented on 12 January 1830. Outlining the basic tenets of English Enlightenment thought, he argued that improved agriculture was at the heart of a well-ordered agrarian society. Therefore, the "importance of a successful and well directed system of agriculture must be obvious to all, because it lays a foundation, a superstructure for all future wealth, prosperity and grandeur. It is the foundation of society." Agricultural societies brought "together people from different parts; and the various means for promoting the grand objects of the whole, are made known to all, so that great good results to the community." Fothergill noted that in his experience, "he had seen very great and beneficial results from such institutions in England and he was satisfied like benefits might be realized from the same means in this country." But in many parts of Upper Canada, he continued, the establishment of these useful organizations was not possible, because "the people are poor, and find themselves unable to carry into effect the object of their wishes." He hoped the adoption of his resolutions would result in the government providing grants of money to farmers in each district of the province for this purpose.[28]

At the time, the County of Northumberland Agricultural Society was the only such organization operating in Upper Canada. Elsewhere, continued support for such organizations was mixed. New Brunswick's agricultural society still existed, while in Nova Scotia only the King's County agricultural society remained active. Prince Edward Island's lieutenant governor, John Ready's call for a Central Agricultural Society in his 1827 throne speech had resulted in its establishment at Charlottetown, and it would spawn some thirteen societies on the island between 1827 and 1842. In Lower Canada, the legislature had again offered one-time funding in 1829 to each district agricultural society in the province while expanding their roles to include oversight of any county agricultural societies that may be founded in their district. It also created honorary memberships in respective district agricultural societies for provincial legislators and clergy. These changes, along with the continued requirement for district agricultural societies to report to government officials as to how the government grant had been spent, bound these societies tightly to the Lower Canadian state. New York's network of agricultural societies that had received Mackenzie's praise in 1824 had seen their public funding cut by the state government in 1826, and without this support, many ceased to exist.[29]

Fothergill claimed there had been no objection to the bill among those he had consulted, only debate as to the amount of money to be granted. But William Warren Baldwin, the reformer representing Norfolk, echoed the words written twenty years earlier by Richard Cartwright. He believed that "the lands were not sufficiently cultivated to require the aid of these societies; nor did he think as much benefit would result from them here, at the present day, as in those old countries which were under a high state of cultivation." This expenditure of government funds would embarrass farmers instead of benefiting them, he claimed, "and thereby be injurious to the country." But

Baldwin's was a minority point of view. Cartwright had written his comments in a pre-war colony of some seventy-seven thousand settlers. By 1830, Upper Canada contained more than two hundred thousand people and was on the cusp of adding some hundred thousand more settlers in the next four years. Representing what was formulating into the majority opinion, James Wilson, a reformer representing the riding of Prince Edward County, "had not any doubt of the propriety of any measure that would call up the country from the slumber into which it had fallen ... the outlay [would] pay the government well ... grant this money to the agricultural societies and it will change the face of the country directly."

How might the proposed agricultural societies be administered? Some looked to the old, failed model of a central board of agriculture. But notably, this was not the only proposal under discussion as it had been in previous years. Others wanted local organizations to be established across the province. As for any standard rules of operation, Fothergill "apprehended it would be impossible to establish any general rule for the regulation of all the societies," and James Lyons, another reformer representing Northumberland, defended the riding's county agricultural society of which he was a director: "[T]he prosperity of all societies and institutions depended on their being established on liberal principles," he argued. To be successful, "they must be untrammeled."

Then came the question of funding. Mackenzie, now the MHA for the Fourth Riding of York, rose to speak in favour of a sum given to each district in accordance with its population. "Lower Canada had adopted this method, and it had proved very beneficial," he noted. But later, he would tell the committee that he had been persuaded that distribution by population "was a mistake." Doing so would deny money to the most recently settled back townships of the districts, thus those in most need of assistance.

The debate over funding settled questions about administration. Fothergill suggested that £100 be set as the amount available to each of Upper Canada's eleven districts on "certain conditions." Not impressed with the idea of a central board or establishing "any general rule for the regulation of all societies," he proposed that the money be placed with Lieutenant Governor Colborne, "whom he knew to be a warm friend to agriculture," and all applicants for funding would have to submit a request "with satisfactory evidence" as to why they were entitled to funding. Primarily, that evidence would be proof of a properly constituted society in the form of a receipt for £50 already raised in membership fees.[30]

At the conclusion of this initial debate, the committee of the whole reported several resolutions in support of government funding to agricultural societies. The first stated the ideology of improvement:

> Resolved, That as a prosperous agriculture is the broadest and firmest base of national strength and wealth, it is highly expedient to promote the welfare of all such institutions as may tend to the accomplishment of that desired object.

The second resolution highlighted the significance of the English, American, and British North American examples:

> Resolved, That the experience of England and other enlightened communities, having shewn that the institution of Agricultural Societies in different sections of the country has greatly promoted the success of the farmer, it is wise and politic in this Legislature to give all the encouragement in its power to such Agricultural Societies as are, or may be, founded on a liberal footing within this Province.

Significantly, the committee asserted the role of the legislature in the leadership of agricultural improvement by offering money to each district once its agricultural society had been established:

> Resolved, That if an appropriation, not exceeding one hundred pounds, in aid of such Societies, was made for each and every district of this Province, it would materially assist and induce the importation of valuable live stock, grain, grass seeds, useful implements, &c. &c. &c. but in no case shall such appropriation be made until the society applying for it shall make it clear they have themselves subscribed and paid into the hands of a regularly appointed treasurer, a sum equal to the sum of fifty pounds.[31]

In sum, the House of Assembly turned away from failed initiatives such as the attempts to encourage the growth and export of hemp and came to an agreement that leadership of agricultural improvement was a role for the Upper Canadian state to play; that agricultural societies should be the semi-public agencies by which that improvement would be promoted and implemented using public funds to assist in the acquisition of good quality seed, better breeds of livestock, and new labour-saving machinery; and that each local society across the expansive province should be provided the independence to determine how the practice of agriculture might best be improved in its specific locale. In the absence of landed gentry with sufficient time and money to form and lead such agricultural societies, the assembly drew upon the New York state and Lower Canada model of employing public funds to establish these new agents of agricultural improvement. The provincial secretary, on behalf of the lieutenant governor, would receive each agricultural society's annual grant application and transfer to it the public funds.

A final resolution requested a select committee be created to draft a bill that incorporated the first three resolutions. The House selected Fothergill, along with Lyons from Northumberland, a director of its agricultural society; and Thomas Dalton, a reformer who had served as the secretary of the FCAS at its inception in 1821. Recently, he had launched *The Patriot and Farmer's Monitor* in Kingston, a newspaper with the motto of "Common Sense" and a front page devoted to the "Farmer's Page."[32] The committee shepherded their bill successfully through the reform-dominated House of Assembly. Funding was not permanent. However, in each of the next four years, a district agricultural society could request £100 from the lieutenant governor after it had collected a total of £50 in membership fees. (The four-year limit on funding was a common approach to the spending of public funds.) In contrast to many other bills generated and approved by the reform-dominated House of Assembly, the Legislative Council approved the bill without delay or amendment, and on 6 March 1830, Lieutenant Governor Colborne provided royal assent to "[a]n Act to encourage the Establishment of Agricultural Societies in the several Districts of this Province."[33] Fothergill, the conservative reformer, embodied the political bridge between tory and reform necessary to secure its passage. The bill incorporated a workable plan that drew on the ideal of the British gentry's agricultural societies and the New York state model of state-sponsored agricultural societies. It managed to address criticisms raised by the reformers, but it did not raise the hackles of tories who were keen to employ colonial institutions to maintain the social and political order of a Georgian agrarian society.

How does one classify these new agricultural societies? They were to be created by private citizens according to no set institutional format but were eligible for annual public funding. There were no legislative requirements as to how a society was to be formed or if it required a constitution, vague details on what it should do throughout the year to encourage agricultural improvement, and little accounting required for it to report how it spent any public funds received. They would not be required to be junior branches of any central agricultural institution established at the capital; they would not even be required to interact or coordinate activities with neighbouring district agricultural societies or those elsewhere in the province. In effect, the provincial government established itself as the central agency with the legislated role of funding the agricultural societies' implementation of improved agriculture with minimal oversight beyond the provincial secretary's role. Thus, district agricultural societies established by the 1830 legislation became semi-public, arm's-length agents of the provincial government charged with directing agricultural improvement. Amplifying their semi-public nature would be those members who filled the executive offices of the agricultural societies who were also elected or appointed members of the Upper Canadian government. Remarkably, in only one instance (explained in chapter 5) would the reform and tory visions clash over who best should found and lead a district agricultural society.

Only one agricultural society in Upper Canada established before 1830 had to be accommodated within the terms of this new legislation. The bill specified that in such districts where agricultural societies had already been formed at the county level, government funds would be divided equally between the district and county societies. Yet, competition between local oligarchies meant this accommodation could not be accomplished that simply. About three months after his bill became law, Fothergill wrote to the provincial secretary, Zachariah Mudge, notifying him that individuals in the Newcastle District were set to organize an agricultural society. However, the County of Northumberland Agricultural Society refused to unite with the Durham County society under formation. Fothergill, as MHA for the Riding of Durham, expressed to Mudge his fears that the Northumberland society, already prepared to claim its government grant, would present itself as the district agricultural society of the Newcastle District, denying the nascent Durham County society the ability to secure any public funding. Lieutenant Governor Colborne provided a very King Solomon–like solution, offering each county £50 of the yearly grant. It highlighted the extent to which it remained difficult to unite local leaders at the district level, let alone unite those of the entire colony into a single organization, and perhaps explains why requirements for interaction between districts had been left out of the legislation.[34]

Colborne certainly understood the task of agricultural improvement. He had arrived in Upper Canada in 1828 having served as the lieutenant governor of Guernsey for seven years. During his tenure on that island, he had aided agricultural improvements, the development of domestic manufacturers, and introduced many infrastructure reforms.[35] Demonstrating his appreciation for the role that agricultural societies might play in the state's attempt to foster improvements, he sought assistance from the newly established societies in November 1830. He asked each agricultural society to provide input on his suggestions concerning the

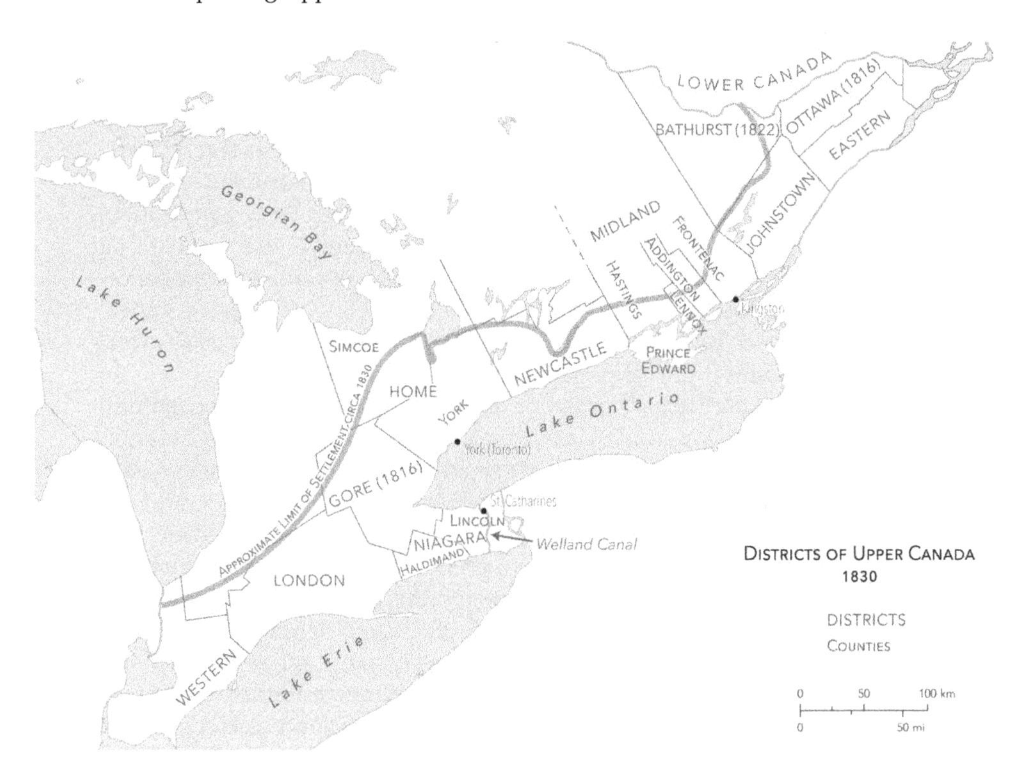


Map 4.1. Map of Upper Canada's districts, 1830, showing counties contained within the Niagara, Home, and Midland Districts. Map produced by Julie Witmer Custom Map Design.

poor state of the province's roads, particularly how they might be better maintained and repaired. He thought the matter to be under the purview agricultural societies because "[t]he system of forming Roads and keeping them in repair by Statute Labor takes the Proprietor or Cultivator from his important occupations to perform work which could be more effectually executed by other means; consumes a considerable sum annually; and produces a result; unfortunately too well known – Roads which destroy the cattle of the Farmer, break his Waggons, and waste his valuable time." Colborne requested each society consider the possibility of having the statute labour commuted by the payment of a rate to each district to be spent on road construction and repairs before mid-January 1831. A report of the House of Assembly's Roads and Bridges Committee indicates that it received responses from eight agricultural societies of the province's eleven districts.[36] But Colborne's request would prove to be a rare example of provincial authorities making direct inquiries to district agricultural societies. Instead, each district agricultural society launched its own activities, principally the hosting of one or two cattle shows per year. While reports were submitted regularly or infrequently to a local newspaper to publish, the only communications with the government, beyond the yearly request for funds, were petitions requesting a continuation of annual funding at such times when the legislated period of funding neared its end.

## Legislative Developments, 1830–1851

The legislative history of Upper Canada's agricultural societies is an overlooked aspect of what Gerald Craig terms a "mania" for internal improvements during the 1830s. Such improvements were conducted using a provincial treasury with minimal resources and, not surprisingly, they generated a sizeable provincial debt. By Douglas McCalla's calculations, the provincial debt underwent a thirty-four-fold increase during the 1825–37 period, and a fiscal crisis in 1837 compounded the province's debt problems, making interest payments on debt almost equal to all other provincial expenses combined.[37] Funding for less tangible improvements such as that provided to district agricultural societies was in competition with funding for construction and maintenance of canals, ports, bridges, roads, and other critical infrastructure for a province witnessing a substantial population growth that stemmed from increased and sustained immigration to Upper Canada.

As the four-year period of annual grants approached its end in 1833, several district agricultural societies petitioned the government for the annual funding legislation to be extended.[38] In December, the conservative-dominated assembly, elected in October 1830, did agree to a first reading of a bill to extend annual funding to district agricultural societies but proceeded no further. Later in the session, it established a committee to examine expiring legislation, which identified the agricultural societies' act. But it made no recommendation for renewal.[39] With the end of public funding, the province's semi-public district agricultural societies were left to continue as private institutions reliant on their own funding to continue operations. Most struggled to do so.

New elections in October 1834 returned the province's second reform-dominated assembly. It was certainly sympathetic to the cause of improvement, adding the largest single increase in the public debt, some £400,000, in 1835.[40] Charles Duncombe, the reform MHA for Oxford, a physician, and a landholder of sizeable acreage, announced in February 1835 that he intended to introduce a bill to revive the expired agricultural societies act. This did not happen, although he did chair the Assembly's committee on expiring and expired laws, which recommended offering district agricultural societies a one-time grant of £1,200 (£100 for each of the twelve districts in the province). A new bill was tabled, and by mid-April 1835, it received support of both assembly and council. The legislation provided just one year of funding and shortened the arm's-length connection between government and each district agricultural society by requiring the latter to become more accountable as to its use of the public grant. Each would have to submit an account of previous expenditures with its petition to qualify for the one-time government grant.[41]

By the time this act expired, Duncombe had gained the reputation as "one of the busiest and most productive members of the house."[42] He again championed the act's renewal by requesting the appointment of a select committee to consider "[a]griculture and the improvement of the breeds of animals and seeds of grains [and] trade and manufactures." Chaired in January 1836 by John Gilchrist, MHA for Northumberland, the nine-member committee was composed of reform members from across the province. Eight were farmers, while three were also merchants. Two members were executives of their district agricultural society.[43]

In the preamble to its recommendations, the committee reiterated why it behoved the provincial government to encourage agricultural improvement, stating the oft-repeated Enlightenment view that agriculture was "the only true and solid basis on which the permanent prosperity and wealth of most nations must rest." In a colony such as Upper Canada, the report argued, "the encouragement of agriculture in its various branches should claim the particular attention of the

Legislature" because "manufactures and the mechanical arts" did not "add much to the general wealth" and probably would not for some time to come. While it admitted that commerce remained of primary importance, the committee recommended that the provincial government secure the easiest and cheapest conveyances for agricultural produce to the best markets. It also asserted that the legislature should diffuse "a knowledge of the best manner of conducting agricultural operations so that the greatest quantity and best quality may be produced by the least labour and expense." This included assisting farmers to improve their livestock, crops, and agricultural equipment.

The committee maintained faith that agricultural societies provided the best leadership for agricultural improvement despite admitting that money given to the district societies had been used primarily for premiums offered at society-sponsored exhibitions and that the recipients of the prizes were usually farmers who resided near the competition. Farmers at the outer edges of the districts were not receiving the benefits of the public funds in equal proportions. Although the "benefits anticipated" from the legislative grants to the district agricultural societies had "not been realized, nor produced that good which might reasonably have been expected to result from such liberal appropriations," Gilchrist's committee argued this was a reason to improve on the model rather than abandon it entirely.[44]

Recognizing how widely scattered the province's farmers were across most districts, the committee recommended agricultural societies be established at the township level. Many farmed at a considerable distance from the principal town of the district and were highly unlikely to travel to an agricultural society meeting or a cattle show located at the district town, let alone pay a membership fee or enter livestock, produce, or domestic manufactures for competition. Therefore, funds should be granted based on the population of each township and the amount that such a local society could raise in subscriptions. Such changes were sensible, for the population of Upper Canada had swelled by some 160,000 inhabitants beyond the population of 213,000 when the original legislation had been passed in 1830.[45] Immigration had pushed the settlement of the province further into the interior, and the number of partially settled townships at the outer edges of a district increased. The province was already responding to this situation by creating new, smaller districts out of the original expansive ones, thus improving local administration by shortening distances between a district's boundaries and its administrative centre.

Gilchrist's committee also recommended that the public funds granted to agricultural societies be used to purchase "the best and most approved seeds of grain, and grass and breeds of livestock," as well as "some practical works or treatises" on agriculture for circulation among agricultural society members. In particular, the committee recommended William Evans's *Treatise on the Theory and Practice of Agriculture*. A strong supporter of agricultural societies, Evans was one of "the most dynamic farmers in the Montreal region" and the secretary of the Agricultural Society of the District of Montreal.[46] It also recommend the *Genesee Farmer and Gardener's Journal*, a periodical that Luther Tucker had been printing in Rochester, New York, since 1831. In the absence of any domestic agricultural periodical, the committee was likely recommending one already being read by Upper Canadian subscribers. Not only did the *Genesee Farmer* contain practical farming advice, but it also advocated government support of agricultural improvement. Tucker and Jesse Buel, editors of *The Cultivator* in Albany – another journal with Upper Canadian subscribers – promoted the establishment of agricultural societies and advocated exhibitions. In 1834, *The Cultivator* had become the official voice of the New York State Agricultural Society (NYSAS), established at an Albany convention two years earlier (although by 1836, the newspaper was struggling to maintain its existence).[47]

After Gilchrist's committee tabled its report, the House agreed to a resolution that the government should provide permanent annual public grants "not to exceed in any instance, the sum of twelve pounds ten shillings to any one township" and subsequently passed a bill "to establish Township Agricultural Societies." The Legislative Council disapproved and rewrote the entire bill. It cancelled the creation of township societies, thereby maintaining the district as the sole administrative unit for agricultural societies in the province, and it cancelled the permanent funding provided by the bill, replacing it with just four years of support. Returned to the House of Assembly for its concurrence, the bill received no further consideration, for there was little opportunity for legislators to do so, a consequence of the ongoing political squabbles within the Upper Canadian government.[48]

The Legislative Council's rejection of the shift to township-based agricultural societies represents a larger effort by the provincial administration and the tory elite that populated its offices to hold on to the structures of a well-structured Georgian agrarian society in the face of reformers pressing for fundamental alterations to the structure of Upper Canada's government and the role it played in fostering the colonial economy and society. Members of the upper chamber were not disinterested in agricultural improvement; they killed the bill because of its broader implications. A shift to township agricultural societies represented a real threat to the central authority of the provincial administration as it would undermine the provincial elite's efforts to create a unified provincial leadership flowing out from the capital to the supportive local oligarchy leading each district. The threat was real, for at the very time that the agricultural societies bill was in the hands of the upper chamber, the House of Assembly issued an address calling for responsible government. To force its demands, the Assembly voted to not pass a supply bill to finance the provincial administration. In response, on the day following the Legislative Council's entire rewriting of the agricultural societies bill, Lieutenant Governor Sir Francis Bond Head prorogued the legislature (he had arrived in January to replace Colborne). A month later, he dissolved parliament for new elections.[49] Such superheated politics meant that, for the second time in three years, district agricultural societies became private institutions dependent on memberships and donations to survive.

The elections of July 1836 returned a majority of tory representatives to the House of Assembly, thanks in large measure to the political campaigning by the lieutenant governor to secure a favourable outcome. Regardless, the campaign for renewed government support for the district agricultural societies passed over to the conservative benches, as the rookie MHA elected for Frontenac, John Bennett Marks, became the tories' champion of this cause. Marks served as a vice-president of the Midland District Agricultural Society (MDAS) by way of his role as president of the Frontenac County Agricultural Society (FCAS). When the legislative session opened in November, he announced his intention to introduce a bill "to provide for the encouragement and improvement of Agriculture in this Province, by establishing under certain regulations, Agricultural Societies in each County."[50]

Several days later, Marks initiated his request that the government grant "a sum of money in aid, and for the encouragement of Agriculture in this Province." The assembly concurred and appointed a Select Committee on Agriculture formed of Marks and three other members. George H. Detlor was the MHA for Lennox and Addington and a director of the MDAS. Edward W. Thomson was the MHA for the Second Riding of York and president of the Home District Agricultural Society (HDAS). John Prince was the MHA for Essex and a lawyer and magistrate who lived on his "Park Farm" estate near Sandwich (there is no evidence that an agricultural society was operating in that district). All were conservative assemblymen, and they drafted a bill more acceptable to the Legislative Council than the one generated by Gilchrist's committee in 1836. The bill "to establish Agricultural Societies and to encourage Agriculture in the several Districts of this Province" announced that district agricultural societies were to be provincial government's chief manner of encouraging agricultural improvement, particularly "for the purpose of importing valuable live stock, grain, grass-seeds, useful implements of husbandry or whatever else might conduce to the improvement of agriculture." It received prompt approval from the assembly, and when sent to the Legislative Council, it received support without amendment.[51]

Whereas the bill promoted by Gilchrist in 1836 had proposed – and, to the tory elite, threatened – to eliminate district agricultural societies in favour of new township societies that might be led by those with little or no connection to the provincial elite, Marks's legislation offered a more suitable compromise. It expanded and deepened the influence of a district agricultural society within a district, while at the same time, it drew these organizations closer into the functions of the Upper Canadian state. The act permitted a single agricultural society in a county, riding, or township, but such local societies could receive government funds only through its district society, which the act provided authority to distribute grants it received in proportion to the total membership fees each local society had raised. In turn, each lower-tier society would have to deliver its financial accounting to the district society's treasurer for submission as part of the district society's annual accounting and application for government funds. To petition for the annual grant, each district agricultural society treasurer was required to complete and submit a standardized form of accounting for the previous year's expenditures along with a standardized certificate stating the amount of funds raised by his agricultural society.

The act also halved the minimum amount that a district agricultural society needed to raise to be eligible for the annual government grant to £25. An eligible society would now receive double the amount raised to a maximum of £200 per year. In the short term, the changes made it easier for a district agricultural society to revive its activities following a year bereft of government funding.[52] The long-term results of Marks's legislation turned out to be quite profound.

By the time the agricultural societies act expired in 1841, Upper Canada had weathered a significant political crisis. Rebellions in Upper and Lower Canada of 1837–8 were quashed, and in their aftermath, imperial authorities united the two colonial legislatures into one that represented a new United Province of Canada, effective 10 February 1841. Facing a complex task of coordinating legislation from each province in a politically heated atmosphere in which few wished to cede their former independent identity, legislators of the inaugural parliament of the Canadas chose to simply extend provisions of numerous pieces of expiring legislation until the end of the parliamentary session that followed 1 November 1844. In fact, separate legislation for each province remained a not uncommon approach to legislation within the united Canadas, for often preparing and securing political support for unified legislation was simply not possible.[53] Upper Canada's 1837 agricultural societies legislation was one such act extended in this manner. Lower Canada had its own system of agricultural societies, and this was not the political climate in which to debate unification or even coordination.[54] As a result, Upper Canadian agricultural societies received, by way of Marks's legislation, an uninterrupted, seven-year stretch of government funding. It was the longest period of stable support agricultural societies had received.

While debate about the future of agricultural societies in both halves of the province was on hold, the civil secretary's office made sporadic use of district agricultural societies as sources of information on the agricultural state of the province. In 1840, the secretary sent a questionnaire to the district sheriffs and presidents of district agricultural societies to obtain information for use in encouraging immigration to the province. His questionnaire was divided into two parts: the first with twenty-eight questions and two tables of prices for agricultural produce and livestock to complete for the purpose of informing "Emigrants with Capital" and the second with thirty-seven questions and two tables on wage rates and retail prices of provisions and clothing to complete for the purpose of informing "Emigrants of the Labouring Classes." The office sent another questionnaire in 1843. Then, in late 1845, the civil secretary distributed a questionnaire to district agricultural societies on behalf of imperial authorities, requesting information about the state of local potato crops between 1843 and 1845.[55] At the time, disease was attacking potato crops in Ireland and elsewhere, and this inquiry was issued just ahead of the repeated failures of potato crops in Ireland that caused high rates of death and displacement, resulting sustained emigration from Ireland to British North America and elsewhere throughout the rest of the decade.

With the establishment of the new government for the United Province of Canada in 1841 came significant changes to the operation of state. Lord Sydenham, the Canadas' first governor general, had adjusted the bureaucracy of government in order to "establish clear and coherent chains of command."[56] Editors of Upper Canada's nascent agricultural press seized upon these developments to generate a sustained and earnest discussion as to how the province's agricultural societies might play a more effective role were they connected more directly to an expanded government administration devoted specifically to leading agricultural improvement.

In April 1841, Kingston's *Chronicle and Gazette* reprinted William Evans's plan for a board of agriculture that he had first outlined in the August 1838 issue of his short-lived agricultural periodical, *The Canadian Quarterly Agricultural and Industrial Magazine*.[57] Kingston was now the capital of the Canadas, and the first session of the united legislature was about to meet in that town. Publishing this call for a board of agriculture in a Kingston newspaper was meant as a spur to legislators who, in Evans's view, needed to do more to encourage agricultural improvement within a union legislature than either Upper or Lower Canada had accomplished individually. Quite rightly, he criticized the "trifling pecuniary aids to Agricultural Societies" that were mostly employed to fund cattle-show prizes for superior examples of livestock. It was not an effective method of improvement because those who competed with fine specimens of livestock were often already good farmers, while those requiring assistance received no aid. His solution was for the government to establish a board of agriculture to directly oversee and coordinate agricultural improvements across the Canadas.[58] A few months later, A.B.E.F. Garfield launched his *Canadian Farmer and Mechanic* at Kingston. He queried: Would not "a more perfect knowledge of the country's resources be obtained, and its wants known and supplied" if an association were formed of delegates from the various districts of the province? Garfield wished only to provoke discussion on the subject and did not publish any specific plans of his own.[59] Unfortunately, there was little opportunity for his readers to respond, for Garfield's newspaper ceased operations in October 1841, and he departed for Syracuse to begin another newspaper there. William Edmundson and John Eastwood of Toronto acquired Garfield's Kingston business and began publishing the *British American Cultivator* in Toronto in January 1842. With William Evans as editor throughout its inaugural year, theirs would become the first agricultural journal launched in either half of the Canadas to continue printing beyond a prospectus or a few issues.[60]

A year after the *Chronicle and Gazette* had published his call for a board of agriculture, Evans published "The Encouragement Which Ought To Be Given By The Government To Agriculture In British America" in which he argued again that the current system of grants to district agricultural societies was inefficient for bringing about improved agriculture and prosperity. He looked to England and the Royal Agricultural Society of England (RASE), which had been established in 1838 following a dinner of the Smithfield Club. With the motto "Practice with

# THE BRITISH AMERICAN CULTIVATOR.

"AGRICULTURE NOT ONLY GIVES RICHES TO A NATION, BUT THE ONLY RICHES SHE CAN CALL HER OWN."—*Dr. Johnson.*

Vol. 1. TORONTO, JANUARY, 1842. No. 1.

## THE CULTIVATOR.

### Introductory.

Since the issue of our Circulars, under date 16th of Nov. last, a circumstance has transpired which has induced us to change the title of our publication. The motives, influencing us to adopt the present title, are expressed at large on page 4. We have issued our first number a few days in advance of its date, in order to give Subscribers to periodicals of a similar character, published in the United States, an opportunity of supporting ours, by transferring their subscriptions, which commence with the New Year to the support of a Canadian paper, which, we flatter ourselves, will be found deserving of their patronage. The circulation of those papers in these Provinces is sufficient to cover all the expenses that would be incurred, on a similar publication issued in our own country. We give a notice in another column, taken from the new *Genesee Farmer*, of the "death of the *Canadian Farmer and Mechanic*, for want of proper care and nourishment." That paper alone has a circulation of 1,500 in the Province of Canada. Our sheet is exactly of the same size, and the difference in price, in favour of theirs, will be only one halfpenny on each number, when the postage is added. We leave the matter to be decided by an intelligent public, whether a publication, devoted exclusively to our local interests, should be supported and *nourished* in preference to one of a foreign character. The immense outlays we must necessarily incur for suitable Engravings, to illustrate the different important subjects that may come under our notice; and the extremely low price of our publication will require an extensive circulation to defray the expenses; confidently anticipating that our spirited Yeoman will use every exertion in their power, to establish a publication in British America, devoted exclusively to their interests,—we have been induced to give them a fair trial—we hope our confidence will not be misplaced. Let but every individual who has any interest in the cultivation of the soil take a prominent part in promoting its circulation; and the scientific and learned become contributors to its columns, the work will then be easily accomplished.

It is a matter of astonishment, as well as regret, while commercial, as well as political papers, may be numbered in our country by the score, that not *one* is established, devoted exclusively to the leading pursuits of five-sixths of the whole population. The result of this neglect, in a great measure, has been the necessity of large importations of the necessaries of life from Great Britain and the United States, whilst we have had but a trifling surplus production to exchange for those commodities: whereas, if a judicious system of husbandry were adopted, throughout every section of these large and fertile Provinces, an annual average surplus would be given, sufficient to meet the demands of our importations. The great truth, that the real source of our wealth, lies in the productive industry of those classes whose welfare it will be our object to further, is beginning to be better understood, and its wide spread agencies more fully appreciated, by every lover of his country.—Our yeomen are, in general, the owners of the soil they cultivate—farm may be added to farm, with the possession of property, a spirit of inquiry is awakened, information of a character that will enhance their interests is demanded; and men of science, experience, and ability, are gladly consulted. Towards lightening the labours of an Editor, in charge of an agricultural periodical, we invite the cordial co-operation of the friends of those interests. Much of the work will necessarily devolve upon them—if each contribute his mite the work will be greatly accelerated. There are many scientific practical Farmers interspersed through these North American Provinces whom, we think, the public interest have claims sufficient to induce them to make known, through our columns, the results of their research. There are, likewise, hundreds of able practical husbandmen who are unaccustomed to write for the public press, and such we wish to give in a plain statement of facts, if they require any dressing, we will place them in a proper form before our readers.

The great advancements which Agriculture has made in Great Britain, within the last half century, furnish a very interesting example of the improvements of which this science is susceptible. We need only notice the amendments introduced into different sections of these provinces within the last fifteen years by emigrants from the British Isles, to show that much improvement may be made in the general practice of Agriculture. Every available exertion shall be used, on our part, to advance the true interests of the cultivators of the soil, by extending an improved system of cultivation throughout every portion of the Provinces, and encouraging the more extensive use of articles, the produce of our domestic manufactures.

We shall address a copy of this number to each Post-master throughout the Provinces, and likewise forward one to many of the most influential Farmers, in the hope of making it generally known; being confident that it is only necessary to bring it to the notice of those classes, for whose benefit it is intended, to induce most of them to become Subscribers and we request those who may receive a copy of it, to use their influence for our publication, if they cannot attend to it personally, we hope they will be kind enough to place the paper in the hands of some individual who will feel interested in extending its circulation. We hope all those who do not feel disposed to subscribe for our paper, or take an interest in its publication, will be kind enough to return the number, (by post,) to the Proprietors.

All Post-Masters, who take an interest in our publication, will be considered authorized Agents—such will please forward their names, and address to the EDITOR of the *British American Cultivator*, without delay. The same premium will be given them as we allow other Agents.

### Cheap Houses.

There has been within the last four years introduced in this District, a style of houses as yet comparatively unknown to other parts of the Province. We feel a pleasure in bringing it into general notice, as it will, no doubt, be brought into general use as soon as its good qualities are fully known. The houses constructed on this style are denominated "the unburnt brick house." The few brief hints we intend to give at this time on the subject, will be more to solicit correspondence than to give a detailed description of the process of building. If those who are more acquainted with the matter than we are, should fail to give the particulars, we will advert to it in our next, and endeavour, by the ensuing spring to give creditable testimonials in their favour, and clearly elucidate the subject to the understanding of all classes who take an interest in reading our Journal. These buildings cost about the same price as a frame, and a farmer who could do much of the work within himself, could erect the walls of such a building nearly as cheap as with logs. The material for the brick is prepared much in the same manner as for common brick, with the exception of its being mixed with straw. The dimensions of the brick are 6 inches thick, 12 inches wide, and 18 inches long. A number of houses have been built this last summer by contract, at the rate of £1. per hundred brick, (including making) containing an area of 75 feet of wall. The walls of a house, 30 feet square and 15 feet high, at that rate would cost only £34. The common practise is to rough-cast, and when built upon a good stone wall, are considered the warmest and most durable house that we have. There are within a circuit of 40 miles of this city, at least 200 of those houses, and the most of them have been built within the last 2 years. We have seen houses, barns, stables, and sheds built upon the same plan. All seem to be well satisfied, and recommend their neighbours "to go and do likewise." Much credit is due to the person who introduced this valuable plan of buildings in our country, and if any are solicitous to hear further on the subject, he would no doubt answer, through our columns, any inquiries that may be made.

### To Correspondents.

WE hope all those who may be kind enough to contribute to the columns of the *Cultivator*, will endeavour to make their articles interesting and useful. We have noticed the speculations of the Multicaulis—the Chinese Tree Corn—the Rohans—the Egyptian—Siberian—and Italian varieties of wheat, and all the other humbugs, which have been practiced among our neighbours, within the last few years; and confidently hope that we will not be the instrument of palming on the public such impositions. We are fully aware that some varieties of seeds, roots, &c., are much better, and more profitable, than others; but, it is quite soon enough to bring them into public notice, when their good qualities, and adaptedness, to our climate, are fully tested—they may, then, be brought safely into notice, and the public in- [illegible]

Image 4.2. The first issue of the monthly, sixteen-page *British American Cultivator*, January 1842. (Private collection.)

Science," the RASE sought to encourage "the application of science to agriculture, the stimulation of agricultural progress and development, and the generation and communication of agricultural information."[61] The society held an annual agricultural show to display the livestock and crops raised by its gentlemen members, which rotated throughout the counties of England. According to Evans, the RASE had nearly six thousand members by 1842. He noted, however, that "[i]n British North America we have no rich and powerful landed proprietors to encourage improvements, or take any active interest in agricultural prosperity." As a result, the Canadas would be unable to establish an organization such as the RASE; however, its government could establish a board as a substitute to accomplish the same goals.[62]

For Evans, England and its more than three hundred agricultural societies provided a convincing reason for Upper Canadian farmers to associate: "If the people of England have thought it necessary to unite all parties in a Society for promoting agricultural improvement and prosperity in a country, where agriculture is already in a higher state of improvement than in any other part of the globe; why should it not be good for us to adopt means that would be likely to produce the same results?" Believing that "the most approved systems of agriculture practiced in the British Isles" were best adapted for Canadian farms, he contacted the RASE in the hope that, as editor, he could tap into the agricultural improvements of society members and publish them in the *British American Cultivator* for the benefit of British North American readers and, of course, the financial growth of his newspaper.[63]

Evans postulated how a "General Board of Agriculture" might be structured. He believed the government's newly established Board of Works to undertake the improvement of the province's infrastructure, primarily the construction of canals and roads, was the model to adopt to "promote the improvement and prosperity of agriculture." He planned for a provincial Board of Agriculture that involved both Upper and Lower Canada, composed of three or five members, all appointed by the Governor, who would retain their positions for at least five years and be paid the same amount as a member of the Legislative Assembly. The board would hold quarterly meetings at Quebec, Montreal, Kingston, and Toronto to coordinate the activities of the district agricultural societies and ensure they were operating under "judicious rules and regulations."[64] No one acted on Evans's proposal, and by the end of the year, his partnership with William Edmundson was finished. Edmundson, the new editor of the *British American Cultivator*, would not look to England and the RASE with the same reverence. Instead, his proposals for agricultural leadership in Upper Canada drew on

New York and its statewide association of agricultural societies that relied heavily on state government support.[65]

New York state did provide a fresh example of government supported agricultural improvement. There, in 1841, the state legislature had endowed the NYSAS with $8,000 per year for five years to be divided among the various counties of the state. Out of this sum, $700 was to be retained by the society for the purposes of hosting a cattle show and fair at Syracuse in 1842. This event would shortly evolve into the widely attended annual New York State Fair. But government funding had not come easily. It was secured only after some nine years of improvers petitioning the legislature and lobbying extensively in the pages of the state's expanding agricultural press.[66]

Edmundson soon employed his newspaper as a forum to frequently promote and discuss the formation of an Upper Canadian Board of Agriculture. In April 1843, he requested the Honourable Adam Fergusson to submit "an illustration of the probable duties and benefits of such an institution."[67] The fifty-eight-year-old Fergusson was one of the preeminent agricultural improvers in Upper Canada. He had been a director of the Highland and Agricultural Society of Scotland which, in 1831, had sent him to the Canadas and the United States to examine the state of agriculture and its potential for emigration. Impressed with Upper Canada, he returned in 1833 with his family, established his farm "Woodhill" near Waterdown and helped found the village of Fergus. Privately at Woodhill, he employed the techniques of improved agriculture, including importing pure-bred cattle to develop his herd. Publicly, he had served the province as a legislative councillor since 1839, being appointed by Lieutenant Governor Sir George Arthur as "a gentleman from Scotland, highly respectable and intelligent."[68]

Before Fergusson could submit his ideas to the *British American Cultivator*, "The Pittsburg Farmer" had responded to Edmundson's comments on the subject. Believing a board of agriculture might be useful agency of government, "especially in the education of the rising generation," he looked to the recently established Boards of Trade in Toronto, Quebec, and Montreal, arguing that a board of agriculture could perform similar tasks; plus, it would generate a uniform view of the needs of the colonies' farms and farmers.[69] The Boards of Trade were not part of the government bureaucracy but were organizations created by local merchants as a forum to discuss matters of commercial interests and as a lobby to raise such issues with elected officials at the municipal and provincial level.[70] When he responded, Fergusson's sentiments would echo those of "The Pittsburg Farmer" in that a board of agriculture should be a government institution and, like the existing

Boards of Trade, should provide statistical information to the government, which might aid both farmers and, possibly, future immigration. He also argued for the necessity of correspondence with the RASE and other such societies on the European continent to gather "discoveries and experiments, useful and applicable to Canada."[71]

Whereas the Canadian Board of Works had influenced Evans's plan for a board of agriculture and whereas Boards of Trade had influenced the thinking of correspondents to the *British American Cultivator*, the district councils – another part of Lord Sydenham's administrative reforms – influenced Edmundson's scheme.[72] Sydenham considered district councils to be his most important administrative reform.[73] Each council was composed of elected township representatives who met quarterly under the authority of a district warden appointed by the provincial government. This new level of municipal government at once freed the central government from issues that were not of a provincial concern by providing the councils authority to handle purely local matters. More importantly to Sydenham, it provided a more effective means by which his centralist reforms could be implemented at the local level.[74] District councils were an attempt to foster local harmony with the central government; they were not a move towards more responsible government. They institutionalized the arm's-length connection between local oligarchies at each district centre and the provincial government. Agriculturists saw the opportunity for closer connections between district agricultural societies and the district councils, considering the potential overlap in membership and interests.

Edmundson proposed restructuring the existing district agricultural societies by creating district boards of agriculture with district councillors as members. This, he believed, would encourage municipal leadership to adopt agricultural improvement into their political mandate. Furthermore, councilors and the district warden could better influence the provincial legislature to commit to financial support. To coordinate agricultural improvement across the province, Edmundson proposed that selected members from each district board would form a "General Board of Agriculture for Upper Canada."[75] Whereas Evans's had believed in a top-down approach in which a government board of agriculture, composed of government appointees, would oversee agricultural improvement in both Canadas, Edmundson looked to grassroots development that built on existing township and county agricultural societies within each district, focused on local agricultural conditions and opportunities for improvement, and a provincial board composed of executives from each of the district agricultural societies across Upper Canada to create a unified provincial voice for improvement practised locally.

Edmundson did more than just plan. He submitted his scheme to the Home District Council for its consideration during the summer of 1843, a move likely facilitated by E.W. Thomson, an executive member of the HDAS of which Edmundson was also a member. Conveniently, Thomson had been appointed as the Home District's first Warden. But a committee of the council rejected Edmundson's plan on the grounds of the "onerous" duties already under the district council's mandate, as well as the short period in which the council had to accomplish them. In fact, the committee recommended that the Home District Council not "interfere in Agricultural affairs" of the province. The council agreed.[76]

By November, Edmundson presented a revised plan for a provincial agricultural association arguing that district agricultural societies were "of trifling service to the country" and that they had "done a vast amount of injury, by introducing stock altogether ill suited to the wants of the country." To best lead agricultural improvement in Upper Canada, he proclaimed: "We must begin at the foot of the ladder." Township agricultural clubs, led by a "board of directors, consisting of the most influential and patriotic farmers in the township," would meet once a month to discuss agricultural topics. At the second level, the most active members of each township club would be selected to form a district board. It would meet once per quarter, assessing and disseminating the information provided by the township clubs. On the next rung of the ladder, a provincial board of agriculture would be formed by the election of one or two representatives from each district board. Communication between the district boards was essential to his plan as it was lacking in the current system. District boards would manage jointly a "Journal of Canadian Agriculture" containing material contributed from the township clubs and district boards. Such a journal would allow "each farmer in the province [to] avail himself of the combined experience of his class." As for the provincial board, it would manage an annual provincial exhibition, to be hosted in a different district of the province from year to year.[77] There is little doubt that Edmundson expected his *British American Cultivator* to become the journal of this new society, much like *The Cultivator* in Albany had been designated the official journal of the NYSAS.[78]

Edmundson pitched his plan to the HDAS during the first week of November 1843.[79] Those in attendance agreed in principle that the matter was worth pursuing. Thomson, the society's president, moved that the plan be presented to a wider group of individuals at another public meeting in Toronto.[80] Two days later, several members of the HDAS, a few district councillors, and other inhabitants of the Home District met at the Toronto courthouse. There, they agreed that "it would materially tend to the prosperity of this Province, if the Agricultural Societies now established were so connected, that an uniform [*sic*] system in their management should be pursued." After agreeing to adopt Edmundson's ambitious scheme for the Home District as well as the promotion of a provincial agricultural association, those present appointed Edmundson and Thomson, along with William Botsford Jarvis, and George Dupont Wells, the society's vice president and its secretary, as a committee to "open a correspondence" with the agricultural societies across the province to publicize the plan widely. In the January 1844 columns of his *British American Cultivator*, Edmundson praised the members of the HDAS for setting "a noble example to their fellow agriculturists of other districts." He commended them for following his suggestion of having the district councillors "exert their influence" in establishing township auxiliary societies and gloated that within three months there would be at least twenty associations in the Home District, each with a membership ranging from forty to three hundred members.[81] His confidence was premature.

The HDAS held its annual meeting in February 1844 at which members agreed to offer a sum of £150 in aid of organizing township auxiliaries in the Home District. It also coordinated itself with the district council by rescheduling the society's quarterly meetings to match the weeks in which the council assembled in Toronto. Yet, as Edmundson reported later, the "clause which has reference to the organization of a Provincial Society [had been] very properly postponed for further consideration."[82] It seems that outside of the Home District, there had been little support for Edmundson's scheme. Evidence suggests only the MDAS and the Gore District Agricultural Society adopted portions of the plan. This was far short of Edmundson's prediction of support from at least six districts by year's end.[83] Optimistically, he declared that his next issue would contain an announcement for "a conventional meeting" to establish the Canada Agricultural Association at either Cobourg or Hamilton. No such a meeting was ever held.

By 1845, Edmundson abandoned his plan and appears to have placed his faith in new legislation before parliament to better govern the district agricultural societies. Once this "sound basis" had been established across the province, he believed a "Grand Provincial Agricultural Society" would be organized.[84] But provincial legislators proved to be little interested in a major overhaul of the district agricultural societies and the county and township societies founded under their administrative umbrella, although the government did provide one significant legislative improvement.

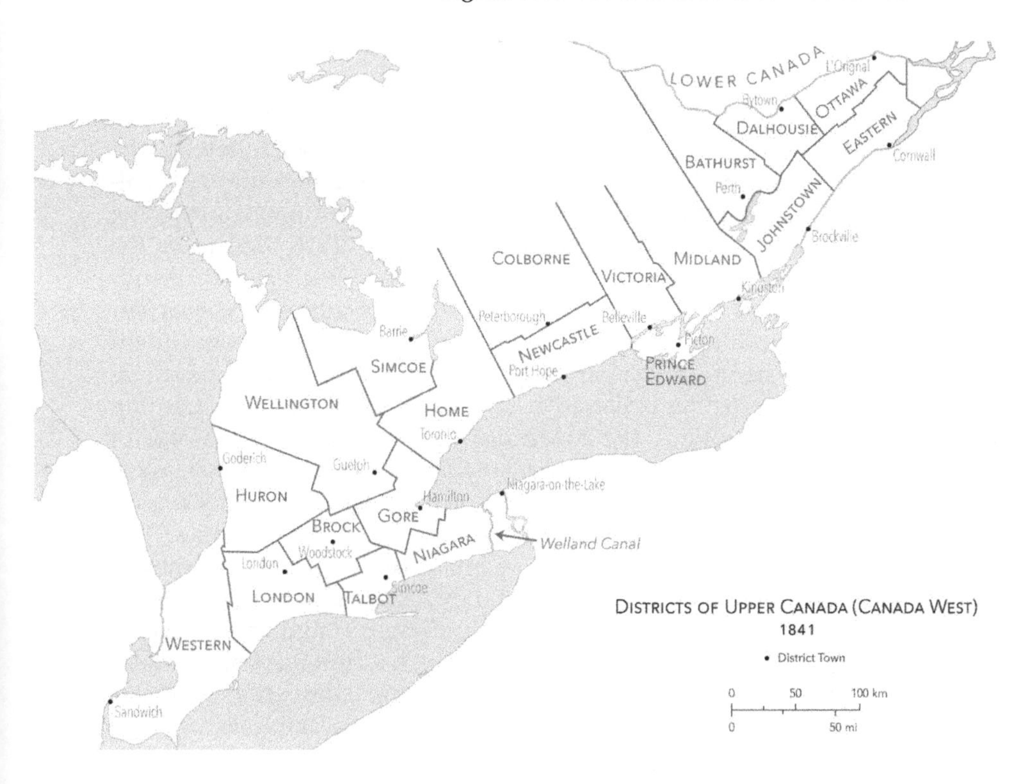


Map 4.2. Districts of Upper Canada (Canada West), 1841. Map produced by Julie Witmer Custom Map Design.

In late October 1843, the Canadian legislature had established a committee "to consider the best mode of granting Legislative aid for the encouragement of Agriculture in this Province," but it did not report, because parliament was dissolved in early December following the resignation of the Louis La Fontaine–Robert Baldwin ministry.[85] Following October 1844 elections, the subject of Upper Canada's expiring agricultural societies legislation was the subject of two debates in the early 1845 session of parliament, resulting in new legislation. Amendments increased the financial commitment to agricultural societies "for the purpose of importing valuable live stock or whatever else might conduce to the improvement of agriculture." Once a district agricultural society treasurer collected and reported the legislated minimum of £25 in subscriptions, the government grant would *treble* instead of double that amount to a new annual maximum of £250. The increase in public funding was again paired with increased accountability requiring each district agricultural society to submit an annual report to "the three branches of the Legislature" within fifteen days of the opening of each session of parliament, whether they sought renewed funding or not. The report was to include an accounting of annual memberships collected, the government funds the society had received, and a statement of expenses. It also had to account for all competitions offered by the society, the names of prizewinners and the amounts awarded to them. Significantly, the act made a first, although indirect, effort to connect the work of agricultural societies from district to district. Each district society was required to submit to the government "all such other observations and information as [the agricultural society's secretary] shall deem likely to tend to the improvement of Agriculture." The act assumed that such accounts and observations would be published as an appendix to the Province of Canada's legislative reports, thus available publicly across Upper Canada (and, in effect, across Lower Canada too).

What set this legislation apart from previous versions was its lack of an expiration date. Fifteen years after the original agricultural societies legislation had been enacted, including two lapses in funding, the government provided permanent public funding for Upper Canada's agricultural societies. As the preamble of the new act asserted: "the science of Agriculture demands encouragement from the Revenues and People of Upper Canada."[86]

By 1846, a provincial agricultural association became a reality, although not by the processes that Edmundson had proposed (see chapter 7), and that association developed into a lobby for agricultural societies in association with the province's agricultural press. By 1850,

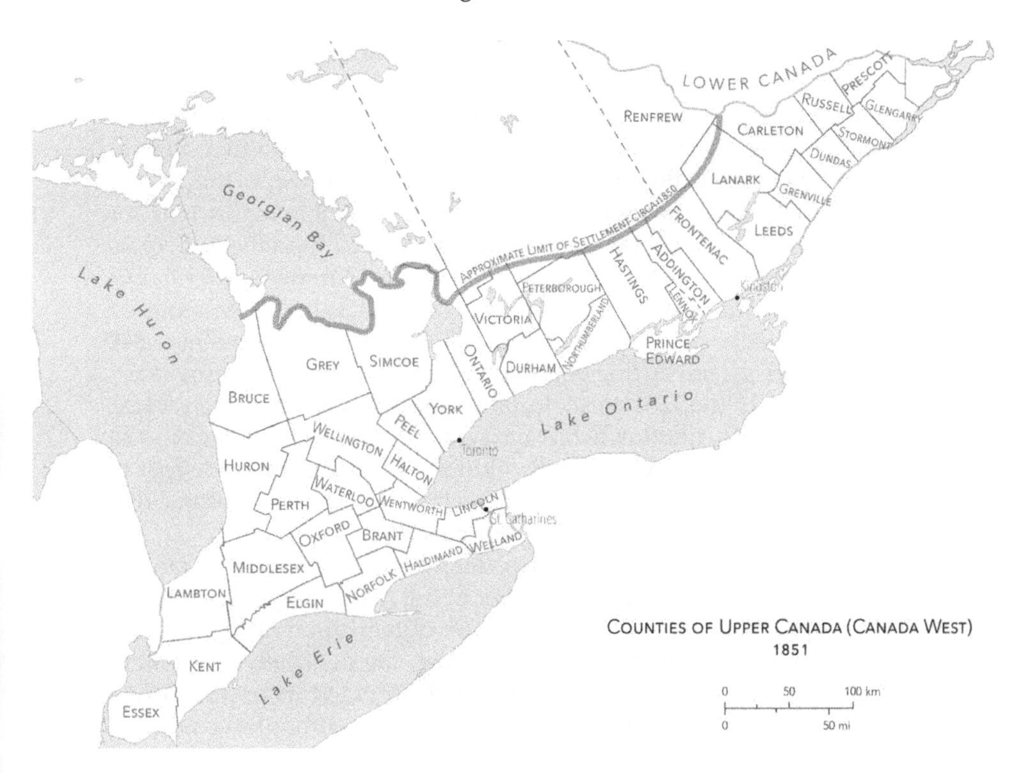


Map 4.3. Counties of Upper Canada (Canada West), 1851. Map produced by Julie Witmer Custom Map Design.

the AAUC presented to the Canadian legislature a bill it had drafted to establish a board of agriculture for Upper Canada and reorganize the province's network of district agricultural societies. Expediencies of the legislative session caused politicians to divide the bill so that it could at least establish the Board of Agriculture (see chapter 8).[87] Thus, one of the first tasks undertaken by the new board was lobbying in 1851 for passage of a bill that contained the clauses trimmed from the draft of its own establishing legislation. This bill sought to eliminate district agricultural societies because Upper Canada's districts had been replaced by counties as the principal municipal division as of 31 December 1849 in response to the advancing populations in townships of the province.[88] Thus, clauses of the 1851 agricultural societies legislation were revised to relate to county and township agricultural societies. In doing so, the legislation largely reflected the devolution of activities from district to county and township levels that had been well underway in some districts since the mid-1840s (see chapters 5 and 6).

On 3 July 1851, James Hervey Price, member of Provincial Parliament (MPP) for York South and life member of the AAUC, introduced to the Canadian legislature a bill "for the better organization of Agricultural Societies in Upper Canada." By 25 August, the Legislative Assembly debated it, made some amendments, and provided its approval two days later. The Legislative Council approved it without amendment and before the end of the month it received royal assent.[89] The act set forth "a uniform system" for a "more efficient working" of the province's agricultural societies. It repealed the 1845 legislation, and the new law standardized the executive of each county agricultural society to a president, two vice presidents, a secretary, a treasurer, and five directors. It also provided for the president of each township agricultural society within a county to be an *ex officio* director of that county's agricultural society. Further standardization included the preparation of an annual report composed of a standard set of information related to the previous year's proceedings. A township society was to produce this report for presentation to its annual meeting in January so that it could be forwarded on to the county society, which had to the responsibility of bundling all township reports along with its own annual report for submission to the secretary of the Board of Agriculture prior to 1 April.

For the first time, the act set a minimum size of an agricultural society. At least fifty members were required for a county society, each paying an annual membership of no less than 5s each. While it set no membership minimum for a township (or branch) agricultural society representing one or more townships within a county, each member had to pay an annual membership of no less than 5s, and the society had to raise a minimum total of £17 10s each year to be officially recognized. Such funds were to be forwarded to the county treasurer for a complete accounting of membership dues raised at both the township and the county level. Schedules attached to the legislation provided the standard form for each county and township society to record individual memberships and the amounts they paid, as well as the form of the county treasurer's financial statement to be sworn in front of a county justice of the peace before submission to the Board of Agriculture.

When the county treasurer filed the annual reports and financial statement, the Board of Agriculture would certify them and forward them to the governor. In return, that county society received an annual public grant equal to three times that raised in membership to a maximum of £250. The grant would be paid to the county secretary who would then disburse funds to the township societies in proportion to the membership funds each had raised. While the act simplified the reporting process for 1851 so that societies could receive funding without having to provide detailed accounting during the initial year of the act's provisions, in future years, any county or township treasurer found to have made a false report as to the funds raised would forfeit the government grant, plus pay a £10 penalty and be guilty of a misdemeanour. Such requirements combined to produce a significant increase in the administrative tasks for the Board of Agriculture and other provincial officials to finance and oversee county agricultural societies. An 1852 report indicated there were at least twenty-two county agricultural societies operating across the province's forty-eight counties. Fourteen of the twenty-two agricultural societies reported as active in 1852 were unions of two or three counties from within their former district. In this manner, the leadership of the old district agricultural societies continued through mid-century under a new administrative guise.[90]

The legislation did much to shorten the government's arm's-length relationship with these semi-public county and township institutions, and it also established county agricultural societies as corporate entities. Doing so permitted them to acquire and own land and other property worth no more than £1,000 "to sell, lease, or otherwise dispose of the same." By owning its own grounds, a county agricultural society could erect permanent buildings, thereby substantially reducing the task of planning and setting up for a cattle show each year. For the first time, rules were set for hosting these annual county exhibitions. If the township in which the county exhibition was to be held had an agricultural society, then the township society would not host its own exhibition that year but merge its funds designated for an exhibition with those of the county to host a single event.[91]

Fothergill's legislation provided an ending as much as it did a beginning. Upper Canada's first government-funded agricultural societies had been created out of concern over the poor state of the province's agriculture in the 1820s, and they had been created by drawing on examples from other provinces and nations. The ideal remained that of the British gentry, but in the absence of a wealthy landed gentry, New York and its provision of public funds and legislative coordination of agricultural societies across the state provided the more practical approach for the North American frontier. This hybrid approach, sponsored by Fothergill, received support from both tories and reformers despite the other issues that generated acrimony between those at each end of the political spectrum. His act set aside old plans. Gone were any ambitions to create a province-wide agricultural society. Instead, the legislation created semi-public organizations in each district of the province invested with the task of leading agricultural improvement in their region of the colony, funded by an annual provincial grant. Yet, little by little these district, then county and township, agricultural societies were drawn closer into the administration of the Upper Canadian state by way of increased requirements for standardized annual reporting in order to claim and disburse the now permanent annual provincial grant. Further changes, which resulted in the creation of a provincial agricultural association and a board of agriculture for Upper Canada are discussed in chapters 8 and 9, while the next three chapters analyse three district agricultural societies to provide comparative examples of local state formation. Each highlights ongoing reasons why local independence remained a feature of Upper Canada's agricultural societies and their governing legislation.

# 5 The Farming Compact: York, 1830

Late afternoon, 6 July 1830, found William Lyon Mackenzie knocking on the locked courtroom door of the York courthouse, having just been ejected from the meeting inside. Suddenly, the door opened, shouts of abuse spewed forth, and someone handed him his hat. Then, the door was slammed shut and relocked. It was the second time in as many months that Mackenzie had been ejected from a meeting at the courthouse. As he put on his hat and stepped into King Street to return to his *Colonial Advocate* office nearby, Mackenzie pondered pressing charges for assault.

The meeting had been scheduled to establish an agricultural society for the Home District. The agricultural societies legislation passed in March provided the tory oligarchy of the provincial capital a new source of government money it could control to help extend its paternalism out from York into the district beyond. The Family Compact, as the oligarchy at the capital and its supporters across the province would come to be known, was not about to allow Mackenzie and his reform supporters interfere with such an opportunity. Common ground between reformers and tories might have been discovered within the provincial legislature during debate and passage of the agricultural societies act, but in the streets of the "political cockpit"[1] of York, the announcement of annual government grants produced yet another public battleground for political and social tensions brewing in the capital. Immediately after the agricultural societies act received royal assent, Mackenzie had seized the initiative to organize a district agricultural society. In the months that followed, York's tory elite launched a counteroffensive aimed at wresting control of the infant institution away from Mackenzie and his supporters.

This chapter analyses events surrounding the creation of the HDAS in spring and summer 1830, for they represent some of the most public, yet mostly overlooked, displays of the York oligarchy's efforts to maintain political and social control of York and its institutions in an overt attempt to entrench and extend its paternalism out into the Home District beyond. The events of spring and summer 1830 were both ungentlemanly and unseemly. Gentlemen of other districts of Upper Canada generated little conflict and certainly no violence while establishing their local agricultural societies in the spring and summer of 1830.[2] But, at York, the response fit a pattern of "legal and illegal intimidation and chicanery in defence of the status quo that stretched back to the persecution of Robert Gourlay."[3] Indeed, here again was the establishment of an agricultural society at York to advance improvement while at the same time squelching reform, broadly defined.

There were many reasons why the York oligarchy felt threatened and were vigilant against social and political change that undermined its authority. York of 1830 remained a small town, and farmland bordered three of its sides. At its fourth was the lakefront where agricultural commerce departed the docks and settlers arrived, ready to head inland to clear their surveyed farm lots. These factors had doubled the town's population from 1,240 residents to 2,860 during the decade of the 1820s,[4] yet the town was still composed of only several hundred families, many connected through marriage. Official circles in York remained even smaller, meaning political opponents had many opportunities for face-to-face battles in their daily lives. Political quarrels were among "York's distinctive entertainments," both on the streets of York and inside the House of Assembly where Mackenzie remained determined to debate to exhaustion the minute details of any issue. It also meant that there were more people to line up on either side of a public outburst.[5] Relations between the tory oligarchy and the Mackenzie had been sour for some years. In 1826, sons of several of the town's elite families destroyed Mackenzie's *Colonial Advocate* print shop and printing press, and after the court compensated Mackenzie substantially for his loss, he employed his new printing press to continue his weekly attacks against abuse of privilege and power at the capital and throughout Upper Canada.[6]

The events of early summer 1830 belie the oligarchy's wider defence of power and influence in the face of the Home District's changing politics. Provincial elections of 1828 and a town by-election of 1830 had demonstrated that within the growing population of York and province there was a significant group of voters who supported candidates opposed to the provincial executive. In 1828, the province's attorney general, John Beverley Robinson, had won the Town of York riding, having run against a radical reform candidate, Dr. Thomas D. Morrison. A year later, Robinson was appointed Upper Canada's chief justice and the replacement tory candidate, William Botsford Jarvis, sheriff of the

Home District, contested the town riding in a by-election. Jarvis was defeated by nine votes to a moderate reform candidate, Robert Baldwin.[7] The results signaled to the tory elite that a new group of professionals was emerging as a political force in York. Chief among its members were merchants whose businesses relied on the farmers of the countryside outside of York and the arrival of new immigrants. In consequence, they shared more political views with the farmers with whom they traded than they did the tory elite.[8]

This had been confirmed by Robinson's somewhat hollow 1828 election victory, which left him the isolated tory victor in the Home District, as voters of the rural ridings elected reformers Mackenzie, Jesse Ketchum, and John Cawthra as their representatives to the provincial assembly.[9] As discussed in the previous chapter, the elections had also tipped the balance of the House of Assembly in favour of the reformers, a loosely organized group of MHAs whose strength, in part, came from the rural ridings on the edge of the capital.[10] In the tempestuous capital, reformers became lightning rods for the ire of an increasingly defensive tory elite. Thus, when Mackenzie attempted to establish an agricultural society for the Home District, a defensive oligarchy lashed out, because it could not condone further loss of control. A district agricultural society, funded with government grants, provided a new foundation for its paternalism and new avenues for its tendrils of authority to extend out into the increasing population of the Home District.

On 8 March 1830, two days after the Agricultural Societies Act had received royal assent, Mackenzie walked the short distance from his newspaper office to York's market square. There, he circulated a petition in support of creating "The Home District Agricultural Society." Thirty-six individuals contributed their support.[11] With this indication of public support, Mackenzie circulated handbills and published a notice in his *Colonial Advocate* calling for a public meeting on 25 March to establish the agricultural society and elect officers for the year.[12]

Poor weather and muddy roads meant that only forty individuals attended. A disappointed Mackenzie thought this support to be insufficient to agree to a constitution; thus, the meeting drafted only a proposed constitution to be considered at a new meeting scheduled for the York courthouse on 8 April. Eight individuals were chosen to solicit subscriptions for the society, and a report of the proceedings was written up for distribution to the six newspaper editors of York. The meeting was then adjourned.[13]

The draft constitution was in no way representative of any reform platform, radical or moderate. In fact, it followed what could be considered an established formula for the structure of an agricultural society; it was little changed from that of the UCAS in 1818. Those present agreed that "members of the Legislative Council and the House of Assembly, residing in the Home District" would be honorary members of the society and they also planned to ask Lieutenant Governor Colborne to be the society's patron.[14] Prestige and securing continued government support were no doubt practical motives behind such invitations. Even Mackenzie realized that the agricultural society needed broad support to be a success, especially if the necessary £50 was to be raised. This meant that both gentleman and common farmer, both tory and reformer needed to unite in support. He warned those planning to attend the next meeting that they would have to "strive to leave their political feelings behind them … and to overcome for a few hours that bitterness of personal animosity which has long made the town of York deservedly a bye-word [*sic*] and a reproach."[15] For York, this was too tall of an order, and his decision to delay the agricultural society's founding robbed Mackenzie of any advantage in his initiative. His opponents organized quickly to remove Mackenzie as leader of what one critic dismissed as an institution organized by "the shopkeeping interest of York under the patronage of Mackenzie."[16] Tellingly, Robert Stanton did not publish the proceedings of Mackenzie's founding meeting in the *Upper Canada Gazette*, despite his earlier statements in support of a new agricultural society for the capital.[17]

The 8 April meeting was not held. Mackenzie was at the York courthouse but for another purpose. A packed public gallery was watching Mackenzie defend himself in a libel trial, the consequence of comments he had made about James Small during a by-election for the riding of York in November 1829. The trial lasted twelve hours, four of which were taken up by Mackenzie's address to the jury. With no meeting possible on that day, James Doyle, a York lawyer who had been elected acting secretary at Mackenzie's 25 March meeting, took it upon himself to postpone the meeting until 8 May.[18] Mackenzie himself later reflected that "so intense was the excitement created thereby and so general the attendance in the courthouse, that no other meeting could have been then held with a reasonable expectation of constituting a society."[19]

One person who did travel to York for the agricultural society meeting had done so for the express purpose of preventing Mackenzie's society from being born. Edward O'Brien was a recently immigrated, retired British military officer who farmed property north of York, near Thornhill. He held the public office of a local magistrate, was an acquaintance of Lieutenant Governor Colborne, and was considered a "natural ally" of the Family Compact.[20] When he arrived at the courthouse and realized no meeting would be held that day, O'Brien and his unnamed travelling companion "formally adjourned the meeting till further notice."[21] How they accomplished the adjournment "formally" is unclear, but O'Brien decided that he and his soon-to-be brother-in-law, Richard Gapper – another retired British officer who farmed property along Yonge Street between Thornhill and Richmond Hill – would take on the project of forming the agricultural society for the district and set about "conducting it properly."[22]

During the month of April, Mackenzie continued to implement his vision for the agricultural society. He wrote to John Neilson, president of the Agricultural Society of the District of Quebec, requesting a copy of that institution's "rules, regulations and system of management." Neilson, the former publisher of the *Quebec Gazette*, member of Lower Canada's House of Assembly, agricultural experimenter, and implement inventor, was one of the most knowledgeable individuals in the Canadas whom Mackenzie could contact regarding forming an agricultural society. He had served as president of his society since 1826 and, earlier, had been its vice president. In his letter, Mackenzie stated he had only newspaper accounts to guide his attempts to create the organizational structure of an agricultural society. These, he noted, were "less full than could be wished for."[23] Along with a form of a constitution, he hoped Neilson could pass along a list of contacts that his Quebec society relied on for acquiring models and plans of new implements, seeds, and other agricultural items to connect the "western folks" into the wider world of agricultural improvers.[24] His efforts were in vain.

The postponed meeting of the agricultural society that Doyle had called for 8 May did not occur. By that date, D'Arcy Boulton, Jr., had circulated his own petition to support "forming an Agricultural Society for the Home District." Then, on behalf of Boulton and the twenty-one "Freeholders of the Home District" who had signed the petition, Sheriff William Botsford Jarvis placed an advertisement in the 6 May issue of *The Courier of Upper Canada*, a newspaper staunchly loyal to the provincial administration, announcing a meeting for the courthouse on 15 May, "for the purpose of forming an Agricultural Society for the Home District."[25]

Although Boulton Jr.'s petition has not survived, one can infer from those in attendance at the 15 May meeting that the collective identity of the signatories as "freeholders" was a major understatement of the influence they wielded. Forty-five-year-old Boulton Jr. was the son of the former attorney general of Upper Canada who Mackenzie would later place at the head of the Family Compact.[26] He had purchased a

Image 5.1. William Botsford Jarvis (1799–1864), ca. 1860 (Toronto Public Library, JRR 899 Cab.)

park lot of some one hundred acres just north-west of York in 1817 on which he promptly constructed a stately brick home known as "The Grange." Boulton Jr. was a gentlemen farmer, but he was also the Master in Chancery and a police court justice, and he held the sinecure of Auditor General of Land Patents. This eldest son of one of the province's most influential men had also married Sarah Ann Robinson, making him a brother-in-law to Chief Justice John Beverley Robinson. His father, D'Arcy Boulton Sr., had been instrumental in the expulsion of Robert Gourlay and the creation of the UCAS, serving as a vice president of that organization in 1819–20. Boulton Jr.'s younger brother Henry John had also served as treasurer in that society. As for the thirty-one-year-old Jarvis, a man characterized as the "Sheriff of the Family Compact," he had purchased his 110-acre estate, "Rosedale," to the north of York in 1824 and by 1830 had improved it into a showpiece by expanding the old farmhouse on the property and by planting orchards, vines, and flower gardens.[27]

Image 5.2. "Rosedale," the estate of William Botsford Jarvis. Watercolour, perhaps by Owen Staples, ca. 1915, apparently based on a watercolour by a "James Hamilton, London, C.W., 1835." (Toronto Public Library, JRR 290R Cab.)

In a later assessment of the events of the spring of 1830, Mackenzie would explain that the meeting that Doyle had rescheduled was not held, for "it was thought by some that it would be better to pass it over for a few days and see what the Sheriff's notice would produce, in order to give all parties a fair chance in a matter wherein the public good alone was professedly sought after."[28] This appears to have been a poor attempt at saving face, for Mackenzie's immediate reaction was less placid. Jarvis had sent his meeting notice to the loyal *Courier of Upper Canada* but not to Mackenzie's paper, which had a much greater circulation throughout the Home District. Not that there was any surprise in Jarvis's snub of the *Advocate*, however, for it had been William's second cousin who had led the vandalism of Mackenzie's printing office in 1826. Undeterred, Mackenzie simply copied the notice from the *Courier* in his 13 May issue. Published just two days before the meeting was to occur, the notice offered Mackenzie a vehicle for his warning as to what Boulton's manoeuvre would produce:

> All former attempts at establishing an efficient Home District Agricultural Society having failed from the non-attendance of the farmers,

> who are the parties to be benefited, the sheriff at the insistence of Mr. D'Arcy Boulton, Jr., and others, has called a meeting in town on Saturday next; and we have now no doubt, but that as an association will be organized by a class of inhabitants who are doubtless both able and willing to spare from their ample official incomes a sum sufficient, with the provincial grant, to confer signal benefits upon the agriculture of the district.[29]

Mackenzie was not only venting his frustration but also admitting things that he knew to be true. He was an educated Scotsman, well versed in the rhetoric of improvement. Both Lord Kames and Sinclair had argued that an agricultural society required the involvement of gentlemen farmers with time, money, and interest to ensure its success. Even Mackenzie's favoured republican model of an agricultural society from New York state drew strength from the improving class more than it did from the yeoman farmers they praised. Most Upper Canadian farmers remained consumed by the never-ending toil on their farms, financially incapable of sparing cash on membership dues. Therefore, it would be easier for a coterie of wealthy gentlemen to raise the £50 in memberships required to secure the public grant than it would be to rouse a large pool of farmers to come forward with their few spare pence and shillings. Regardless, Boulton's action drew clear battle lines against Mackenzie's efforts, and the York courthouse hosted the battle to rob Mackenzie of his "nursling."[30]

There exist two differing accounts of Boulton's meeting of 15 May, the first by "Humphrey Clod, Esq., Major 7th Regiment, York Militia," resident of Thornhill. Clod was one of several pseudonyms invented by Mackenzie and employed to record the events of York society in a third-party voice.[31] In Clod's opinion, the meeting was "indeed confused and disorderly." It had been called for noon, but when he arrived at one o'clock, only two other persons were present. He left for half an hour and returned to find Boulton Jr. along with twenty-two other "freeholders" who had signed his petition. While he ridiculed this small gathering for making the courtroom look like "a Brobdingnagian pantry or an Ogre's cheese-closet." Clod found it most obnoxious that these individuals had sat themselves within the bar, "as if to represent the aristocracy."[32]

In colonial society, the courthouse occupied a unique place, both physically and socially. Even on days when court was not in session, its chambers were more than just public meeting rooms. As historian Rhys Isaac argues in his study of colonial Virginia, "[t]he court was central to the organization of society. Its functions went deeper than the conduct of business and the distribution of patronage. The court was the guardian of the Law, and the Law defined rights and obligations."[33] Thus, even though court was not in session, the elites had drawn on all the authority the courtroom offered by seating themselves "within the bar," meaning the area directly in front of the judge's bench, separated from the public gallery by the "bar" that divided the courtroom space. They might have waited in the public galleries but, instead, assembled right under the shadow of the throne and other symbols of the Crown mounted above it to display their control to all those who stepped through the courtroom door. This was not a meeting of mere freeholders; it was a display of who held, and who should hold, authority in the Home District.[34]

Among those whom Clod identified as being assembled within the bar were the meeting's organizer, D'Arcy Boulton Jr. and John Elmsley, a recently arrived gentleman farmer of York with deep ties to the town's elite, for he was the son of the late chief justice. John had retired from the Royal Navy at the rank of Captain in 1825 and returned to Upper Canada to set about administering his late father's sizeable land holdings in the province. As part of his welcome back to the province, his mother had given him a gift of sixty acres up Yonge Street to the north-west of the town. Here in 1829, Elmsley began improving his Clover Hill estate. Representing a younger element of the York elite, Elmsley's parentage, landholdings, and status as one of the capital's wealthiest and most eligible bachelors opened all doors to the inner social and political circles of the capital. He was a director of the Bank of Upper Canada as well as one its largest shareholders. A justice of the peace, he would be appointed to the Executive Council later in 1830, an indication of his fast-rising status in the capital.[35]

The Jarvis family was also well represented within the bar. Joining the sheriff was his father Stephen, the former Home District Registrar who had secured his son's position as Home District Sheriff three years earlier. The sheriff's cousin Samuel P., eldest son of William Jarvis, Upper Canada's first provincial secretary and registrar, was also present. The ringleader of the vandalism of Mackenzie's printing shop, Samuel had inherited his father's park lot on the northern edge of York in 1798 and built "Hazel Burn," a sizeable and well-appointed brick house in 1824. There, he developed orchards, gardens, and kept an extensive fowl house and a large rabbit warren in the southern fifty acres of the lot. Like a country gentlemen, he was able to hunt snipe and deer for pleasure in the forest at the north end of his estate. Having married Mary Boyles Powell, daughter of the former chief justice

Image 5.3. "Moss Park," the estate of William Allan (ca.1770–1853), viewed from the southwest, 1834. John George Howard (Courtesy of the Royal Ontario Museum, 995.35.2.)

William Dummer Powell, Samuel was well connected to York society and influence.[36] Samuel's eastern neighbour, William Allan, also sat within the bar. Among the earliest residents of York, Allan remained the pre-eminent York merchant. A magistrate and legislative councillor, he improved his "Moss Park" estate, as a landscaped portion along the southern quarter of his two-hundred-acre lot. Allan was certainly no friend of Mackenzie's, having suffered the brunt of several attacks in the columns of the *Colonial Advocate*.[37]

There were other gentlemen of rank within the bar who detested Mackenzie. James Small, a York lawyer, was the man whose unsuccessful libel trial against Mackenzie had filled the courthouse in April. His brother, Charles C., was also there. Both were sons of Major John Small, who had arrived with Simcoe to assume the position of clerk of the Executive Council. Charles had taken over his father's other role as Clerk of the Crown and Common Pleas of the King's Bench and was the owner of Berkeley House, his private home and well-known centre for the gatherings of York's elite society. The Parish Clerk of St. James Church, John Fenton, was loosely connected to the Family Compact by his association with the religious head of the oligarchy, Archdeacon Strachan. Fenton drew an annual salary as well as a percentage of the pew rents he collected from the church where most Family Compact members worshipped. He held several other positions, too, including messenger for the Bank of Upper Canada, clerk of the police office at York, and an assistant teacher at York's Central School and Market Lane School. The War of 1812 hero Lieutenant James FitzGibbon also attended. He had retired from the military in 1825 and shortly thereafter had accepted the appointment of militia colonel along with several other minor government appointments. When appointed clerk of the House of Assembly in 1828, Mackenzie targeted FitzGibbon as one of the worst examples of government patronage, claiming that FitzGibbon was being rewarded for having collected a fund so that the vandals could pay the £650 in damages that a jury had awarded Mackenzie for the destruction of his printing shop. Robert Stanton was also within the bar. A magistrate, and a lieutenant colonel of the militia who lived in a substantial brick home in York, Stanton voiced the views of the Family Compact through his *U.E. Loyalist* section of the official *Upper Canada Gazette*. Other associates of the York elite who were present included: Robert G. Anderson, chief teller of the Bank of Upper Canada; Francis T. Billings, the Home District treasurer; and John W. Gamble, a brother-in-law of William Allan, magistrate and gentleman farmer who managed his properties and mill on Mimico Creek in Etobicoke. Edward O'Brien had also returned to York for the meeting.[38]

When he entered the courthouse, Mackenzie joined quite a different crowd standing in the public space "usually filled by the farmers in court-time." Among them were John Macfarlane of Etobicoke, the man who had been elected chair at Mackenzie's March meeting, and Michael Whitmore, owner of the Yonge Street pottery. Shortly thereafter, John Elmsley reached across the dividing line formed by the courthouse bar to request that those standing with Mackenzie pay their membership to the proposed agricultural society. Mackenzie took the request as a clear affront, for Elmsley was speaking to those who had already met in March to organize a Home District agricultural society.

At this point, three more individuals joined Mackenzie's group. "Mr. Law," a veterinarian and "practical agriculturist," arrived with the shoemaker George Nichol, and Edward W. Thomson, a militia captain and brother of Hugh, the MHA for the Frontenac riding surrounding Kingston and editor of the *Upper Canada Herald*. During the previous few years, Thomson had been completing contracts on the construction of the Rideau Canal along with similar contracts on the St. Lawrence River. Only recently arrived in York, Thomson appears to have been sorting his place within York society.[39] He had attended Mackenzie's meeting in March and had been charged with raising subscriptions for the proposed agricultural society, but in the years ahead, Thomson would be elected a conservative MHA. Moreover, he and William B. Jarvis would become close and constant partners in leading agricultural societies in the Home District and establishing and leading a provincial agricultural association.

As Clod reported, Boulton and his associates had not thought out their strategy beyond blocking Mackenzie's efforts. No one had prepared an agenda or draft constitution for consideration. In fact, the meeting was delayed while those assembled awaited the return of an individual who had been dispatched to Mackenzie's *Colonial Advocate* several blocks away to retrieve a copy of the resolutions agreed on at the original society's March meeting. During this delay, those within the bar looked anxiously at the door in vain hopes that more farmers would appear so that the meeting might "carry an appearance of a farmers' meeting." Francis Billings, who had tried to find recruits among the farmers in the market square, was sent outside numerous times to call farmers in from the streets to attend the meeting. But, in Clod's words, the response he received was "if the meeting was called by D'Arcy Boulton and the big-bugs, our safety lies in steering clear of it – burnt children dread fire."[40]

If Clod's reporting is in any way accurate, the paternalistic tone of the resumed meeting – "We give and they receive" – was not well received by those in the public gallery. When one individual within the bar presented his membership as a gift, "from the gentlemen of the town of York to the farmers in the country," and when another suggested that society officers should be chosen only from among the gentlemen within the bar, the meeting erupted into chaos. "Some of the speakers recommended an adjournment and a meeting to be called in the country – others were for doing what they did quickly – 'we are no farmers' said one – 'it's their own fault if the country-folks won't attend when we offer them our aid and the use of this building for nothing' quoth a second – 'this can scarcely be termed a public meeting,' added a third, 'for the public *are not with us*.'" York's MHA, Robert Baldwin, entered the courtroom, "looked at the extraordinary scene before him with unfeigned astonishment for about 90 seconds" and departed. Clod, too, left the meeting when the uproar became too great. He did not report what role Mackenzie might have played in generating or sustaining the mêlée.[41]

The version recorded from within the bar was much less dramatic and suggests that order was soon restored after Clod's departure. That evening, O'Brien's new wife, Mary, related in her diary that Edward and her brother Richard had been elected as two of the society's directors for the ensuing year. The meeting had agreed upon a set of resolutions on which "The Home District Agricultural Society," should be based and raised the necessary £50 through memberships and donations. Membership was 5s per year (the same rate as had been set by the UCAS in 1819).[42] The society's executive would be composed of a president, twelve directors, a secretary, and a treasurer. It would meet with the membership at four general meetings per year to be held on the same days as the General Quarter Sessions for the district.[43]

George Crookshank was elected the society's inaugural president. At age fifty-seven, he was very much the quintessential Upper Canadian gentleman farmer. A brother-in-law of another early settler of York, James McGill, the commissary of stores and provisions, Crookshank had been appointed by Lieutenant Governor Simcoe as deputy commissary general but had retired his post following the War of 1812. He had served as a director of the former UCAS, and in the 1820s, he had also served as the Bank of Upper Canada's second president. He lived on "an opulent scale" in his large home on Toronto Harbour at the front of an estate of some 330 acres. By 1830, the legislative councillor owned many other properties in York, New York state, and the Home District, including a farm near Thornhill of about the same size as his property on Toronto Harbour, another farm to the town's north-west, and one at Newmarket.[44]

William B. Jarvis and John Elmsley filled the offices of treasurer and secretary, respectively. The directors of the new society included many of the men assembled within the bar at the courthouse: William Allan, John Elmsley, D'Arcy Boulton Jr., Edward O'Brien, Richard Gapper, John W. Gamble, Charles C. Small, Robert Stanton, James FitzGibbon, and Robert Anderson. Also elected a director was Peter Robinson, who had served with Crookshank as a director in the UCAS. The brother of the chief justice of the province and known for his settling of Irish immigrants into Upper Canada in the early 1820s, he had developed farming, milling, and merchant interests centred in Holland Landing at the north end of the district. Robinson wielded considerable influence as an Executive and Legislative Councillor, not to mention his post as the commissioner of Crown lands and surveyor general of woods, a position invested with the responsibility of opening Crown reserves for new settlement. Rounding out the dozen directors was Alexander Wood, once one of York's prominent merchants and a close friend of Crookshank. Retired from his business activities for a decade, he served as either a director or an executive member of many York organizations, including the Bank of Upper Canada, and acted as a representative for several of York's absentee landowners and other businessmen.[45]

Before concluding the meeting, the provisional executive drafted its petition to Lieutenant Governor Colborne for the annual government grant.[46] The directors were to draft a constitution to put to members at a first general meeting of the agricultural society, scheduled for the same location on 6 July. New elections for offices would be held at that time.

In a second report of the May meeting, Humphrey Clod expressed his disgust at the behaviour of "'the gentlemen,' (as the worthies of the Court House are pleased to style themselves *par excellence*)." They had commandeered the HDAS, and by petitioning the government for the grant "took the *public* money under their own especial control." Clod warned that the York elites needed to remember that they were "located in the midst of *ten thousand* North American Freemen, *owners of the soil*, and jealous of their liberties."[47] Yet, Clod did not broach the reasons why those thousands of farmers had failed to show at either Mackenzie's March meeting or Boulton's May meeting to offer their support.

Five days after the meeting, Mackenzie (not Clod) published a brief antagonistic editorial in which he bitterly attacked York's "official junto." Although he hoped that "the Agricultural Society (such as it is) may flourish, become useful to the country and promote its happiness," he claimed the meeting had contravened the government's purpose for establishing agricultural societies on a district-wide basis. The £50 had been subscribed very quickly from a mere twenty individuals at the courthouse, and a hasty election of officers had followed. This, he declared, did not follow "the Spirit of the Act," and he warned readers that this junto intended to use the £100 government grant to finance their own private club at the capital. It troubled Mackenzie that he had counted only four farmers present at the courtroom meeting, and they were not among the eighteen gentlemen arranged within the bar. Thirty or forty individuals had attended his March meeting, he claimed, and nearly a dozen were "respectable farmers," but, Mackenzie expressed in disgust, "*not one of the faction* would grace a meeting of real farmers by their presence."[48]

Mackenzie (and Clod) raised some fundamental differences of opinion as to who constituted "farmers" or "the public." The twenty subscribers to Boulton's HDAS may not have been the individuals Mackenzie wished to see form the society, but they had not contravened the law. Nothing in the legislation indicated there were restrictions on who could and who could not form an agricultural society in any district.

It offered no definitions of "farmer" or "public meeting." In fact, many of the "official junto" he criticized lived on estates that they farmed. As noted, such vagueness had likely been instrumental in facilitating the passage of the 1830 legislation in the provincial parliament. But the clash that the legislation had tried to avoid in the legislature was now being played out at York.

Mackenzie was not alone in expressing his disgust at the actions of the York elite and their all-too-familiar tactics. William Buell Jr., a reform MHA for Leeds and editor of the *Brockville Recorder*, concurred with Mackenzie's opinion of what had transpired. A champion of Brockville's ability to compete with York and Kingston, Buell was appalled by the actions of the tory oligarchy at York and provided a succinct assessment of social leadership in Upper Canada, an opinion that Mackenzie gladly reprinted for readers of his *Colonial Advocate*:

> It now only remains for the Farmers to enter and humbly compete for the prizes under these great men and all will be complete. Truly the manner in which things of this nature are generally conducted in this Province is calculated to excite disgust in a thinking mind. A certain set of men must manage all public matters in their own way or they will withdraw their support from useful public objects; while a no less respectable and worthy class, who do not wish to assume so much, see that they can have no chance of enjoying equal privileges, quietly attend their private occupations. Thus, important and useful public institutions often fail, for want of a proper amalgamation of the individuals which should have a share in their management. An Agricultural Society is of this description, and is it in vain to think of a society operating to advantage without placing its management proportionably [*sic*] within the power of those most interested in its success.[49]

Even the York elite would soon discover the validity of Buell's opinion. Although the hastily selected executive was eligible to petition the government for £50 by means of its own largesse, efforts to secure memberships throughout the district elicited a cool response.

Before the May meeting, O'Brien and Gapper had already enlisted their wives' help with gathering memberships and after the meeting, the two new society directors returned up Yonge Street to solicit more memberships. O'Brien thought the annual militia muster in June would provide a good opportunity, but his efforts produced few results. Writing to his brother-in-law Anthony Gapper in England in early July, O'Brien suggested the society was "badly in want of some person, you for instance, sufficiently active and idle and somewhat scientific withal to take an interest in the thing and to induce others to do the same, for I am much in dread that notwithstanding all that has been done, the Ag'l Soc'y will die, not a natural death but what they call in Ireland of a 'decay.'"[50] Edward's assessment was correct: There were few Upper Canadians wealthy enough to be completely independent of government commission or professional role to devote their time and finances to the scientific improvement of agriculture. In fact, Edward's comment belied a fundamental problem for societies like the agricultural society in development for the Home District. Gentlemen amateurs such as Anthony[51] who toured the province (often part of a wider tour of British North America or the United States) mostly returned to England. Upper Canada did not have a university at which they could conduct research and instruct students; it possessed few printing presses and provided a minuscule market for learned publications. Thus, in the absence of gentlemen of leisure, such as Britain's gentry, Boulton's offensive faced a rocky beginning. The 6 July meeting descended into chaos.

Once again, we know of events from two sources, Mary O'Brien's recording of her husband's version in her journal, and Mackenzie's reporting of "The Official Riot" or "The Official Outrage" as he termed it in two separate issues of the *Colonial Advocate*.[52] According to O'Brien, the general meeting had progressed much better than her husband had anticipated. Some farmers were taking an interest, having "come forward in tolerable numbers and with their support." The only major problem, claimed Edward O'Brien, had been the attendance of Mackenzie. The society's provisional treasurer, Sheriff Jarvis, made an offer of membership to Mackenzie at the meeting, but Mackenzie refused to subscribe. Procedurally, this was a problem. The meeting was in a public place, and all were free to witness the society's proceedings, but only paid members possessed the right to offer their opinions regarding the creation of the society's constitution and bylaws.[53] Keeping Mackenzie silent proved to be difficult.

O'Brien's version of events describes how, shortly after the meeting began, Mackenzie rose to speak with "his accustomed impudence." He insisted that members of the public still had a right to speak because the May meeting had produced only resolutions, not a constitution. Secretary Elmsley and Treasurer Jarvis promptly tried to persuade him to be quiet, but Mackenzie persisted. To allow the dispute to be resolved, Chairman Crookshank was asked to leave the chair. Elmsley then grabbed hold of Mackenzie but let him go because others suggested Mackenzie should be allowed the opportunity to remain quiet. But Mackenzie would not do so. In O'Brien's view, "it became quite evident he came for the express purpose of interrupting the proceedings and causing a row." O'Brien and Gapper seized Mackenzie and forcibly removed him from the courthouse, locking the door behind him so that he could not return. According to O'Brien, the "little blackguard then like a spoilt and ill-behaved baby, kept thumping at the door." Sometime later, George Taylor Denison, resident of his Belle Vue estate west of York, suggested that the door should be unbolted as there were farmers who might wish to come in. He offered to personally see that Mackenzie remained silent. When asked how he planned to do this, he replied, "[B]y giving him a slap on the chops to be sure." The door was then unlocked, but Mackenzie had gone. Recalling this incident, Edward O'Brien suggested, "had Mackenzie come in a second time, a tumble from the windows of the Grand Jury room would most certainly be his fate."[54]

Mackenzie recalled events differently. At about 3 p.m. on that Tuesday, someone had dropped into his newspaper office and told him that the meeting had commenced at the York courthouse. An hour later, Mackenzie wandered over to the Grand Jury Room to find some twenty "country people" and about thirty of the "others" in attendance. Once again, those among the provisional executive of the new society assembled themselves within the bar. Samuel Jarvis, Elmsley, O'Brien, Crookshank, Gapper, Wood, and William B. Jarvis were all in attendance, along with the latter Jarvis' brother Frederick.[55] Secretary Elmsley was reading the names of the fifty-some farmers who had signed on as members, although Mackenzie remained sceptical that the list had not been padded. President Crookshank took the chair, and Secretary Elmsley began to lead the meeting through the clauses of the draft constitution. The proposed constitution called for a board of directors of sixty members to manage the affairs of the society. Sixty? Had Mackenzie heard correctly? He could not resist commenting.

What Mackenzie actually said at the meeting is unclear, for in his report to the readers of the *Colonial Advocate*, Mackenzie claimed that, when trying to raise his objections, it was not "possible for any one to offer a remark in a more unassuming and unobtrusive manner than he did on that occasion." Anyone who knew Mackenzie's bluster and tenacity must have smirked when they read this claim.[56] Mackenzie stated something about the folly of choosing sixty individuals to be directors of a society with less than one hundred members. Not even the East India Company or the Bank of England had that many directors, he reasoned. It seemed to him that the suggested number belied the fact that the temporary executive was very anxious to demonstrate that "party spirit, monopoly of office, and exclusion were by them discarded." How very liberal, Mackenzie commented, that they were prepared to make "not only members but also directors, of men who had neither subscribed one shilling nor probably ever given the society a thought."[57] He made a valid point; this was a ridiculous number of directors for a society of this size, and it was impossible to believe that sixty members or even a proportional quorum from across the district would gather at York each time a meeting was called.

His objections were not well received. When Mackenzie tried to forge ahead with his views, members of the provisional executive seated within the bar hissed, clapped, and stamped their feet to silence him, and more voices added to the chorus raised against him. Elmsley declared Mackenzie's views to be "obnoxious to the majority" and threatened to eject him. Shortly thereafter, O'Brien, Gapper, and a Mr. Young, who had also been seated within the bar, laid "violent hands" on him. After a struggle, the men let Mackenzie go. Chairman Crookshank resumed the meeting, and Mackenzie resumed his criticisms. The howling and stamping from those within the bar recommenced, and Elmsley, O'Brien, Gapper, and Young grabbed Mackenzie a second time and tossed him out of the courtroom, locking the door behind him. Mackenzie claimed that he knocked "peaceably" on the door, and it was opened partially for a moment so that Elmsley could hand Mackenzie his hat, while others hurled more abuse at him from within. Elmsley closed the courtroom door and locked it once again. Mackenzie walked back to his office planning to press assault charges, for he recalled being struck once, by Richard Gapper he believed.[58] This was the second time in as many months that Mackenzie had been expelled from a meeting at York. At the beginning of June, the stockholders of the Bank of Upper Canada had met to elect a board for the ensuing year. Not a stockholder but acting as a proxy on behalf of two others, Mackenzie insisted that this meeting, too, was public and all had a right to attend and speak. The meeting was suspended until Mackenzie departed. There was a considerable overlap of attendees at each meeting and few objected to Mackenzie being expelled again, for they had little appetite for an encore performance.[59]

Mackenzie's second editorial about the incident drew heavily on the symbolism within the courtroom. He – an elected member of the provincial assembly, representing the farmers of the Home District – had been "collared and insulted" by Elmsley and his "majesterial compeers" right in front of "the King's throne." He claimed the attack he had suffered at the hands of O'Brien and Gapper to be deeply troubling because they were appointed magistrates, yet they chose to employ violence in front of the symbol of the very Crown they had sworn an oath to judiciously defend. All the while, the other magistrates in the courtroom, plus the sheriff, watched "and evidently enjoyed the riot." He had a point: this was particularly ungentlemanly behaviour. Moreover, as Rhys Isaac notes, the coat of arms on what Mackenzie termed "the King's throne" at the front of the courtroom represented the very symbol of the "descent of authority from above."[60] Mackenzie concluded that if such abuses were allowed to occur to an elected representative of the provincial legislature at a meeting held only to establish an agricultural society for the Home District "the liberty of the subject and the rights of the many must be at a very low ebb."[61] Mackenzie vowed that "a jury of the country" would soon have the opportunity to decide if the conduct of Elmsley, O'Brien, Gapper, and Young had been proper, or whether it deserved "public reprobation." But Mackenzie did not press charges, for he was able to secure political advantage from the incident. News of George IV's June death reached Upper Canada in September 1830, consequently dissolving the provincial legislature with immediate effect and initiating new elections. Voters in Mackenzie's riding re-elected him, a result he announced to be a satisfactory verdict on the matter.[62]

The vituperative nature of the HDAS's birth highlights the intensity with which politics divided the provincial capital of York in 1830, spurred on by the fact that reformers dominated the House of Assembly and, of course, the columns of Mackenzie's *Colonial Advocate*. York's tory oligarchy may have wrested the agricultural society away from Mackenzie (who would take no further role in the organization), but it faced the challenge of extending its paternalism out into the Home District at a time when the old patron–client relationships that once underpinned its control were disappearing. Thousands of immigrants were settling the district each year with no connection to these established elites. Chapters 6 and 7 analyse the development and struggles of the HDAS from 1830 through to mid-century alongside parallel analyses of agricultural societies founded in the Midland District and the Niagara District. Doing so provides further evidence that faith in the power of paternalism to operate a successful agricultural society and to encourage agricultural improvements among a district's farmers was similarly misplaced by oligarchies of other Upper Canadian district centres. It also highlights how the 1830 agricultural societies legislation spawned semi-public agents of agricultural improvement that developed local variations in form and function.

# 6 The Home, Midland, and Niagara District Agricultural Societies, 1830–1850

In the summer of 1831, George Gurnett opined to readers of his *Courier of Upper Canada* that there was "every reason to hope that, the Province is likely to reap a good many of the advantages" from the district agricultural societies that had been created during the previous year. In several districts, the societies had been "warmly supported by all the most respectable, and best practical agriculturists in their respective neighbourhood." In others, however, the public had not "come forward to support them in the way that was expected," and in no district was any society "so universally, and so liberally patronized, as could be desired." Nonetheless, he had been assured "the greatest advantages, are beginning to be derived – particularly in the breed of stock."[1] Gurnett was being generous. In 1831, most district agricultural societies remained clubs of local oligarchies and were struggling to survive. Furthermore, the *Courier*'s editor was devoutly loyal to the Family Compact, and his comments whitewashed that oligarchy's defensive attacks in 1830 for control of the Home District agricultural society. A year on, it was similar to all other district agricultural societies that had organized in the province, demonstrating that its paternalistic leadership of agricultural improvement produced neither universal nor liberal support.

This chapter explores the leadership and development of district agricultural societies in the Home, Midland, and Niagara Districts from their establishment in 1830 through to 1850. In those twenty years, each society provides a window into state formation at the local level, including specific consequences that resulted from temporary lapses in public funding and the legislated changes in organization and operation discussed in chapter 4. Lapses in funding hobbled a district agricultural society, while permanent funding provided stability and consistency to its operations. But Upper Canada's district agricultural societies were shaped by other provincial legislation, too. A district was an administrative unit of municipal government undergoing its own transformations during the 1830–50 period. The original four districts, established in 1788, had been subdivided into eleven by 1830, each still containing massive tracts of territory. More subdivisions lay ahead. Throughout the 1830s and 1840s, immigrants flowed into the province at unprecedented rates, increasing the provincial population from some 213,000 to more than 725,000.[2] Eleven districts increased to twenty by 1849, at which time all districts were eliminated in favour of counties as the top level of municipal governance.

In 1830, the York oligarchy demonstrated how control of a district-wide institution funded with annual public grants was an attractive means to extend and strengthen its paternalism throughout a district. A culture of clientelism had long influenced Upper Canada's society and politics,[3] and the assumptions of clientelism at the heart of the agricultural societies legislation likely aided its passage through the legislature in 1830. But local oligarchies in the Home, Midland, and Niagara Districts soon faced a pressing need to devise new ways to connect with those they wished to be their clients in the out-townships of the district. Lapses in funding during the decade only exacerbated that need. Waves of immigration during the 1830–50 period created new communities across the outlying counties and townships of each district and gave rise to local leaders therein, over which members of the oligarchy of a district centre might have no direct or indirect control. Maintaining the primacy of the district agricultural society in each update to the governing legislation during the 1830s and 1840s provided a conservative means to sustain the leadership of the oligarchy-controlled district institution. But, by the late 1840s, primacy meant only that a district agricultural society served as the umbrella organization that coordinated the financing and activities of the township or county societies under its authority. Then, in 1849, when the province dispensed with the district unit of municipal governance, the district agricultural society was legislatively dispensed with shortly thereafter, replaced by county agricultural societies. The examples of district agricultural societies established in the Home, Midland, and Niagara Districts, presented in this chapter and the next, demonstrate variations in development in response to the peculiarities of their local oligarchy, geography, politics, and patterns of settlement, despite each being governed by the same provincial legislation.

## The HDAS

We know little of the early days of the HDAS in the immediate aftermath of the July 1830 battles. No copy of the original constitution has survived, but it appears that the society's executive tried to secure support throughout the district by appointing local patrons as directors. Richard Gapper's brother was elected to this role, as was Thomas S. Smyth, a farmer from Etobicoke, and Benjamin Thorne, whose store and milling complex founded the village of Thornhill along Yonge Street, not far from the Gappers and O'Briens. Sixty directors were not appointed. Despite the fierce reaction to Mackenzie's criticisms, it appears that the HDAS rejected the proposal. In fact, the HDAS would struggle to achieve quorum in the months ahead. A late September meeting to organize the first HDAS cattle show drew only Treasurer W.B. Jarvis, Secretary John Elmsley, and seven directors. Nevertheless, the society did host its first cattle show on 4 October 1830 at York's market square on the first day of a six-day "public fair and mart," proclaimed by Sheriff Jarvis, the steward of the fair. It was a small competition featuring a few "very fine cattle" on show. Only members of the society were able to submit entries, and the society awarded just a single prize each for the best male and female specimens of horse, cattle, sheep, and swine.[4] Its second cattle show, hosted in conjunction with the York Fair of May 1831, was a slightly expanded exhibition. Although attendance improved, it was not abundant. The show of stud horses was good, few cattle were exhibited "and nothing to boast of."[5]

In 1831, the HDAS elected one of its directors, Alexander Wood, as its second president.[6] The rather poor competition at the HDAS's October 1831 cattle show held in front of the York courthouse suggests his leadership did little to invigorate the society. Egerton Ryerson argued in his *Christian Guardian* that many who might have brought their livestock to the show had not been informed of the event because the HDAS had advertised only in newspapers with few subscribers throughout the Home District. He also suggested that individuals were circulating a petition requesting that a separate agricultural society be formed in the Home District at Newmarket and that it be allowed to apply independently for a provincial grant. It was a plan, he claimed, that had the support of Lieutenant Governor Colborne.[7] Although the plan did not come to fruition, neither did the HDAS make any effort to alter its character as a tory and York-centric organization. The following May, it hosted its cattle show on the eastern edge of town. Few attended on this rainy day, and the reform editor of the *Canadian Freeman*, Francis

Collins, complained that the HDAS hosted its show too early for the cattle of the "ordinary farmer … fed partly on winter fodder, and partly on spring grass." They could not compete "with the owners of old rich towns parks in the vicinity of York, where there are good opportunities of forcing early vegetation, even tho' he had a superior breed of cattle." A June show, he concluded, would permit "the poor to compete, in some degree, with the rich."[8] The tory HDAS executive did not act on their critic's suggestion and the society's fall cattle show appears to have been another casualty of the cholera epidemic that swept into the town that summer. The agricultural society revived and hosted its cattle show at Market Square in conjunction with a fair again in the spring and fall of 1833.[9]

Wood served just one year as HDAS president, as the society elected John Elmsley to be his replacement.[10] He was an interesting choice. Yes, the thirty-two-year-old gentleman was busy improving his Clover Hill estate, but in 1833, he was also creating deep rifts within elite circles by his conversion to Catholicism and his bitter and public debate about doing so with his friend John Strachan, the Anglican archdeacon. He also resigned his seat on the Executive Council over a policy disagreement with the lieutenant governor.[11] Not only was Elmsley unable to increase support for the HDAS across the district, as he would promise, his abrupt resignation as HDAS president in 1834 was its own scandal.

A May 1834 letter submitted to the *Colonial Advocate* by "An Old Canadian Farmer" reported the collapse of Elmsley's ambitious plans for the HDAS. Apparently, in October 1833, the president had arranged for the society to host a "Grand Ploughing Match" and had pledged to the society's executive that he would secure £30 in new memberships, sufficient to host a ploughing match for Canadians and those who had left "the Old Country" prior to the age of ten. To do so, the society would need to provide him £20 from its funds to be spent hosting another ploughing match open to all entrants. Despite Elmsley's efforts to raise memberships, and although the event was advertised with a substantial list of twelve prizes, including three different ploughs to be awarded to the top competitors, the event had to be delayed until spring because dry weather had made the soil too hard to plough.

On 23 April 1834, competitors and spectators assembled on John Elmsley's farm for the ploughing match among Canadian competitors but were soon informed that the contest would not be held and that Elmsley had abruptly resigned as HDAS president. He had just lost a libel suit lodged against George Gurnett and an indignant Elmsley declared that because "a Jury of the freeholders of the County had given a verdict in favour of the Editor of the *Courier*, and against him … he did not consider that the county had any further claim on him to forward their interests." But "An Old Canadian Farmer" claimed there was a more direct reason why Elmsley resigned. Rumours suggested that the ploughing matches could have gone ahead in October, but Elmsley had not been able to raise his promised £30 in memberships. Also, the only reason Elmsley had advertised the rescheduled event in April 1834 was to show off his benevolence to the district in the days immediately prior to his libel trial. If "An Old Canadian Farmer" is to be believed, Elmsley also left a local implement maker holding £30 worth of ploughs that the departed president had ordered as prizes.[12] The HDAS did manage to host a cattle show in October 1834, a commendable feat considering its disorganization and that cholera had swept through Toronto again that summer.[13] That year's expiration of government funding for district agricultural societies may have caused the HDAS to not host a spring cattle show in 1835, and while an August notice called for a meeting to plan a fall cattle show, no prize list announcement or report of the event has been found.[14]

President Elmsley's resignation was just the most prominent of the number of changes within the executive of the HDAS after 1834. None of the first three presidents would hold any future position within the society and only two gentlemen from the original 1830 executive continued to be involved. Just one of them served with any consistency. William B. Jarvis, the HDAS secretary since 1832, emerged as one of three longstanding leaders of the organization. From 1834 through mid-century, Jarvis would alternate the office of president and vice-president with Edward W. Thomson, while the young George Dupont Wells regularly served as secretary until his early death in 1854. William Atkinson would be the society's treasurer until the mid-1840s.[15] Jarvis continued to improve his Rosedale estate throughout the 1830s and 1840s, which became a showcase with its flower gardens, orchard, and vinery.[16] During this time, he maintained his post as sheriff of the Home District and was a leading member of many of its benevolent societies. Thomson would compete against Mackenzie for the Second Riding of York County in the provincial elections in 1834 and 1836, being successful in his second campaign.[17] Thomson possessed some five hundred acres throughout the district, and at Aikenshaw, the first brick house on Dundas Street between Toronto and the Humber River, he established himself as a borderline "fanatical" farmer, spending much of his time and money importing and breeding livestock.[18] George Dupont Wells was the son of Lieutenant-Colonel Joseph Wells, the legislative councillor, who had purchased his two-hundred-acre Davenport estate five miles north of York in 1821, and who had served as a director

Image 6.1. Edward William Thomson (1794–1865), ca. 1860 (Toronto Public Library, JRR 465 Cab.)

of the UCAS. George, just entering his twenties in the mid-1830s, spent his days on the estate like a quintessential gentleman farmer.[19] In contrast, Treasurer William Atkinson, a saddler from London who operated his shop in Market Square, represented the early 1830s rise of leading Toronto merchants who achieved "symbolic admission ... to a junior membership of the elite."[20]

Neither the provincial capital of York nor the Home District remained the same small settlements they had been in the spring of 1830 when the tory elite took control of the HDAS. York soon transformed from a small town of 2,860 individuals into the incorporated City of Toronto in 1834, which contained 9,252 residents.[21] The Home District's population stood at 28,565 in 1830, an early marker in a 128 per cent increase of the district's population between 1829 and 1835 alone. By 1841, the district's population was 51,043 and Toronto's population was 14,249. In response to this rapid population growth, the provincial government altered the boundaries of the Home District in significant ways. Prior to 1837, the HDAS had within its mandate

Image 6.2. "Aikenshaw," the estate of E.W. Thomson, c. 189–? Watercolour attributed to Owen Staples, 191–? after a pen and ink drawing by W.J. Thomson (Toronto Public Library, JRR 928 Cab.)

the rapidly developing townships contained within the two counties of York and Simcoe. Then, in that year, new legislation granted Simcoe County district status as soon as a courthouse and jail were built in Barrie. Following the elimination of districts in 1849, the Home District population, now over 77,000, would be divided among the Counties of York, Peel, and Ontario. By then, Toronto's population surpassed 25,000 residents (see maps 4.1, 4.2, and 4.3).[22]

Reflecting the rapid expansion of the Home District in the 1830s, the HDAS elections for executive offices in 1834 attempted to draw new local leaders of the district into the organization. Members elected David Gibson to replace Jarvis as secretary, following the latter's election to the office of president. Gibson was "a prosperous farmer" and cattle breeder on land he had purchased in 1829 at Willowdale on Yonge Street. Following his election to the HDAS executive, voters of the First Riding of York would elect him as their MHA in October. Regarded as a "moderate and sensible" reformer, Gibson was nonetheless a "reasonable but forceful proponent of radical reform." He served as HDAS secretary for just one year and remained a director until late 1837.[23]

The HDAS made a similar effort to extend leadership offices across the political spectrum in 1835. William B. Robinson, President Jarvis's son-in-law, and the Tory MHA who shared the riding of Simcoe with another vice-president, Samuel Lount, joined Vice-Presidents Thomson and Gapper. Lount, a blacksmith and farmer who was well known throughout the Holland Landing area as "one of the most highly respected settlers in the area," had won his first and only election as a reform candidate for the Simcoe riding in 1834. Bridging the political spectrum between Lount and the three tories, was the "conservative reformer" Charles Fothergill, now resident of Pickering Township. The other vice-president, Thomas Mair, represented the upper reaches of Simcoe County. An early settler of the Penetanguishene Road, he was an improving farmer who imported Durham cattle from England and, with his brother, operated the first set of saw and gristmills established north of Lake Simcoe. This 1835 increase in vice-presidents (from one to six) presumably meant that each vice-president was to represent two of the twelve townships within the Home District.[24]

While the HDAS's engagement of support from local patrons across the political spectrum might appear generous and welcoming, in fact, they were tinged with desperation. The legislation funding district agricultural societies had expired in 1834 and was revived for one year in 1835. Funding would not be renewed again until 1837. Thus, the HDAS needed all the help it could get to solicit memberships from the farmers of the district. In the polarizing politics of 1835–7, it was likely becoming evident to the elite at Toronto that they had little hope of attracting newly arrived settlers or established farmers of the Home District without the assistance of local patrons, even if their politics clashed with high tory politics of President Jarvis and other executives. This is also likely why the HDAS increased its number of directors to twenty-five.[25]

For similar reasons, the HDAS began to sponsor the creation of township agricultural societies in 1835, two years ahead of provincial legislation that sanctioned such societies at the township, county, or riding level. In fact, this move may show the HDAS acting in defence, for an "Agricultural Society for the East Riding of the County of York" in Whitby Township had begun operating sometime before March 1835. In the face of this show of independence, it appears that the best the HDAS could do was to provide the opportunity for "branch societies" of no less than twenty-five members to be created in each township of the district and appoint HDAS members as agents for soliciting subscriptions in thirteen of the district's more populous townships. The hope, of course, was that each branch society would see the benefit of maintaining ties with the HDAS at the capital. If an annual government grant was restored, the township societies could help the HDAS raise the money required to submit a request for provincial funding and might receive a portion of such funds in return. Yet, such efforts met little response. An "Agricultural Society for the Second Riding of the County of York," formed in Toronto Township in 1836 accepted branch status to the HDAS. But this was E.W. Thomson's influence at work. The existing society at Whitby maintained its independence.[26] Despite the transformations happening throughout the Home District, the HDAS made no attempt to host a cattle show or any of its meetings anywhere other than Toronto.[27]

Expansion of the HDAS was up against larger political, social, and economic forces that had caused some of the local patrons drawn into the society's executive offices to become more radical in their politics. First, the House of Assembly elections of 1836 witnessed more than twice the number of tory victories than reform across the province's ridings, a result partly engineered by the political machinations of the new lieutenant governor, Sir Francis Bond Head. Second, in the acrimonious political atmosphere that followed the July elections, Mackenzie, freed from his duties as a legislator by his electoral defeat, increasingly fanned the flames of discontent by gathering grassroots support for the list of grievances he had been collecting against the colonial administration. Third, an economic collapse in 1836 caused a shortage of cash in Upper Canada by the following year when low crop yields in the province made foodstuffs short in supply and high in demand.[28] As a result, the Home District became an economically and politically turbulent place. Although the tory-dominated legislature had renewed funding for agricultural societies in the spring of 1837, the remainder of the year was hardly the climate in which to grow membership in the HDAS.

Rebellion erupted on Yonge Street north of Toronto in early December 1837, and executives of the HDAS were among its instigators. Director David Gibson, one of the few remaining reform members of the provincial legislature, was a close associate of Mackenzie's in planning the uprising. Although Gibson would protect loyalist prisoners during the rebellion and deliver them to government forces once the rebel headquarters of Montgomery's Tavern to the south of his farm came under attack, Lieutenant Governor Head ordered Gibson's farm to be set ablaze. Gibson managed to hide from authorities for a month and escaped across Lake Ontario to New York state, where he would remain until 1848, despite having obtained a pardon for treason five years earlier.[29] Samuel Lount, who had been a vice-president of the HDAS in 1835, was one Mackenzie's strongest supporters of an attack on the capital, believing that the rebellion would be bloodless. Using his prominence in the Holland Landing area and the wider area south of Lake Simcoe, Lount was one of the first to arrive at Montgomery's Tavern on 4 December with a contingent of supporters. The following day, both Lount and Gibson prevented Mackenzie from setting fire to the Rosedale estate of fellow HDAS executive W.B. Jarvis. (And when Mackenzie finally did march his men down Yonge Street that evening, it would be Jarvis's picket that his men would face – and flee.) When loyalist forces routed the rebels on 7 December, Lount tried to escape to the United States but was captured. Convicted of treason, he was sentenced to hang. Despite the widely supported petitions seeking mercy for him, none other than Home District Sheriff Jarvis led Lount and another convicted traitor, Peter Matthews, to the gallows at the Toronto jail on 12 April 1838. That the office-holding world of the Home District remained a small and interconnected one is highlighted by John Ryerson's note to his brother Egerton, relating how "Sheriff Jarvis burst into tears when he entered the room to prepare them for execution. They said to him very calmly, 'Mr Jarvis, do your duty; we are prepared to meet death and our Judge.' They then, both of them, put their arms around his neck and kissed him."[30]

In the rebellion's aftermath, the HDAS drew back from the local patrons of the district to its deep roots among the conservative elite of Toronto.[31] This was partly due to a reduction in the Home District's territory with the creation of the Simcoe District. But the city was changing rapidly, too, and the Georgian gentlemen of old York would need to adapt to the dawning Victorian era of Toronto and its character as a growing commercial city. Upon the union of Upper and Lower Canada into the Province of Canada in February 1841, Toronto was no longer the provincial capital. To maintain some sense of prestige, the HDAS secured the patronage of Sir George Arthur, shortly thereafter. Arthur had served as the final lieutenant governor of Upper Canada and at Lord Sydenham's request had stayed in Toronto to serve as a deputy governor. But he remained only until March 1841 when he left for England, unimpressed with the job and suffering from ill health. Previously, the HDAS had secured the patronage of Lieutenant Governor Sir Francis Bond Head, and it would secure the patronage of successive governors general throughout the 1840s, from Sir Charles Bagot, Sir Charles Metcalfe, and Lord Elgin.[32]

By 1839, the HDAS could report a balance of nearly £300, with a Toronto Township branch and another formed in Markham Township helping raise the necessary membership fees to receive the provincial grant of £200.[33] A Newmarket Agricultural Society was also formed in 1839, fulfilling a demand first expressed eight years earlier. Another branch would be formed in Etobicoke Township in 1841, and Vaughan, York, Scarborough, and Whitby Township branches existed by 1844.[34] In that year, the HDAS promised to designate £150 of its funds to finance individual township auxiliaries in proportion to the amount they raised in subscriptions. To provide a closer connection with these societies beyond the distribution of public funds, the HDAS also amended its constitution to automatically consider presidents of the township auxiliary societies as *ex officio* directors of the HDAS. By April, new branch societies were founded in the townships of West Gwillumbury and Albion.[35] At a February 1845 meeting, the HDAS amended its constitution again to allow each township branch to elect two directors to represent the society on the district executive, and in return, the district society would elect one director for each township branch so that the district organization was represented in each township. If all the townships in the district were represented on the HDAS board of directors, it would swell to some sixty members.[36]

A rare, complete list of executives and directors of the HDAS for 1846 provides a useful glimpse into the spectrum of gentlemen holding office within the society at mid-decade. While the top offices remained the preserve of Jarvis, Thomson, and Wells, the HDAS relied on the support of gentlemen of the city and the Home District, whose interests reflected the commercial opportunities of the Victorian city and who were moderate conservatives or reform in their politics. Thomson, now warden of the Home District, was president, while Jarvis took his turn as vice-president. Joining him in this role was John W. Gamble, a tory district councillor, miller, manufacturer, and self-proclaimed squire of Pine Grove in Vaughan Township. Gamble owned a milling complex which included a grist and flour mill, a sawmill, a distillery, and a cloth factory.[37] Wells retained his position as secretary and was aided by an assistant secretary, the Toronto Township councillor, William B. Crew. Replacing William Atkinson as treasurer was a vice-president of the York Township Agricultural Society, Franklin Jacques. Wells was also a director, along with fifteen others, including William Henry Boulton, resident of the The Grange, mayor of Toronto, and tory MPP for Toronto; James Hervey Price, resident of Castlefield in York Township and reform MPP for the First Riding of York; George Skeffington Connor, a lawyer and "excellent Irish gentlemen in speech and manner"; J.P.

De la Haye, the French master at Upper Canada College; Justice Peter Lawrence; William Augustus Baldwin, son of William Warren Baldwin and brother of Robert; Richard L. Denison, the gentleman cattle and horse breeder at his Dover Court estate to the northwest of Toronto; George Miller, former vice-president of the HDAS and livestock importer and breeder at his Riggfoot farm near Markham; William Atkinson, the Toronto saddler and former HDAS secretary; and two Toronto butchers Jonathan Scott and Jonathan Dunn, whose involvement demonstrated the society's interest in supporting the fat cattle trade required to feed the growing city.[38] The list is deceptive, however, for the HDAS was struggling to keep united its members who held diverse political views and economic interests.

In February 1844, the HDAS announced plans to host its largest cattle show to date, and William Edmundson told readers of his *British American Cultivator* that this fall show would "be by far the largest and most splendid thing of the kind that ever took place in British America." The society planned to offer £200 in prizes in a wide range of competitions for its "Grand Autumn Fair and Cattle Show" to be held at the St. Leger Race Course to the north-west of Toronto on 2–3 October. The HDAS promoted the prize list throughout the summer, but Toronto newspapers suddenly announced in the last week of September that the society would hold only a single-day cattle show at its usual location on 9 October.[39] The prizes on offer for a much-reduced list of categories amounted to just £84.

President Jarvis had been in England for some months, and it seems that, in his absence, other officers of the society mismanaged the planning for a larger event. For his part, Secretary Wells had not attended any HDAS meetings for the past twelve months due to illness. A disappointed and upset Edmundson and M.P. Empey of Newmarket, Edmundson's fellow member in the Fourth Riding Branch Agricultural Society, complained to their MPP, Robert Baldwin, that the HDAS accounts of Treasurer Atkinson, were "so discreditably kept" that Atkinson could not account for "at least £1000." Although Empey demanded an investigation into the accounts of the HDAS, no such investigation occurred and the society re-elected Wells and Atkinson to their offices in 1845.[40]

The failed grand cattle show was just a symptom of a more fundamental malaise affecting the HDAS. Soon after Edmundson and Empey had penned their complaints, Secretary Wells appended a warning notice to the March 1845 advertisement for the agricultural society's spring cattle show: "N. B. – No politics !!"[41] The source of these "politics" is not entirely clear, although one can readily identify the main tensions of the day. In Montreal, the government under Governor General Charles Metcalfe was disintegrating, and the United Province of Canada was shifting to responsible government in all but name. Furthermore, a significant part of the administration's demise was a failed bill for a provincial university in Toronto that had attempted to strike a compromise between an Anglican-controlled institution and a secular one. One can reasonably imagine that it was the intensity of these two issues that generated "politics" in Toronto and the Home District, further enflamed by the accelerated decline of high toryism in the face of moderate conservatives and reformers, plus the partisan Toronto newspaper editors who fought and fanned the flames of such political battles with their editorials. Regardless, Wells's attempt to stifle the disgruntled within the society stirred up more political debate than it suppressed.

When W.A.C. Myers, editor of the tory *Patriot*, reminded his readers of the upcoming May cattle show, he suggested that "[i]f people should transfer a tithe of their attention and enterprise from politics to agriculture, Canada would be a far happier and better country."[42] In his report of the cattle show in the reform *Examiner*, James Lesslie (a friend and former business partner of Mackenzie), presumed Wells's warning had been mostly aimed at preventing political subjects being raised in speeches at the members' dinner after the cattle show. The warning was pointless, Lesslie argued, because everyone knew the "political predilections" of the agricultural society's tory executive. He claimed that the sizeable advertisement for the exhibition, "published for months" in four tory papers in Toronto, and "carefully excluded from every one of the Reform Journals" read by farmers, was a most obvious way that politics were at work within the HDAS. Leslie blamed this fact alone for the "falling off" of attendance at the society's cattle shows.[43] Inserting a notice "No politics!!" was "the sure way to create political party distinctions in the Society." If partisanship were used to announce the exhibitions, he argued, the "same spirit [would] pervade all its operations, preferences would be given to the competitors of a party, and the prizes [would] be awarded on the same principle."[44] Myers concluded that the *Examiner*'s attack boiled down to Lesslie's displeasure at not being supplied with an advertisement from the agricultural society.[45] This was perhaps true, but the exchange belied the fact that the HDAS was incapable of eradicated politics from its proceedings.

In his report of the speeches that followed the society's dinner to close its cattle show, Lesslie claimed that those delivered by high tories president Jarvis and Toronto mayor William H. Boulton were well received by the audience. However, when William Edmundson attempted to speak, he was "*coughed down*." Edmundson, Lesslie admitted, was not "a fluent speaker" and "tedious in his remarks," but he was a devoted member of the HDAS and "was entitled to the courtesy and respect of a gentleman."[46] Although the HDAS employed Edmundson's journal to publish its proceedings, Edmundson had complained publicly about the management of the HDAS and its finances. Moreover, in this moment of heightened political tension between the old tories and reformers, Edmundson's imprisonment on suspicions of treason following the Upper Canadian rebellion of 1837 was likely an added reason for the coughing. Among the insular and vehemently tory members of Toronto's post-rebellion elite, the fact that he had been implicated in the rebellion at all remained a permanent stain on his character.[47]

"Politics" continued to dog the HDAS. In a move that either stemmed from Wells's continued illness or the HDAS's petulance at having been the subject of Lesslie's editorial attacks, Wells sent no full announcement to any Toronto newspaper to advertise the society's October 1845 cattle show. Instead, he sent a short notice from his Davenport estate to the *Toronto Patriot* in late September directing the public's attention to handbills being circulated that contained complete details about prize categories. To this short notice, Secretary Wells again appended the warning "No Politics!" amplified by "GOD SAVE THE QUEEN!"[48]

Over the next five years, several transformations addressed the "politics" problem for the HDAS. By the end of the 1840s, nine township societies were active in the district;[49] thus, the HDAS became increasingly an umbrella organization, requesting, receiving, and distributing funds to these organizations. The HDAS did continue hosting its spring and fall district cattle shows, and for the first time since 1830, it decided to host the occasional cattle show outside of Toronto, with Richmond Hill the location in October 1849. In an all-too-late recognition of the district beyond Toronto, the HDAS declared: "The District Society belongs to the district, and not to the people in the neighbourhood of Toronto, who may very naturally wish to have its meetings held, and its money distributed among themselves." The experiment returned an increase in attendance, but at the end of that year, districts were eliminated as a municipal unit and, soon after, a revised agricultural societies bill replaced district agricultural societies with county societies.[50] The HDAS disbanded in the following year, leaving its long-time executive members to lead the County of York Agricultural Society, which held both its spring and fall cattle shows at Toronto and served as an umbrella organization for the township societies active in York County.[51]

## The MDAS

News of the public funding available for agricultural societies did not take long to reach Kingston. At the close of the parliamentary session in March 1830, Hugh C. Thomson, the MHA for the riding of Frontenac on the outskirts of the town, announced the passage of the agricultural societies' legislation to readers of his *Upper Canada Herald* and recommended that the farmers of the Midland District meet soon to establish such an institution. The district town of Kingston was far from the geographic heart of the sprawling Midland District; thus, Thomson suggested that the meeting be held "as near the centre of the District as possible," perhaps Adolphustown or Napanee. No one followed up on Thomson's useful suggestion. Instead, a public meeting to establish an agricultural society was announced for Kingston, although it planned to coincide with the spring Quarter Sessions in late April when many of the leading men from across the district would be in town.[52] By 21 April, unidentified individuals had prepared a draft constitution for a district agricultural society and published it in Kingston's *Upper Canada Herald*. Attendees to a subsequent meeting on the 27th at the Kingston courthouse considered the draft and returned the following evening to adopt the constitution and formally establish a "Midland District Agricultural Society."[53]

Despite its aspirations to be a district-wide organization, the three central executive members of the new society were unabashed Kingston gentlemen. Thomson chaired the first meeting of the society, and once the constitution had been ratified, he was elected the society's first secretary. It was an office he had held with the old Frontenac Agricultural Society between 1823 and 1825. A long-time gentleman resident of Kingston, David J. Smith, was elected Treasurer. To fill the position of president, members elected John Macaulay, who, during his time as editor of the *Kingston Chronicle*, had been a staunch supporter of agricultural improvement and the late ASMD and FCAS. By 1830, this Kingston gentleman with Loyalist roots reaching back to the beginnings of the town had become one of the town's most prominent businessmen. He would remain president to 1836, during which time he was at the height of his professional career and maintained strong connections to the government at the capital, including several government appointments.[54]

The MDAS constitution reflected the format of the old ASMD in that each of the district's five counties would organize a county agricultural society and be represented in the district society by a vice-president and a director. John Bennett Marks was elected as the vice-president for Frontenac County. A Royal Navy veteran who had served under Lord Nelson at the battles of Copenhagen, the Nile, and Trafalgar, he arrived in Upper Canada during the War of 1812 for service on the Great Lakes. Following the war, he assumed the role of clerk in charge of the Kingston Royal Naval Dockyard, later serving as secretary to Commodore

Image 6.3. John Bennett Marks (1777–1872), ca. 1860s (Courtesy of St. Mark's Anglican Church, Barriefield, Ontario.)

Robert Barrie, the Canadas' senior naval officer. Marks also farmed his property in Pittsburgh Township, outside Barriefield, not far from the dockyard.[55] Four other county patrons of the Midland District were chosen on 28 April to organize meetings in their respective counties for the purpose of establishing a county agricultural society and collect the subscriptions required for the MDAS to apply for the government grant. These gentlemen included: Isaac Fraser of Addington County, registrar and former MHA for the riding of Lennox and Addington, who operated a woolen mill in Ernestown Township; Allan Macpherson of Lennox County, miller, merchant, and patron of the developing town of Napanee; Asa Worden (Werden) of Prince Edward County, a farmer, miller, lumberman, and land speculator from Athol Township, who would be elected to the House of Assembly in the October 1830 elections; and William Bell of Hastings County, a justice of the peace and coroner from Belleville.[56]

The constitution of the MDAS stated the organization's mandate to be "the importation of live stock, grain, grass seeds, useful implements, and whatever else may conduce to the improvement of agriculture within the limits of the Society's influence." To accomplish this objective, the MDAS offered a certain degree of independence to each county society. Both the district and county boards were free to "frame by-laws and regulations, for their own guidance and conduct as they may seem fit," provided there was no infringement by one organization on another. For its part, the district society would synchronize the county and district activities and it would ensure that relevant matters discussed in the various counties were tabled at the district meetings.[57]

But the district agricultural society was born into an era of significant transformation for the Midland District, producing barriers to the Kingston-area improvers who wished to exert their leadership across the district. In the spring of 1830, the Midland District consisted of five counties, Hastings, Prince Edward, Lennox, Addington, and Frontenac, containing a population of some 34,000 inhabitants. By 1841, the district had been reduced to the County of Frontenac and the United Counties of Lennox and Addington with a population of nearly 29,000 persons distributed among Kingston – the first capital of the Province of Canada from 1841–4 – and the district's thirteen townships. A decade later, the Midland District was abolished, leaving its population of 45,423 divided among the counties of Frontenac (16,914) and the United Counties of Lennox (6,412) and Addington (11,820), plus the 10,097 residents of the City of Kingston (see maps 4.1, 4.2, and 4.3).[58]

The MDAS soon abandoned its original intention of organizing five county societies. At a July 1830 meeting, Frontenac and Lennox were the only two counties to report any activity. A little more than £36 had been collected within the former county; notably, more than £26 of that amount had been raised by town residents of Kingston and by those in Pittsburg Township on the edge of town. In his report, Frontenac vice-president J.B. Marks, offered the discouraging and telling news that many of the people approached in his county had stated they would gladly join an agricultural society if one were formed in their county and not for the district. Lennox County had recently held a meeting to elect six directors and Allan Macpherson as the county vice-president. But there was no word from the counties of Addington, Hastings, or Prince Edward. This indifference and reluctance prevented the MDAS from raising the £50 required to apply for the provincial grant. Thus, the July meeting ended with members encouraged to drum up new memberships and remit fees collected to the society's treasurer as soon as possible.[59] By November, additional subscriptions raised in Lennox and Addington Counties brought to the fledgling society just enough funding to petition the government for the annual grant. In doing so, President Macaulay had to admit that financial support had come from monies raised in Frontenac, Lennox, and Addington Counties alone.[60]

The western-most counties of the district were the most vocal about their dissatisfaction with their place in the Midland District. In June 1830, a meeting had been held at Hallowell, Prince Edward County (present-day Picton), for the purpose of electing a vice-president and directors for the MDAS. Yet, after some discussion, those present expressed their concern about "the insular situation" of Prince Edward County "and the disadvantage therefore under which the farmers would labour" if the funds they raised were to be controlled by the district society. As a result, the meeting adjourned with plans to hold another soon to establish a Prince Edward Agricultural Society. This occurred several days later, and the new society wrote to the MDAS stating the reasons why a separate society had been formed and of its intentions to cooperate for the general interests of the district.[61] The outcome was not surprising. At that same time, local leaders of Prince Edward County were lobbying the provincial government to grant district status to their county for similar reasons. The peninsula was a community unto itself, operating apart from the distant district town of Kingston. In 1831, the provincial government would grant Prince Edward County its wish. The new district of Prince Edward was proclaimed into effect in 1834 when a proper jail and courthouse were completed in the new district town of Picton.[62]

Separate status did not immediately translate into success, however. The Prince Edward Agricultural Society was unable to collect the £50 necessary to petition the government for an annual grant in 1830, plus there was an important legal question as to whether the society was even eligible for provincial funds. By July 1831, sufficient funds had been collected and the society informed Lieutenant Governor Colborne of the reasons for its independence, mainly because members were "unable to see a prospect of any good arising from continuing themselves with the Midland District Agricultural Society on account of the distance from Kingston" and the limited funding they would receive through that connection. Their petition gained them only £25, because Colborne, supported by the opinion of the attorney general, determined that the Prince Edward County Agricultural Society was still under the mandate of the MDAS and entitled only to such funds as it might have received as a county branch.[63]

In Hastings County, too, "it was found that the people were generally averse to any connections" with the MDAS; thus, no county branch was formed.[64] In his 1831 petition for the government grant, President John Macaulay seems to have been overly honest in his submission.

By having raised £50, the MDAS was entitled to £100, and the act directed only that funds be divided equally among county agricultural societies "in the event of there being" such organizations.[65] Yet, Macaulay requested only £75, three-quarters of the total bounty, because Hastings County had not yet established a society. Once again, in Macaulay's 1834 petition for the government grant, he felt obliged to note that no agricultural society had ever existed in Hastings County. And, in his petition for the following year's grant, he stated emphatically that all subscriptions to his society had been made by residents of the counties of Frontenac, Lennox, and Addington. "Nothing," he assured the provincial secretary, "has been subscribed in the County of Hastings, which has never taken any interest in the concern, and has no independent Society." On this occasion, Macaulay argued his society was eligible for the full sum, pledging as he had in 1830, that should Hastings County form a society before the end of the year – "a most improbable event" – to offer that county's agricultural society its due share of the provincial grant. This was the last mention of Hasting County by the MDAS, and in 1839, the issue would be settled in the same manner as it had been in Prince Edward County. New legislation elevated Hastings County into the Victoria District with Belleville as the district town.[66]

The counties of Lennox and Addington were also cause for debate while the MDAS struggled to establish its influence across the district. The society's constitution recognized Lennox and Addington as separate counties because it was commonly believed that the two counties had been combined only for the purpose of forming an electoral riding for the House of Assembly. But provincial authorities considered them to be also a single municipal entity. Learning of this in 1831, President Macaulay informed the MDAS that its constitution required amendment to recognize that the society was composed of only two counties: Frontenac and Lennox and Addington. Members expressed "great surprise" at the news. Nevertheless, in January 1832, the society amended its constitution, scaling back its executive offices to two vice-presidents and twenty directors representing each of the townships contained in the two counties.[67]

Such efforts aimed at determining what defined the Midland District impeded the agricultural society's early activities. Unable to petition the provincial government for a grant until November 1830, the society's first real exertions, beyond gathering memberships, were made at a February 1831 meeting at which society members discussed and reported on Lieutenant Governor Colborne's query as to the improvement of the province's roads.[68]

In contrast to the HDAS, the MDAS planned to have an annual cattle show and ploughing match held within each county during October, followed by a district show. Competitors and their livestock resident in either Frontenac County or the counties of Lennox and Addington could enter whether they were members of the agricultural society, and in the interests of promoting agricultural improvement, the society's constitution contained a clause stating that "prizes shall always be awarded in live stock, grain, grass seeds, and useful implements of husbandry, or works on Agriculture of established merit to the extent of the annual grant from the public funds." But in 1831, the society had spent most of its funds elsewhere, and "owing to the impracticability of procuring grain, live stock, or proper implements of husbandry," the MDAS offered £30 to each county society to be used as prizes at their cattle show. There would be no district event.[69]

Those in attendance at the Frontenac County's first cattle show at Waterloo in October 1831 assembled to hear Vice-President Marks acknowledge the less than auspicious beginning for MDAS. "Although many public spirited men residing in Kingston, who are not connected with farming pursuits, will remain subscribers," he pointed out, "it must rest with the Farmers to meet the other subscribers with a corresponding liberality, and it must ultimately depend on the Farmer, whether this Agricultural Society, formed under so many advantages, shall stand or fall." He commended those who "begged money from house to house for its support." Secretary Thomson expressed similar sentiments at a society meeting in April 1832, noting that the progress of the society had not been as great as "friends of agriculture might desire."[70]

The MDAS determined that the county cattle shows of 1831 would be replaced by a first district-wide exhibition scheduled for October 1832. Doing so, members hoped, would increase the amount and number of premiums offered, and "by throwing open a wider field of competitions than mere County Shows can afford, induce the farmers to take a deeper interest in the proceedings of the institution than has yet been exhibited." Only £124 remained in the district society's account, demonstrating its reliance on the annual government grant. Nevertheless, the MDAS planned to offer £120 in prizes at an October cattle show and ploughing match to be held at Bath. The prizes for this event would be offered in the form of money, as the society had given up on its original plans for prizes to assist agricultural improvement.[71]

Success continued to elude the MDAS. The cholera epidemic that killed thousands in Lower Canada and Upper Canada in 1832 prevented any planning meetings to be held, and by autumn, cholera continued to spread throughout the district resulting in the event's cancellation. Thus, the only activity for the year was an October ploughing match hosted by the Frontenac County Society at Waterloo.[72] Perhaps this was a mixed blessing for the fledgling society. The MDAS had planned to award all but about £4 of its finances on premiums at the show, on the assumption that the district society would be able to refresh its accounts through annual subscription fees and the annual government grant. As it turned out, the society was unable to collect its yearly membership fees in 1832 and subsequently failed to qualify for the annual public grant. The MDAS relied on the unused prize money to help it struggle along into the next year.[73]

The MDAS requested the establishment of chartered fairs for the district. At the directors' meeting of January 1832, it was thought two such fairs should be held in Kingston each year, plus two at other places in the district. The matter was taken up again at a public meeting in Kingston in March, at which Thomson acted as secretary. A petition was drafted and sent to Lieutenant Governor Colborne citing the increasing population, the success of such institutions in Great Britain and Europe, and "the advantage to the commercial and agricultural interests of the whole Province of Upper Canada." He concurred and ordered a Kingston Fair be established.[74] At the annual meeting of MDAS in April 1833, Vice-President Marks recommended that the society's cattle shows and other meetings should take place in conjunction with these periodic fairs as a means to increase support of the society and effectiveness.[75] Later that year, the FCAS scheduled its fall cattle show for the second day of the Kingston Fair "with the view of obtaining a greater attendance of Farmers." The hosting of both events on the commons in front of St. Andrew's Presbyterian Church was a relationship that lasted until 1841. The lieutenant governor also authorized a fair in Napanee in April 1834, and the Lennox and Addington County Agricultural Society decided it would hold its fall cattle show in conjunction with the September fair. But reports of the fair make no mention of the agricultural society's cattle show, and September 1838 is the first recorded instance when agricultural society the coordinated its cattle show with that chartered fair.[76]

Considering the failed 1832 district show and the fragile existence of the MDAS, Marks's report to the society's April 1833 annual meeting delivered the accepted view of his FCAS that the district agricultural society had been established "on too large a scale to reap all the advantages intended by the legislature." No matter where any district event might be held, the distance would be too great for farmers to bring their livestock, manufactures, or ploughs to compete. Three years' experience, Marks argued, had taught his branch that separate county agricultural societies for the district would be more convenient and beneficial.[77] With his FCAS being the only source of activity for the district,[78] Marks again expressed his frustration about the society's mediocre support and successes to those attending its October cattle show: "If we have hitherto failed in establishing a general District Society let us endeavour to keep the Institution in existence on a smaller scale, by inviting persons to join in a County Society, who are willing to take an active part in its management, when by perseverance no doubt but your labours will be crowned with success."[79] Soon, Marks would receive his wish, but not in the manner he expected.

As public funding for agricultural societies came to an end, the MDAS disbanded at its annual meeting of April 1834. In its stead, two new county societies were formed, with the former district society's remaining £79 divided between Marks, representing Frontenac County, and Peter Davey, representing Lennox and Addington Counties. In a final tribute to the old society, thanks were offered to Marks and the former MDAS executive for their efforts on behalf of the society and the district's agriculture.[80] Significantly, the society's two principal supporters, President John Macaulay and Secretary Hugh C. Thomson were not present at the final moments of the organization that they had helped form. Macaulay had arrived for the start of the meeting, but soon left in haste for the home of Thomson, who was suffering through his last moments of life. He had suffered a haemorrhage of the lungs in York in December and had only just returned to Kingston with the opening of navigation on Lake Ontario. Two hours after the disbandment of the MDAS, the forty-three-year-old Thomson died with his younger brother Edward (the vice-president of the HDAS) at his side.[81]

Did the MDAS disband completely? The actions of President Macaulay muddy the waters. The only activities conducted during 1834 were the FCAS's hosting of a cattle show in conjunction with the autumn Kingston Fair and a ploughing match held during the following week.[82] Yet, at the close of the year, Macaulay petitioned the government as president of the MDAS for the annual bounty. This is rather curious for two reasons. First, the funding legislation had expired; new legislation would not come into effect until April 1835. Second, Macaulay submitted to the provincial secretary a certificate from the society's treasurer for little more than £39, far short of the £50 required under the now-expired legislation. Macaulay wondered if his society might be eligible for an amount of public funds in proportion to what had been raised, something the defunct legislation had never allowed. He wrote to the provincial secretary again on the last day of 1834, transmitting a new certificate as the required £50 had been raised. This time he requested the full grant of £100, and Lieutenant Governor Colborne agreed. It is unclear why Colborne considered the MDAS to be eligible for funding at that time.[83]

Resuscitated by this government grant, an "annual meeting" of the MDAS was held in April 1835 in Kingston. The society's treasurer reported funds of nearly £150 on hand, and with the possibility of receiving another £100 public grant for 1835, members agreed to form committees to solicit memberships in each township of the district as well as in the Town of Kingston. When the committees met at Bath in July, they had collected almost £70, more than enough to petition the government for the bounty. This was not entirely good news, however. No subscriptions had been collected in three of the townships, and once again, the greatest number of memberships by far had been purchased by gentlemen resident in Kingston.[84]

Those attending the meeting at Bath believed it necessary to create a new constitution for a new "Midland District Agricultural Society." As of 1 September 1835, both county societies would be folded back into an MDAS, although they could act as "separate Societies to choose their own Directors and Secretary and make such other arrangements as may be necessary for Cattle Shows and premiums." The bounty collected from the government, plus money raised in Kingston, would be distributed by the MDAS to the county societies in proportion to the amount of subscription fees raised therein. With Macaulay restored as president and a new secretary chosen to replace the deceased Thomson, the Bath meeting's chairman, J.B. Marks, expressed his hopes that the society's "bad times" were behind it and that its business could, in future, be supported "by cheerful subscriptions, instead of the present unpleasant practice of begging from door to door."[85]

The renewal was short-lived. With the end of government support in 1836, the MDAS kept its executives in office for the coming year and appointed committees for each township and for Kingston to conduct a membership drive. No further business would be transacted by the society until the committees congregated in June with the funds they had been able to raise.[86] Two small cattle shows, one at the Kingston Fair in October for Frontenac County and one at Bath in November for Lennox and Addington Counties, plus a new rule that competitors had to have their yearly subscription paid by 1 October 1836 to be considered as members of the society suggest the society was once again struggling to survive.[87]

Several developments affected the MDAS that year. First, the organization lost its first and only president when Macaulay moved to Toronto to assume his appointment as a legislative councillor and surveyor general.[88] At the same time, Marks won the election for the Frontenac riding in the provincial elections of July 1836. It was a victory for the tory candidate that added to his assortment of various public offices held in the Midland District. He had been appointed a justice of the peace in 1825, and in the years ahead, he would serve as colonel of the militia to defend the province in the fallout of the December 1837 rebellion. He would also be chosen warden of the Midland District upon the creation of such a position in 1841. By the time of Marks's retirement from the dockyard in 1845, a notice published in the *British Whig* stated that "no gentleman has so well deserved, and at the same time so universally received, the esteem and approbation of all classes of the Canadian community."[89] As we saw in chapter 4, the rookie MHA would play an instrumental role during his first session of the provincial legislature in shepherding legislation through parliament that renewed funding to district agricultural societies. Once the new agricultural societies legislation became law in early 1837, Marks would replace Macaulay as president of the restored MDAS. He would retain this office until his resignation in 1850.

Buoyed by renewed funding for the next four years, the MDAS under Marks's leadership appointed a committee at its April 1837 annual meeting to refashion the society's constitution for yet another rebirth. Reported to a July meeting in Bath, the revised constitution provided for new "County Boards" for Frontenac and for Lennox and Addington to be composed of two vice-presidents and seven directors with a mandate of governing county agricultural societies, coordinating cattle shows, ploughing matches, and other activities within their respective county.[90] As for the MDAS, it would now function as a general board, composed of a president, a recording secretary, a corresponding secretary, a treasurer, and the vice-presidents and directors of the newly named county boards. Responsible for all money received from the government, as well as membership fees in the Town of Kingston, the general board would also organize an annual district cattle show.

The new constitution doubled membership rates to 10s a year and placed new requirements on those entering competitions at the county or district shows. Competitors had to be members of the society and were required to submit to the society advanced written notice of their intentions to compete. Furthermore, anyone entering an animal for competition would have to produce an authenticated statement of its pedigree and particulars as to how it had been raised. Anyone competing for prizes in crop categories would have to supply a written statement as to how the crops had been raised.[91] The assumptions informing these rules did not square with the realities of the Midland District and likely hindered memberships from those who could not afford the additional cost. It limited entries into competitions from those not literate enough to write and submit a report before the deadline.

In fact, the MDAS sponsored little activity over the next two years.[92] The Frontenac County Society's October cattle shows in 1837, 1838, 1839, and 1840 continued to be hosted in conjunction with the Kingston Fair, the first two held at Selma Park and the latter two held in front of St. Andrew's Church. As mentioned earlier, the Lennox and Addington society appears to have held one cattle show during the Napanee Fair in September 1838.[93] Not until June 1839 did the MDAS host a district cattle show but, this time, at Blake's Tavern, Links Mills, Addington County, with prizes of more than £94 as well.[94] The MDAS hosted a second district-wide cattle show at that same location in June 1840,[95] but the premiums awarded, plus the cost of purchasing a stallion for breeding purposes (see chapter 7), prohibited the society from hosting its annual event in 1841. Instead, all funds were directed to the county cattle shows and ploughing matches, the Lennox and Addington event being held at Bath and not the Napanee Fair.[96] The MDAS rallied in 1842, hosting a district cattle show held at Waterloo in June, with £100 in prizes.[97]

During the early 1840s, activity within the district became increasingly centred on the county and township branches. For example, after the June 1840 militia muster on Wolfe Island, across the harbour from Kingston, several men met to form the "Wolfe Island Branch Agricultural Society of the County of Frontenac." They did so because residents of the island had been "disbarred from the benefits" provided by government funding "by the great inconvenience under which they labour of attending meetings, and competing with farmers on the main land."[98] In response, the MDAS reformed its constitution to define more clearly the roles of its various executive members and branches. County Boards would now operate at arm's length from the district board; representing them on the general board would be a single vice president. In turn, the county boards were reorganized to better represent the townships within their borders. The county boards for Frontenac and Lennox and Addington (the latter two being considered separate) would consist of two vice-presidents, four directors from each township, a secretary, and a treasurer. The Township of Wolfe Island would also adhere to this structure. County boards would continue to be responsible for scheduling their own meetings, raising their own subscriptions, and managing their own cattle shows and ploughing matches. At each quarterly meeting, they were to prepare a report on the state of agriculture in their region and were to submit it to the corresponding secretary of the general board, who would then prepare and publish a general report on the state of agriculture in the Midland District "for the general information of the country."[99] For its part, the general board of the MDAS would, in accordance with the provincial legislation, collect the subscriptions submitted to it by the county boards and petition the government for the annual bounty. It would control the funds it received and distribute them in a proportion double to the amount subscribed by each board to a maximum of £20. The general board had the authority to purchase livestock, grain, seeds, and implements of husbandry and direct them to be sold at public auction at the district or county cattle shows or to members of the society.[100]

In 1843 and 1844, the MDAS agreed to several more constitutional amendments to provide for the growth of township agricultural societies. First, the MDAS offered any agricultural society that might be established on Amherst Island the same privileges as those enjoyed by the Wolfe Island Agricultural Society.[101] Second, the MDAS responded to William Edmundson's plan published in his *British American Cultivator* for establishing a provincial agricultural association. It replaced the two county agricultural societies with the possibility of a "Branch Agricultural Society" to be formed in each township of the district. In response, eight township societies were formed in the Midland District during 1844, each hosting its own cattle show. To ensure the branches received sufficient public funding, the MDAS committed to reserving only £50 from the government grant, plus any membership fees collected from Kingston residents.[102] To formalize these changes, the MDAS was dissolved at its annual meeting in July 1845. That same meeting approved a new "Agricultural Society of the Midland District" based on a constitution with a very simple structure of nine articles constructed around the township branches created in the previous year. To encourage remaining townships to create a branch society, subscription rates for membership were halved to a minimum of 5s.[103] By the end of 1845, ten of the thirteen townships in the district had active agricultural societies, and as a consequence of the new constitution, there was improved attendance at the MDAS annual meeting of May 1846, providing for more vigorous discussion.[104]

In early 1847, the MDAS revised its constitution yet again, permitting each township branch to set its own membership rate and any branch of at least one hundred members who each paid 20s in membership fees to host a district cattle show and ploughing match in their township. At its meeting in August of that year, the MDAS refined details of financial arrangements with its branch societies. First, it set £20 as the amount distributed to each township society, so long as they provided the MDAS a minimum of £12 10s. Second, any township that could not attain that target, would be provided government funds in proportion to the amount they could provide the MDAS. Clearly, the district society was devolving into an umbrella organization for its township branches, for a February 1848 financial report showed that the MDAS held a little more than £60 on account (just slightly less than what was reported in the previous two years) but had expended over £404 in sponsoring activities of its township branches. That same month, the MDAS announced it would host a district cattle show at Mill Creek (present-day Odessa) in early June, should it collect £50 before 1 May. There is no evidence of such support from its townships or that the event materialized.[105] A last moment of prominence for the MDAS came during 1849, when Kingston was chosen to host the annual Provincial Agricultural Exhibition (see chapter 8). Yet, the MDAS failed to host its regular meetings or cattle shows that year because hosting the September exhibition required Marks and his directors to form a local committee to plan and manage the event. Another wave of cholera in the Midland District that spring and summer also limited the society's activity.[106]

John B. Marks resigned as president of the MDAS early the following year. Now in his early seventies, Marks expressed his pride at the success of Kingston's provincial exhibition, the fact that the MDAS was debt-free and, most importantly, that there were many within the society who might be chosen to fill the office he was vacating. But shortly thereafter, the MDAS was no more, for the elimination of the districts in favour of counties transformed the district agricultural society into the Agricultural Society of the United Counties of Frontenac, Lennox and Addington.[107]

## The Niagara District Agricultural Society

A meeting of local gentlemen founded the Niagara District Agricultural Society (NDAS) at a St. Catharines hotel in June 1830. No copy of the original constitution exists but there are certain aspects of the NDAS that can be gleaned from a list of its founders. Six individuals were chosen to draft a constitution for the society: George Adams, the agricultural improver, magistrate, and Welland Canal investor who farmed near St. Catharines; Samuel Wood, the Niagara District coroner who lived at Grantham; Adam Stull, a farmer from Grantham Township; Johnson Butler, a gentleman merchant from St. Catharines; Cyrus Sumner, the well-known physician of the Niagara communities who had been a member of the NAS; David William Smith, a lawyer from St. Catharines; and John Gibson "a well-known Englishman," a "successful farmer ... an enterprising wool carder ... cloth dresser," and justice of the peace.[108] Several of these gentlemen knew each other from service in the Upper Canadian militia during the War of 1812 and through Freemasonry. Notably, the committee added a clause to the proposed constitution stating that only farmers were eligible for election to an office of the new society.[109] The June meeting elected directors for each township of the district who were then assigned the task of soliciting memberships so that the NDAS might raise the £50 necessary to apply for the provincial grant. This took the NDAS six months to accomplish and only upon receipt of a donation by the wealthy merchant Samuel Street Jr., who, for his generosity, was "constituted a member for life."[110]

The NDAS aimed to represent a Niagara District with a population of almost 21,000 individuals in 1830, divided into eighteen townships contained within the counties of Haldimand and Lincoln, plus the Town of Niagara. The district's population increased to 32,504 by 1841, and correspondingly the number of townships had been expanded to twenty-three by that date. Then, in 1845, some of the townships of Haldimand and Lincoln Counties were expropriated to create a new County of Welland. By 1851, after the elimination of districts, a population of 53,029 was divided among the County of Haldimand (27,378), the United Counties of Lincoln (16,473) and Welland (14,749), plus the Town of Niagara (3,282).[111] From its very start, the NDAS was infused with support from both agricultural and commercial aspects of one of the longest-settled districts of the province, which now featured the Welland Canal cutting across its peninsula to circumvent Niagara Falls' impediment to Great Lakes shipping. St. Catharines was not the administrative centre of the Niagara District – the Town of Niagara maintained this role; however, St. Catharines was the district's emerging commercial centre with the opening, then improvement and expansion, of the Welland Canal. A village of only 384 persons in 1827, its population would swell to 1,130 by 1835. In 1845, St. Catharines was incorporated as a town with a population of some 3,500 persons (see maps 4.1, 4.2, and 4.3).[112]

George Adams was the inaugural and long-serving president of the NDAS, who had become well known throughout the district through his government appointments and personal connections with other Niagara patrons. Upon his death in 1844, the *St. Catharines Journal* remembered Adams as "one of the oldest and most prominent inhabitants" of the Niagara District, whose "energy and example" had been the "chief cause" of the "triumphant success" of the NDAS. Another

Image 6.4. George Adams (1771–1844), ca. 1820. Artist unknown (John Ross Robertson, *The History of Freemasonry in Canada*, vol. 1, [Toronto: 1899], 453. Toronto Public Library)

commentator noted that in the St. Catharines area, "none was more *loved* and *beloved*." Throughout the period under study, he served as a magistrate on the bench of the Court of Request; thus, residents of the district remembered him as, "as their general arbiter – settling all their differences, and zealously promoting harmony and peace."

Born in Londonderry, Ireland in 1771, Adams had immigrated as a child to Canandaigua, New York, where he had learned the currier and tanner trade. He arrived at the Niagara settlements during the American Revolution and by the mid-1790s had set up shop as a tanner in Queenston. Significantly, the young Adams found himself a valuable patron in Robert Hamilton, who established him as a partner in a milling operation on the Niagara River with another local entrepreneur, Benjamin Canby. Adams purchased two hundred acres of Robert Hamilton's land near Twelve Mile Creek (later St. Catharines) next to that of Thomas Merritt, the father of William Hamilton Merritt, the future promoter of the Welland Canal. Adams and the younger Merritt would forge an association with each other as promoters of the canal project, (they also served together as magistrates), although Adams's support would wane due to decisions about the route of the canal.[113] Adams began his ascent into the gentlemen circles of Niagara society in 1796 by joining the Freemasons, acting at times as secretary of the local St. John's Lodge. By 1820, he had attained the status of provincial Grand Master of the Upper Canadian Lodge, a position he would hold for two years.[114] Adams, like many settlers of the Niagara District was a veteran of the War of 1812. He had been appointed lieutenant of the 2nd Flank Company of the 1st Lincoln Regiment and was severely wounded during the defence of Fort George in April 1813.[115]

Yet, it seems the wounds that Adams suffered during the war did not hinder him from agricultural improvement. One author noted that when Adams purchased his farm, it was "in the worst possible state, completely overrun with woods." He used the off-farm income that he received from his public appointment, as well as his investments in the Welland Canal, to employ a gardener and experiment with new crops and breeding stock. As a result, Adams's farm was a showpiece of improvement by 1841. He was, perhaps, the first to introduce the cultivation of rutabaga, mangel-wurzel, and sugar beet into the district. He grew record-sized onions, and like most of his contemporaries he tended to an orchard, which by the 1840s spread to several acres around his house. As for Adams' cattle breeding, one observer suggested that no herd in Canada surpassed his "in blood and symmetry." In 1842, W.H. Merritt, by then MPP for Lincoln North, complimented the President of the NDAS, as "the most direct and appropriate channel through which any communications to the public, on any subject relating to [agriculture], can be made."[116]

The NDAS struggled through a difficult first year and a half, although it was able to host a spring cattle show in Chippewa and a fall cattle show at Clinton in 1831.[117] To expand support, the society revised its constitution in January 1832 by increasing the number of directors per township from one to five, presumably to reach out the local patrons of the district's townships while at the same time reducing the burden on directors of scouring townships to gather memberships. The society also hoped to attract more prominent gentlemen residing in the district by providing a life membership to those who subscribed £2 10s.[118] That spring, the society hosted its cattle show in St. Catharines.[119]

The NDAS received considerable assistance in the promotion of its activities from Hiram Leavenworth, editor of *The Farmers' Journal and Welland Canal Intelligencer*. An American who had once worked with

Mackenzie on his *Colonial Advocate*, in St. Catharines in 1826, Leavenworth began publishing his own newspaper devoted to "the agricultural interests of this fertile and growing land."[120] In January 1832, he supported the NDAS by requesting that farmers of the district aid "the unwearied efforts of those now struggling not only for a measure eminently calculated to encourage the agriculture of the district, but to promote the general prosperity of the whole country." He also sent copies of the issue to several gentlemen throughout the Niagara District who were not subscribers of his paper but whom he hoped, as patrons, would lend their influence and support to the agricultural society. It had little effect. Early in the following year, Secretary Wood was pleading to the officers of the NDAS "to use diligence in obtaining funds" and turn over to him all monies raised so that the society might secure the government grant.[121]

The NDAS held its spring cattle shows in St. David's in 1833 and St. Catharines in 1834. From the scant information available, one cattle show per year appears to be all the activity the NDAS could afford to conduct. Perhaps in response to the end of public funding, the NDAS amended its constitution at its June 1834 annual meeting by removing the clause: "none but actual *Farmers* can serve as officers of the society" and replaced it with an allowance for "any person … considered competent" to fill any position in the society, "no matter what his profession or calling may be."[122] With the renewal of public funding in 1835 the NDAS was able to host a spring cattle show at St. Catharines and a fall cattle show at Niagara.[123]

Early the following March, President Adams and several members of the society, including three of the area's MHAs, William Hamilton Merritt, David Thorburn, and George Rykert, petitioned Lieutenant Governor Bond Head for a semi-annual fair to be established in St. Catharines on the last Thursday in May and the first Thursday in November.[124] By May, Bond Head had approved the plan, and the NDAS organized its next cattle show to coincide with the first scheduled fair on 26 May. After the event, Leavenworth proudly declared that "[t]his holiday of our Farmers" had proved to be a well-attended success. The NDAS hosted its fall cattle show at Niagara again,[125] but Leavenworth sensed that the society was again struggling for support, noting, "the praiseworthy exertions of a number of individuals in this vicinity, and a very few residing in other parts of the district, in keeping alive the almost dormant energies of our Farmers generally, in regard to this Institution." He hoped the active members would continue to sustain and expand the society.[126] But the NDAS fell dormant until 1838 when an urgent circular was issued for officers and members of the NDAS to meet on the 31 May "for the purpose of devising some plan to fulfil [*sic*] the required conditions ... to obtain the annual grant of Parliament." The society planned to host a cattle show in St. Catharines in late June.[127] It proved to be an important moment in the public life of the district, for significant events transpired that month.

Following the Upper Canadian Rebellion of December 1837, the Niagara District had been the scene of a variety of "Patriot" raids and border skirmishes involving escapees from the failed Toronto and Western District uprisings, aided by American sympathizers. One of the last such raids began on 11 June 1838 when approximately thirty rebels attacked the Short Hills area outside of St. Catharines. This area was known for its political radicalism and, on 20 June, rebels executed a raid on the village of St. Johns. Between these two attacks, Leavenworth wrote an editorial proclaiming that the NDAS's upcoming cattle show would be a timely demonstration of the goodness of man rather than "the rude arts of war that lately burst on our slumbers." Mourning "the evils entailed on our nature," he explained how he was "sickened at the sight of our brother imbruing his hands in the blood of his brother." In sharp contrast, Leavenworth anticipated the upcoming cattle show, noting "with pleasure, the efforts that are made to meliorate society by uniting its members in stronger and sweeter ties of unity and love." He declared: "Let the formation of a Niagara District Agricultural Society, be a rallying point to bring forward the sturdy yeomanry of the country to a knowledge of the duties and privileges set before them, in this favoured and fertile colony: and as time fades away from our grasp, we shall be more and more an united, loyal, happy and triumphant people."[128]

Leavenworth anticipated more from a struggling agricultural society with limited funds than it could be expected to produce. Nevertheless, by the 28 June cattle show, the patriot raids had ended and many of the rebels captured. Therefore, the agricultural society's event at St. Catharines reaped the benefits of sudden popularity, for it served as the first opportunity for area residents to celebrate victory, and it was weighted with the additional significance of being the day of the young Queen Victoria's coronation. On the evening of the cattle show, an illumination took place in St. Catharines, "accompanied by the usual demonstrations of joy, on such occasions, such as the throwing of fireballs, the firing of musketry, and the lighting of bonfires." In the week that followed, Leavenworth was able to note that "notwithstanding the recent insurrectionary disturbances at the Short Hills, and the consequent unsettled state of publick feeling, the concourse of Farmers and others interested, was much larger than usual."[129]

The significance of the agricultural society's timely revival was not lost on its executive of the NDAS either. In "consideration of the particular interest taken in the success" of the institution, a list was published in the *St. Catharines Journal* of those gentlemen who had voluntarily subscribed £1 each to permit a resumption of its operations. As a reward for their subscription, each of the forty-one individuals was appointed a director of the society.[130] However, these one-time supporters could not be relied on for continued support. Less than a year later, Thomas Sewell, editor of the *Niagara Reporter*, grew impatient with the inactivity of the NDAS. He had heard rumours of a forthcoming spring cattle show but had no details to pass along to subscribers. Chastising the society's secretary for not having submitted notice of the cattle show to his office, Sewell wondered if the secretary was sleeping on the job and suggested that the President and Treasurer "give him a shake." The society did rouse itself to host a poorly attended cattle show in Niagara in June but proved unable to host another in fall.[131] Increased and sustained funding to district agricultural societies permitted the NDAS to host spring and fall cattle shows at different towns of the district as of 1840. St. Catharines and Drummondville were the most common locations.[132]

In 1842, George Adams tendered his resignation as president of the NDAS, but the membership refused to accept it. Instead, at the "unanimous request of the members," Adams reconsidered and was later officially re-elected as president.[133] Adams's term lasted just two more years, for he died in April 1844 at the age of seventy-three. At a general meeting following his death, the NDAS offered its "late venerable President" this heartfelt tribute:

> Resolved, That in Mr. Adams, this Society feels it has lost one to whom, above all others, it is deeply indebted for its first formation, for a long continuance of unremitted care and attention – one, who, as its President since its formation (now a period of 14 years) has, by his personal exertions in its behalf, by the admirable pattern which as a practical Farmer he always presented to his brother agriculturists, and by his unwearied attention to the duties of his important office, left behind a bright example which every member of the Society would do well to emulate.[134]

Upon Adams's death, Walter H. Dickson, a vice-president of the NDAS and the MPP for the Town of Niagara, acted as the interim president. Dickson was the son of the Legislative Councillor William Dickson, who had been a member of the NAS in the 1790s. The society elected Samuel Wood as its second president in April 1845, after having served regularly as the society's secretary since 1830.[135] Little biographical evidence exists about Wood, but along with his appointment as district coroner, he was a member of various local boards, committees, and benevolent societies. One late nineteenth-century local historian recalled the pride taken by Wood in the success of the society's cattle shows.[136]

The NDAS may have been Adams's and Wood's pride and joy, along with another equally long-serving executive member, Treasurer John Gibson, but the society appears to have grown little beyond a club of St. Catharines' area gentlemen during Adam's fourteen-year presidency. Unlike the HDAS and the MDAS, the NDAS did little to sponsor the formation of county or township agricultural societies despite the provision for them in the revised agricultural societies legislation of 1837. Branch agricultural societies in Haldimand County and in Grimsby and Clinton Townships were the first organized in the Niagara District by the time Wood began his presidency. Even on that occasion, he expressed ambivalence about supporting additional county and township branch societies. While he claimed that nothing was "more conclusive as evidence in the improvement of the soil, and attention to agricultural interests," than the recent formation of branch agricultural societies in the Niagara District, he worried that if societies continued to be formed in every township, "the funds would be so limited for each, that they would avail but little." The district would be better served if societies were formed out of every three or four townships, he believed. If township branches held their cattle shows on the same day as the district cattle show or those in other townships, it would cause all to suffer from a lack competitors and spectators. Wood did try to remedy this situation, hosting a meeting of delegates from all the township societies at Port Robinson in October 1846 "to make such arrangements as will tend to the general improvement" of the agricultural society's operation. The results of that meeting are unclear.[137] Regardless, only one county branch and ten township branches were operating throughout the district by the end of 1847.[138] No copy of the NDAS constitution has survived, so the precise connection between the NDAS and its new branch societies is unknown. Moreover, information about the NDAS in the late 1840s is difficult to ascertain, because the chief activities were being carried out by the township branches and their reports were published infrequently.

The NDAS continued operations through 1850, gathering reports from each individual branch, petitioning the government for its annual funding, and apportioning those funds among the branches when it organized and hosted the Provincial Agricultural Exhibition at Niagara in September.[139] In fact, the NDAS appears to have persisted long after the elimination of Upper Canada's districts, for a notice of its April 1851 annual meeting records the men elected to its executive offices. By then, those elected were an entirely different set individuals from the agricultural society's leadership of the 1830s and 1840s and there was no longer any provision in provincial legislation for its continued existence.[140]

The examples presented by the agricultural societies established in the Home, Midland, and Niagara Districts demonstrate that the 1830 agricultural society legislation presented practical challenges to members of the local oligarchies who founded them. An inherent assumption proved largely incorrect that gentlemen improvers of some leisure would come forward with their time and money to lead a society that would naturally attract lesser agricultural improvers from across a district through paternalistic authority and patron–client relationships. Yes, a few devoted individuals in and near a district or commercial town came forward in each district to provide their leadership throughout some or all the 1830–50 period, but enthusiasm for improvement itself proved insufficient in the face of a province undergoing political, social, and economic transformations. The reliance on paternalism that seemed possible in 1830 became much less so as hundreds of thousands of immigrants arrived in the province during the 1830s and 1840s, with little need for deference to the elite of the capital of Toronto, a district seat like Kingston, or a commercial town such as St. Catharines. As a response to the rapidly growing population, the provincial government altered district boundaries throughout the 1830s, carving new districts out of the old with their own administrative towns led by their own local elites. Moreover, each society studied in this chapter struggled to survive when public funding disappeared, indicating that securing an annual grant was critical for their existence; it was not just a public handout to supplement generous private donations and lifetime membership fees supplied by gentlemen improvers. Significantly, when permanent yearly funding was established, the focus shifted to fostering county and township societies within each district. Hence, during the 1840s, members of local oligarchies who had initially stepped up to form a district agricultural society witnessed their presumed authority further diminished to the perfunctory role of requesting and distributing funding for the county and township societies of their district while perhaps hosting a district-wide cattle show or two each year. When municipal reforms eliminated districts in favour of counties in 1849, the district agricultural societies disappeared too.

Yet, the devolution of district agricultural societies after twenty years of existence represented a strengthening of agricultural improvement's place within the developing Upper Canadian state. The provincial government's increased and permanent financial commitment caused a shortening of the arm's-length connection between government and agricultural societies, not just ones led by gentlemen elites at a district seat or other commercial centre but specifically those also led by locally prominent men at the county and township levels. In turn, the more local agricultural societies were required to be increasingly accountable for their use of public funds to improve agricultural practices in their locality. By mid-century, this produced a network of publicly funded, semi-public agents of agricultural improvement operating in most counties of Upper Canada. As chapter 8 relates, a collection of district agricultural society executives would create a provincial agricultural association in 1846. But "association" was a misnomer in its early years, for independence remained a key characteristic of district, county, and township agricultural societies. Although bound together by the same governing legislation and funding opportunities, no controls were ever legislated as to what forms of agriculture they were required or recommended to improve. No cooperation among district societies was ever legislated to share knowledge of improvements across the province. Along with the independent nature of local leadership, the different state of clearance from forest to farmland and the varied nature of soils found in those cleared fields from one township to the next underscored the need for local decisions about recommended agricultural improvements. But what results might the provincial government have expected to witness as it handed out large annual sums of money from the treasury to Upper Canada's agricultural societies? Chapter 7 attempts to provide some assessment of the provincial government's return on investment by analysing how the Home, Midland, and Niagara District agricultural societies spent their public grants to improve agriculture in these three regions of the province.

# 7 District Agricultural Societies and Their Improvements, 1830–1850

In 1846, William Henry Smith assessed the state of Upper Canadian agriculture in his *Canadian Gazetteer*, noting every district of the province had its own agricultural society, the direct result of government legislation and funding, which gave "impetus to the progress of improvement." From its once "very low standard," the province's agriculture had recently made "great advancements," he observed, and was "beginning to keep pace with the improvements introduced into England and Scotland." Specifically, Smith posited: "The emigration into the country of scientific agriculturists, with the establishment of agricultural societies, have been mainly instrumental in producing this great change; stock of a different and better description has been imported, and much land that was previously considered by the old proprietors worn out, has been improved and brought back, by means of judicious treatment, to its old capabilities."[1] Other observers were less enthusiastic. In that same year, James Taylor, an English traveller to Upper Canada, acknowledged the existence of agricultural societies in the province but claimed "their scientific labours do not, at present, seem much regarded by one-fourth of the province." Nevertheless, he commented on his attendance at cattle shows in Toronto, hosted "to encourage improvement in the breed of stock; and a spirited competition appears to be the result."[2]

On their own, these impressionistic observations provide little real measure of the spread of agricultural improvement in Upper Canada, but they do provide a rare alternate point of view on the subject beyond the boosterish reports of agricultural society meetings and cattle shows, often submitted to a newspaper or agricultural journal by the agricultural society itself or written by an overly enthusiastic and overly optimistic editor. As no minute books exist for any of the three district agricultural societies studied here (and, apparently, for any Upper

Canadian district agricultural society of the period), it is not possible to produce a detailed and comprehensive assessment of the improvements introduced by a district agricultural society or that district's farmers' level of engagement in improvement. Furthermore, newspaper reporting of district agricultural society activities was inconsistent across the twenty years examined here and, similarly, there is little consistency in that era's newspapers preserved in the historical record. Some relevant newspaper issues are missing, while entire runs of newspapers that apparently published regular advertisements and reports were not preserved. Despite the emergence of an agricultural press in the 1840s, reporting on district agricultural societies decreased due to the shift in activity to county and township societies that had fewer resources or less inclination than district societies to submit regular reports to the press. This leaves the historian with available newspaper reports and editorials from the 1830–50 period as the primary source of information, and they, like the reports of Smith and Taylor, must be viewed with a judicious eye. Was an exhibition *that* well attended? Were the society's efforts appreciated *that* warmly by the residents of the district? Were the entries in its cattle show competitions *that* improved over the previous exhibition? As inferred by the previous chapter's analysis, these limited records make it difficult to identify the membership of a district agricultural society and their residence in the district. Biographical information exists only for the most prominent agricultural society executives, meaning it can be difficult to identify even a full set of directors, let alone the society's membership, and few such lists exist. The same is true with attempts to identify prize winners to assess who competed at spring and fall cattle shows and from where in a district they brought their livestock and produce for competition.

Acknowledging these significant limitations, this chapter employs available newspaper advertisements, reports, and editorials to generate some level of answer to the question: What return on investment did Upper Canada receive from the system of annual public grants to support the activities of district agricultural societies operating across the province between 1830 and 1850? It continues with an analysis of the efforts by the HDAS, the MDAS, and the NDAS to introduce agricultural improvements in their districts by examining what competitions the societies offered at their cattle shows for the purpose of understanding what agricultural improvements received priority. It does so in combination with an analysis of other initiatives undertaken by each of the three district agricultural societies, grouped in several categories – livestock, crops, domestic manufactures, ploughing matches, agricultural education, horticulture, and commerce – to compare the initiatives undertaken by the HDAS, the MDAS, and the NDAS and to assess what level of success each achieved where the historical record permits.

The agricultural societies legislation of 1830 made no explicit requirement or suggestion that a district agricultural society should host annual or semi-annual agricultural exhibitions. It specified only that societies be established "for the purpose of Importing valuable Live Stock, Grain, Grass-Seeds, useful Implements, or whatever else might conduce to the Improvement of Agriculture in this Province."[3] Yet, the HDAS, the MDAS, and the NDAS (and similar district societies across the province) adopted the cattle show as their chief means by which to employ public funds to encourage agricultural improvement. In terms of form, all continued with the Berkshire model introduced to the province in 1819. The ideology of improvement had long rested on the notions of "spirited competition" and "emulation." Visitors to a cattle show were expected to recognize the superiority of the prize-winning examples entered for competition and return to their farms to work on raising livestock and crops of equal or better quality than what they had seen on display. Additionally, the agricultural societies hoped that any farmers who believed their livestock or produce to be superior to that which they saw at a cattle show would become members of the agricultural society and enter their superior specimens in future competitions, thus spreading the improving creed across the district.[4] But the practical challenge of spreading agricultural improvement by emulation across the frontier world of Upper Canada is perhaps encapsulated best by William Canniff's blunt 1870s assessment of its slow acceptance by most provincial farmers. He argued:

> [I]t must be admitted that a vast number [of farmers] were content to follow in the footsteps of their fathers so long as food and enough were yielded by the soil. The land was plentiful, and productive … He saw no other mode of tilling the soil, and with no reason sought not a change, so no innovations by scientific agriculturalists disturbed the quiet repose of many of the steady going plodders … They wanted no new-fangled notions … But the establishment of agricultural associations and the occasional coming of a new man upon an old farm gradually, and frequently very gradually, dispelled the old man's ideas.[5]

Across the 1830–50 period, the farming world of Upper Canada was a fluid one, simultaneously existing and operating at several states of improvement. Many farmers continued to grow wheat (and barley and oats) for their own use and for off-farm sale. Many had increased or were increasing their crop output and contributed to the province's flourishing wheat trade. But livestock husbandry was an increasing part of farming and the local agricultural economy, more so as mid-century approached. What any specific Upper Canadian farmer chose to plant or raise reflected numerous factors of his locality: the state of clearance of the farm and the number of acres under cultivation, the soil types and drainage of the farm, and the farmer's access to roads and markets to engage in off-farm commerce.[6] Across an expansive colony such as Upper Canada, all these factors varied widely. The same could be said for just the territory within the boundaries of the Home, Midland, and Niagara Districts.

In the new townships of a district were newly arrived settlers clearing forests and farming small acreages between stumps on what was recently the forest floor until they could clear the stumps to produce the open fields being cultivated in longer settled townships. But those open fields could be a deceiving sign of improvement. Perhaps the first owner or two had exhausted the soil completely by poor ploughing and growing wheat crop after wheat crop without crop rotation or manuring to replenish the original soil nutrients. When that occurred, perhaps that owner sold the farm to move to a new farm in Upper Canada not yet similarly exhausted or left the province for the US frontier. To reap any successful returns from labour invested, the next owner of such a farm would need to undertake additional improvements to restore the fields to fertility. In doing so, they could take cues from farmers in the district practising mixed farming, whose practices improvers lauded in the press and within their agricultural societies. Raising livestock produced manure to spread on the field. Sowing grasses on a field to pasture livestock similarly restored nutrients. Thus, the challenge for a district agricultural society was to praise the mixed-farm practices and livestock but support the former types of farming in the district, too, so that all might improve to achieve maximum production.

The HDAS, the MDAS, and the NDAS and their cattle shows prized mixed-farming and its livestock. A district cattle show's emphasis on livestock, specifically cattle and horses (and lesser emphasis on swine and sheep), was a choice that reflected the interests and farms of the gentlemen improvers who occupied the executive offices of a district agricultural society. Yet, the struggling existence of district agricultural societies in the 1830s and the poor results of several initiatives described in the following discussion highlight the problems with this approach. It offered little to the farmers most short on additional farm labour and without access to family wealth or off-farm income from government appointments to purchase livestock and construct barns for their shelter, let alone import pure-bred stock.

During the 1830–50 period, the longer-settled farmer might be making some strides in growing higher-yielding crops and raising some livestock, but did he have specimens of either sort that he believed worthy to compete for prizes? Did he have the resources to import purebred or otherwise improved livestock from Britain or the United States? Did he and his family have time to raise livestock or cultivate a crop for the specific intention of entering it into competition? If a farmer from the most distant township of a district drove his livestock across broken-up roads to compete at a spring or fall cattle show, would his livestock arrive in show-quality shape to compete directly with a gentleman improver's entry brought from a long-settled farm on the edge of the town hosting the competition? Did the straightness and consistent depth of his ploughing in a variety of soil types matter to him, believing himself a strong competitor for a ploughing match? Or did he just get the job done the best he could with the model of plough, draught animal, and time available to him? Could that farmer afford the luxury of a "farmers' holiday,"[7] travelling kilometres across a district to attend a society's cattle show or ploughing match when he needed to complete farm work at home? As for the newly arrived settlers to unimproved lots, the ordeal of wrestling some return from crops planted among the stumps was all too time-consuming, and what little they harvested was unlikely to produce prize-worthy specimens, as were any livestock that grazed on the forest floor or the leaves of freshly felled trees. And what rudimentary elements of agriculture might such a farmer emulate having attended a cattle show? With only broad language in the provincial legislation to guide them, the executive and directors of a district agricultural society produced cattle show prize lists that responded to published ideals of agricultural improvement highlighted in books and journals published in Britain and the United States. Only rarely did they adjust competitions to match more basic levels of agriculture practised by many farmers in their district who needed the most assistance to improve. The agricultural society also appointed the judges from among their gentleman-improver ranks. Hence, any improvements selected as priorities for a district reflected the vision of an improved future held in common by the gentlemen-improver leadership of the agricultural society.

## Livestock

Throughout the twenty-year period under study, livestock received top billing in any cattle show prize list and report, particularly cattle and horses. Additionally, each district agricultural society made some effort to introduce improved bloodlines into the livestock of its district, including the importation of pure-bred cattle. These importations by agricultural societies and by individual cattle breeders of Upper Canada are quite early efforts at establishing pure-bred herds. To underscore that point, it is worthwhile noting that Margaret Derry begins her study of Ontario pure-bred cattle breeders in 1870 when dual-purpose cattle (dairy and beef) "were bestowing significant financial benefits on Ontario farm families." By then, she argues, mixed farming had become profitable because Ontario cattle products were being sold on international markets, particularly the north-eastern United States.[8] In the early 1830s, attempts to purchase pure-bred cattle and other livestock consumed a considerable amount of a society's annual government grant, if not its entire funds, and opinions differ as to the immediate and long-term value of such imports. Occasionally, the society hosted auctions of livestock or amended rules to ensure prizewinners remained in the district for breeding purposes to improve the calibre of the district's herds.

The NDAS declared its intentions to purchase a bull "of the best breed that can be obtained" in January 1831, just weeks after it had secured the £50 required to petition the government for the public grant. The society allotted up to £40 for this animal, plus another £20 for the purchase of a ram with six ewes. But it made no such purchases that year. In 1832, the NDAS agreed to use half of its funds for the purchase of "approved breeds of Bulls, Cows and Rams, for the use of the Society." To ensure the quality of the animals purchased, the society appointed "proper persons" – President George Adams and Vice-President Cyrus Sumner – to travel to Canandaigua, New York, where they purchased four bulls, one cow, and two heifers. All were auctioned off in March 1832 with the hopes of disseminating them as widely as possible among cattle breeders in the district. Unfortunately for the society, the sale returned just half of the cattle's original cost.[9]

The poor sale was not a reflection of the quality of livestock selected; rather, it was likely a gauge of the relatively low value placed on livestock among farmers of the Niagara District. The NDAS would claim that the best bull and best heifer in the auction were of better quality than any that had been previously introduced into Upper Canada and were "equal to any in the State of New York." One bull was a pure-bred short-horned Durham that had several ancestors of show cattle in both Massachusetts and England. The other bulls and cows were crosses of Durham, Holderness, and Devon. Adams later purchased a half interest in the Durham bull and set about breeding one of the district's finest cattle herds. Yet, when Adams and the other co-owner chose to sell the bull, no farmer in Upper Canada could be found to purchase him, and he was sold back into the United States for $100.[10]

The end of government funding in 1834 stopped the NDAS's buying and selling of breeding stock for the district, but prior to that date, it sponsored an auction in June 1832 for a thoroughbred bay stallion "Duroc" that had won first prize at the May 1831 cattle show. At its 1833 spring cattle show the NDAS sold "a number of half-blood Durham and Devonshire breed of bulls."[11] A decade later, an unnamed long-term member of the society told a newspaper that "[i]n Horned Cattle, Sheep, and Swine, we are steadily and rapidly improving, but in Horses we seem to be almost stationary." In fact, the best horse on display at that spring cattle show in St. David's was a stallion brought from Lewiston, New York, having been imported from Yorkshire, England, the previous year.[12] Others would challenge the claim that cattle stocks were "rapidly improving" given that, in 1842, Hiram Leavenworth had complained of the indifference of "so many of our wealthy farmers" to the value of pure-bred shorthorn cattle. He cited a recent sale of cattle in the Niagara District at which someone from Buffalo, New York, purchased nearly all cattle and took them to the United States. "We considered it a public loss," he lamented.[13]

At its February 1831 meeting in Bath, the MDAS decided its funds for the year should "be principally appropriated to the importation of good breeds of domestic animals, and good seeds." It established a committee, funded with £120 from the society's account of just over £150, to purchase one bull, six boars, three rams, and one hundred bushels of potatoes "or such seeds in any other proportions they may deem advisable." The latter were to be disposed of at private sale in equal proportions for each county of the Midland District, under the direction of the vice-presidents and at a cost of no more than the purchase price. The bull was to be circulated among the counties, residing at a place determined best by each county board. MDAS members would be able to use the bull for breeding purposes upon paying a fee per cow serviced. Non-members would pay a higher fee, equal to annual membership in the society. As for the boars and rams, they were to be divided equally among Lennox, Addington, and Frontenac Counties for sale by the respective county boards in an auction restricted to residents of each county. Money raised would replenish the society's funds.

Despite the detailed planning, the committee decided instead to purchase three bulls in the United States, because they were "of the best kinds that could be procured" and all had been imported recently from England. Each of the three counties was assigned one bull in March 1831: Lennox County received a full-blood Durham;

Addington County, a half-blood Durham; and Frontenac County, a three-quarter-blood Durham. After a period of four months (with no indication of how many members made use of the bulls), the society decided to offer them for sale to residents of the district at a public auction to be held at Bath in late July. The society reasoned that, by this method, the bulls could be used for breeding more cows than if they remained the property of the MDAS at its expense. Only the two bulls designated for Frontenac and Addington Counties were sold, "at prices below their original cost and actual value." The third and most valuable bull, the full-blooded Devon, met an untimely demise; it had drowned "in consequence of some mismanagement while being ferried in a scow across Hay Bay."[14]

With government funding again available in 1837, the MDAS used £50 of its funds to purchase another bull that members could use to breed their cows. "Independence" was a stout and "neatly framed" five-year-old mixed Durham, the offspring of a Durham–Lancashire cross bull that had been imported in 1830 by Commodore Barrie.[15] This renewed interest in promoting cattle breeding may have been in response to the rather low state of cattle stocks in the district by the late 1830s, according to the commentaries of two Kingston editors. The *Kingston Chronicle and Gazette* reported in January 1835 that, recently, an individual had purchased thirty to forty head of cattle from the Bay of Quinte area, presumably taking them to Jefferson County, New York (directly across the border from Kingston), where, the editor opined, they would be used to improve the herds of that location. The fact that beef from Jefferson County furnished the Kingston market "almost exclusively" should "occasion some reflection among our Canadian farmers," he argued. In June of the following year, the editor of the *Upper Canada Herald* also commented on the number of cattle that had been purchased in the district and taken to the United States, as many as sixty at a time. Claiming that Americans were busily "changing their farms from arable to grazing farms," cultivating less land, raising more stock, and producing cheese and butter – in other words, adopting the improving ideal of mixed farming – he argued it was certainly time for the Midland District farmers to follow their lead and keep their cattle for dairy production and sale to local markets.[16] By the time of the district cattle show at Links Mills in 1840, winning competitors were required to adhere to a new rule stipulating that all stallions, bulls, boars, and rams would remain in the district for the remainder of the year in order to help improve the quality of the district's livestock.[17]

That fall, the *Herald* reflected on the recent cattle show in Kingston and delivered its opinion that the competition demonstrated a "visible improvement taking place in the science of Agriculture generally; and in the rearing of stock particularly." He believed that it must be "gratifying to the members of the Midland District Agricultural Society to see their labours crowned with such success" and hoped that the progress would continue so that contractors and butchers would no longer have "to pay money out of the country to foreigners for bread, meat, flour, cheese, &c." Notably, he gave credit to the "beneficial influence" of gentlemen who paid great attention to breeding, "considering that they are not farmers exclusively."[18]

Again, in late 1841 or early 1842, the society purchased a pure-bred Durham bull "Union," which it kept for breeding purposes until the society sold it at auction to a member of the society from Loughborough Township in July 1843.[19] Later, the MDAS spent £275 to purchase a stallion that had been imported from England in 1836 by H.P. Simmons of the Ancaster area at a cost of £840. The society kept "Somonocodrom" through 1841 and 1842, allowing members to have their mares bred for a fee, with nonmembers paying more. Having seen "Union" and "Somonocodrom" exhibited at the June 1841 MDAS cattle show at Waterloo, the editor of the *Chronicle and Gazette* suggested that "the best Breeds that money could procure" were now within reach of all horse and cattle breeders of the district. "Such very superior stock," he argued, "could not have been procured by the Society without the aid of the Government bounty which has been so liberally been provided for the encouragement of agriculture."[20] Despite such praise, the MDAS planned to sell the horse in the United States until Allan McDonell, sheriff of the Midland District, intervened. He purchased Somonocodrom and made the stallion available for breeding purposes by stabling him at a racetrack near Kingston.[21]

In 1833, the HDAS made its singular attempt to import cattle into the province for breeding purposes, using its entire annual public grant to purchase bulls from the United States. An unidentified director of the society travelled to Canandaigua, New York, with instructions to select "either 3 full blood – or 2 full blood and 2 three quarter blood bulls." The director returned with four shorthorn bulls that had no pedigree and the HDAS sold them all by auction at its May 1834 cattle show. It is not clear if the agricultural society had always planned their sale to spread improved bloodlines throughout the district or if the HDAS was selling off its assets because the yearly government grant had ended.[22] Each received £6 to £7, well below their value. A contemporary observer thought the bulls were inferior quality; however, in 1867 the Board of Agriculture of Upper Canada claimed that these bulls "bred from imported stock ... became the sires of very good grades, and thus tended to improve the general character of the cattle in the district." Decades later, the Dominion Short-horn Breeders expressed yet another opinion, recalling that R.L. Denison had purchased one of the bulls and another was found later "at some place on Yonge Street nearly starved." An unnamed farmer purchased the latter and returned home to Cobourg, taking "care to reach home at night for fear of ridicule." The animal, the report continued, "proved to be a grand stock getter," once nursed back to health, and revolutionized the quality of cattle in that neighbourhood.[23]

Following this singular attempt, the HDAS left the importation of cattle to private individuals in the Home District, of which there is some record. A rare, detailed report of an HDAS cattle show held in October 1836 noted that seven of the forty-five sheep exhibited at the show were thoroughbred English rams, three of which had been imported since the autumn 1835 cattle show. In addition, ten were thoroughbred English ewes, six of which had been imported, three within the past year.[24] As a means to encourage farmers who imported livestock into the province to enter their livestock for competition at an HDAS cattle show, the society added a new rule for its fall 1838 competition. Any animal awarded first in its competition would receive a prize double the advertised amount upon presentation of adequate proof that it had been imported from Great Britain or Ireland within the previous year.[25] As of 1839, the HDAS also hosted auctions of improved breeds of cattle following its competitions to spread improved bloodlines, primarily Durham, beyond the herds of agricultural society members.[26]

Beginning in 1837, the HDAS changed the name of its spring and fall events to "Fair and Fat Cattle Show." Indicative of the society's efforts to encourage farmers of the district to provision the expanding city of Toronto with quality beef and mutton, the society ruled that for its spring 1839 event, all prize-winning fat cattle and sheep had to be sold at auction before leaving the show grounds. To further ensure that this occurred, an 1841 rule ordered that cattle be offered for sale to the butchers prior to any award being presented.[27] These changes indicate a progression from improving the quality of individual cattle in the Home District for the purpose of elevating cattle show competitions to encouraging improved herds capable of supplying the beef market for the city. Hugh Scobie, editor of Toronto's *British Colonist*, remarked with gratification in 1841 on "the steady improvement in the breed of different kinds of stock, which is gradually taking place in the district, in consequence of the importation of thoroughbred horses, cattle and sheep from England," although he did wish "that the rivalry for improving the stock of the country was more general."[28]

## Crops

The 1831 committee of the MDAS that decided to spend £120 on livestock and one hundred bushels of potatoes "or such seeds in any other proportions they may deem advisable" did not purchase any potatoes. Instead, it spent the society's funds on clover seed to be divided among members. But due to the lateness of its arrival in the district, not all the seed could be distributed that spring. Thus, it was suggested in July that the clover seed be given away as premiums for the agricultural society's upcoming competitions.[29] That fall, the society made a second attempt to procure improved varieties of seed for its members, sending Lennox and Addington director Peter Davy to Genesee County, New York, with instructions to purchase the best quality seed wheat available, and upon his return, divide it equally among the three vice presidents of the society. He returned empty-handed, reporting there was no good seed wheat for sale in that area of the state. Regardless, the MDAS had to pay Davey his expenses for his effort, draining some £3 from its dwindling funds.[30] The society tried to import seed wheat again in early 1832. Those present at the society's quarterly meeting in January listened to President Macaulay deliver several papers on agricultural subjects, one of which concerned a wheat variety called Tea Wheat that was being offered for sale in the United States. The MDAS resolved to purchase a barrel of this variety to be sold to members of the society.[31] There is no mention of this occurring, though an April meeting of the society reaffirmed its desire to import some because there was "a general wish" among the farmers of the district "to obtain the best seed wheat." At that same meeting, the MDAS acknowledged the donation of fifteen types of turnip seed provided by Adam Fergusson, the Scottish improver recently arrived in Upper Canada. The seeds were immediately distributed to members, with the recognition that "the advantage of cultivating the Turnip on a great scale is not fully appreciated in the present infant state of Canadian Agriculture."[32] The turnip seeds were, in fact, encouraging livestock farming because turnip, along with rutabaga and mangelwurzel, were root crops that European and North American agricultural improvers encouraged farmers to grow for cattle feed.

In the spring of 1833, the FCAS imported one hundred bushels of seed wheat, half of which was purchased and sown by farmers of the county. But the other half was sold to a merchant at a substantially discounted price after seeding time had ended.[33] Five years later, the MDAS recommended that seventy-five bushels of clover seed, sixty pounds of Swedish turnip seed, sixty pounds of onion seed, and one hundred pounds of hemp seed be purchased and divided equally between the three counties. But it appears that the society spent a sizeable amount of £192, plus £22 in import duty, on clover seed alone.[34]

Early in 1832, the MDAS responded to a theme of agricultural improvement promoted by writers who praised "neat" farms whose appearance alone showcased the results of improvement. It established a committee "to make a tour of the Counties and examine Farms offered for inspection." It was to report "on the general arrangement and management of said farms, the nature and state of the fences, the situation and plan of barns and outhouses, the quantity of land sown with grain &c. last year, as well as the present year, with any other remarks they may see fit to offer." As a consequence of the cholera outbreaks of that year, the tour did not occur. The epidemic, which had cancelled what was to be the MDAS's first district cattle show, also ended plans to award prizes at that event for the individuals who had grown the most wheat on one acre, two acres, or five acres of land manured with lime and to those who had grown the greatest quantity of best wheat, corn, and potatoes "in the ordinary manner" on one acre of land.[35] In 1835, both the Frontenac and Lennox and Addington county agricultural societies offered competitions for crops judged in the field. And again, in 1843, the district society instructed four individuals per township to examine the crops, gardens, fences, and homesteads so that it could award cash prizes to those deemed worthy.[36]

Competitions for samples of field crops at MDAS fall cattle shows were rather sporadic. The Frontenac County branch first awarded prizes for the best wheat, barley, and oats grown in the county during 1833, but no similar competition appeared again until 1838 when the Lennox and Addington sponsored competitions at its October cattle show for best fall wheat, spring wheat, barley, and oats. Again in 1839 and 1840, the Frontenac County branch sponsored a competition for the best rutabaga, and its counterpart in Lennox and Addington hosted 1839 competitions for the best-quality fall wheat, spring wheat, barley, and oats. Crops were not a prize category at any district show until 1840.[37] Occasionally, individuals were awarded a special premium for bringing prime examples of new varieties of fruits and plants to be displayed at a cattle show. Examples included cabbage, Ox apples, rutabaga, Norfolk and Globe turnips, English thorn plants, coconut squash, and hops.[38]

At a September 1830 meeting of the HDAS, its executive ordered the secretary "to procure sundry kinds of seeds" for use of the society. There is evidence of the HDAS doing this in April 1833 when it distributed seeds received "from the coast of Africa" by a Dr. William Rees of York, as well as grass seeds it obtained from an unnamed gentleman recently arrived from England.[39] In that same year, the society advertised prizes for the best three-bushel samples of oats, barley, peas, and potatoes to be competed at the society's May cattle show and announced plans to use another £50 to award prizes for those members who could grow the greatest amount of the best quality wheat, barley, oats, peas, potatoes, or Indian corn on one acre during the current growing season. To compete, farmers were required to submit a statement of "the mode of culture pursued together with the nature of the soil." There is no evidence that anyone competed in these categories or that any prizes were awarded.[40]

At its 1834 fall cattle show, the HDAS added prizes for the best examples of at least fifty turnips and at least twenty mangelwurzels. That same year, the society also invited competition for fifty-pound samples of dressed flax and one-bushel samples of flax seed.[41] Its 1836 competition for root crops and grains, displayed in City Hall, was repeated in 1837 and 1838, with the seed category expanded to include red wheat, white wheat, black oats, white oats, and the root category expanded to include carrots, rutabaga, white turnips, yellow turnips, and potatoes.[42] However, with its shift in attention to fat cattle and livestock in the late 1830s and early 1840s, the HDAS hosted no competition for crops again until its fall 1843 cattle show.[43]

In contrast to societies in the Midland and Home District, the NDAS did not sponsor any competition for crops (wheat, barley, oats, peas, hops) until its 1844 fall cattle show. Only two other prize reports of an 1840s district cattle show exist, showing a similar list of crop prize categories offered in the fall of 1845 and 1846.[44]

**Domestic Manufactures**

The NDAS also stands apart from its contemporaries in the Home and Midland District for the early and consistent attention it paid to the promotion of domestic manufacturing to foster a domestic supply of manufactured goods for the province, an effort applauded several times by local newspaper editors.[45] Its first cattle show in 1831 offered prizes for the best piece of woollen and linen cloth as well as the best-made plough for sward. Afterwards, competitions expanded to include a wide array of household items and farm machinery: leather, cheese, factory-manufactured cloth, potter's ware, axes, shoes, wagons, forks and edge tools, calfskins, coverlets, flannel, carriages and harness, corn harrows, horse rakes, butter, drill barrows, hay-cutting machines, satinett, and even window blinds. In 1845, the exhibition of agricultural machinery was "unusually large and very satisfactory" and included various ploughs, fanning mills, straw cutters, and cultivators. The show was also complemented by the display of a cooking stove, parlour stove, and portable boiler, all manufactured in the district.[46] Occasionally, women were the recipients of prizes for their entries. For example, in 1836, a Mrs. Doddy won first place for her thirty pounds of butter, and a Miss McFarlane won an award for her reticule.[47]

In contrast, the HDAS offered little encouragement to domestic manufactures during the 1830s, beyond the ambitious prize lists of May 1831 for the best examples of agricultural implements manufactured within the Home District. Prizes were offered for the following categories: a double mouldboard plough, seed harrow, common brake harrow, drill harrow, common swing plough, and scuffler or horse hoe. Only entries in the first three categories were awarded prizes.[48] Adjusting its competitions to better acknowledge the primitive state of much of the district's farming practices, the HDAS offered a competition at its October 1831 and May 1832 cattle shows for the best plough, the best seed harrow for ground clear of stumps, and the best such example for stumpy ground.[49] By the mid-1840s, prize lists of the HDAS showed more interest in domestic manufactures, reflecting Toronto's increasing manufacturing and commercial economy. Prize categories for farm implements and domestic manufactures, including dairy products, were offered at the spring 1845 cattle show, and the HDAS continued offering similar competitions at its cattle shows on through to mid-century.[50]

In the Midland District, J.B. Marks, then the Frontenac County society vice-president, assured those assembled at its 1832 cattle show that the district society had "not overlooked the great advantage of encouraging domestic manufactory." His comments followed the Lennox and Addington County Society having offered prizes for butter, cheese, and flannel at its first cattle show in October 1831. Along with dairy products, he stated, the district executive would consider "some plan for encouraging the industry, and abilities of our fair country women, in other branches of rural economy." The MDAS did so in 1832, planning to award prizes for domestic manufactures at its first district-wide cattle show. However, the cholera outbreak of that year meant that competitions were never held for items such as cheese, twenty yards or more of fulled cloth, ten yards or more of linen, pair of wool socks, pair of stockings, best straw or grass hat, best woollen yarn, best thread, and best barrel of cider.[51]

Although Marks pronounced ambitious plans for his FCAS's promotion of domestic manufactures in 1833, it was not until the following October that the society offered a few competitions for cheese, fulled cloth, and woollen socks. Four years later, the Lennox and Addington Agricultural Society offered those same categories of prizes, plus a competition for the best factory-made cloth. It sponsored categories in butter, cheese, and factory cloth again in 1841.[52] In recognition of women's labour and skill, each prize awarded in these categories at the 1835 competition was given to a wife of a society member who had entered the samples for competition.[53] In that same year, the Frontenac County society first offered an open class for individuals to enter an "Improved Plough or any improved Implement of Husbandry or Farming Tool" for the consideration of the judges, although only one individual was awarded a prize for the hay rake he had made.[54] From 1835 onwards, similar categories of domestic manufacture were generally advertised for competition in either the district, county, or township cattle shows.

## Ploughing Matches

In his speech at a Frontenac County Ploughing Match in the fall of 1832, Vice-President Marks asserted, "Ploughing is one of the most important operations in Agriculture, and ought to be executed with the greatest attention ... Ploughing Matches have always been considered the source of much improvement of this necessary operation of Agriculture." Habit, he concluded, "has bound us to a poor system of farming and the sooner this bad system is abandoned the better."[55] Upper Canadian ploughing matches were, perhaps, the aspect of a district agricultural society's activities that were most reflective of the varied nature of agriculture practiced across each district. For example, they featured a variety of competitions for ploughing different types of soil, ploughing using horses or oxen, and competitions that separated the Canadian ploughman from those who had recently arrived in the province. Such competitions indicate an acknowledged need to adjust improved skills in ploughing to the varied conditions and challenges an Upper Canadian field might present.

The MDAS paid regular attention to ploughing and ploughing matches. In most every year from 1830 to 1850, a match was hosted at the county or township level in the district. The HDAS hosted its first ploughing match in conjunction with its inaugural cattle show on 4 October 1830 and another in the following year. The society would not return to hosting another competition until 1842 and again in 1844, at which time its interest in fat cattle and domestic manufactures took precedence.[56] It is likely that the Toronto-based HDAS left its county and township agricultural branches to host their own ploughing matches rather than the district society hosting a cattle show in downtown Toronto as well as a ploughing match at a location beyond the edge of the city. Similarly, the NDAS hosted a ploughing match at its very first cattle show but did not hold another until 1841. The limited records of NDAS cattle shows in the 1840s identify that district ploughing matches were held again in 1842, 1844, and 1846.[57]

## Agricultural Education

During the winter of 1833–4, the HDAS hosted a series of free monthly lectures at York, delivered by William Sibbald, who had arrived with his family in Upper Canada in 1832 to accept a land grant available to retired British military officers. Keenly interested in agricultural improvement, particularly ploughs and ploughing, Sibbald met with local farmers during 1832 and attended the few ploughing matches hosted in the district (not HDAS-sponsored events). The result was his essay on plough improvements, published in the first issue of his short-lived *Canadian Magazine* in January 1833. Following his stint as editor of this failed journal, Sibbald wrote the lieutenant governor and the colonial secretary requesting an appointment as professor of agriculture, noting that the HDAS had done little to advance the science of agriculture "theoretically or experimentally," because its founders did not have the means to do so. Sibbald's request was not granted, although he continued to deliver lectures for the HDAS until the spring of 1834 when he moved north to settle and improve his farm on Lake Simcoe's Jackson's Point.[58]

There is no indication that the MDAS was able to gather books for the library provided for in its first constitution, or again in 1837 when the new constitution expressed the hope that "as soon as it may be found practical" the society would form a library "of the most approved and useful works on agriculture and domestic economy."[59] That said, the MDAS supported the nascent agricultural press of the United States and, later, of Upper Canada. The society must have purchased, or have been donated, copies of the *New England Farmer* (published in Boston), because the society spent funds to have its copies bound in 1832. Three years later, the society ordered the purchase of twelve copies of *The Cultivator* (published in Albany, New York) "in order to give the members of the Society information on the American system of Farming." In 1840, the MDAS again looked to New York state for an agricultural journal for its members, when it ordered forty-four copies of the *New*

*Genesee Farmer* (published in Rochester) to be deposited with the treasurer for the use of members. With payment of "an extra quarter of a dollar," members could receive a year's subscription.[60]

The MDAS become a regular supporter of William Edmundson's monthly *British American Cultivator*, ordering fifty copies of the January 1843 issue to be distributed among members. The following year, it ordered twenty-five copies be supplied annually to each township in the district, and in September, the MDAS extended "[a] strong expression of approbation ... in favor of the *British American Cultivator*, and it was regretted by the members that this useful work had not been more extensively patronized, especially as the price was so low."[61] Again in 1847, the MDAS ordered 250 copies of "an agricultural paper" so that 25 copies could be sent to each township branch agricultural society. A similar decision was made the following year, with an order of £25 worth of the *British American Cultivator* placed for equal distribution of copies across the townships.[62]

The NDAS also attempted to make agricultural publications available to its members on several occasions. In 1831, the district society used £2 of its funds to purchase four copies of the *Genesee Farmer*. Years later, it purchased some sixty-five copies of the *British American Cultivator* for distribution to members in 1845 and again in 1846.[63] Its most significant support of agricultural education, however, was the sponsorship of a publication on agriculture for use in the province's schools. *The Canadian Agricultural Reader*, written anonymously by "a Vice-President of the Niagara District Agricultural Society and Township Superintendent of Common Schools," was launched at the society's 1845 fall cattle show dinner.[64] The textbook offered youth of the province an opportunity to learn improved agricultural techniques approved by its gentleman author, the Reverend Thomas Brock Fuller. The godson of Sir Isaac Brock and former student of John Strachan, Fuller had moved from Chatham to Thorold in 1840 to assume the position of rural dean of Niagara, and within five years, he had been elected a vice-president of the NDAS.[65] Before the end of 1845, the Niagara Municipal District Council considered the appropriateness of *The Canadian Agricultural Reader* for use in the district's schools. Fuller's work was, essentially, a three-hundred-plus-page compendium of agricultural tips gleaned from British, American, and British North American sources. Local improvers and provincial politicians heaped praise on the book, considering it to be "eminently useful." Lincoln North MPP W.H. Merritt praised it for containing "the most useful, practical information," while Chippewa MPP James Cummings ordered a dozen copies for distribution to schools in his township. The book also received praise from several newspapers, and William B. Jarvis, then vice-president of the HDAS, promised to recommend it to his fellow members. In Kingston, the MDAS presented the reader for the consideration of its members at its 1846 annual meeting.[66]

**Horticulture**

Unique among the three societies studied, the HDAS announced it would sponsor a horticultural society for the City of Toronto as a part of its 1843 plan to foster township branch societies. The horticultural society would "have nearly the same relation to the District Society as the Auxiliary Branches of the Townships." It was a plan that stretched the very purpose of the semi-public district agricultural societies and the public money that funded them, for the Toronto Horticultural Society was not founded to promote market gardens or kitchen gardens for factory workers. Instead, it embraced the city's gentlemen – and gentlewomen – who tended their flower gardens and cultivated exotic varieties of fruits and vegetables they imported or purchased from local nurseries.[67] All HDAS members living in Toronto were to be considered members of the horticultural society, and their membership fees would be divided equally, with one half given to the horticultural branch society to be spent on prizes for horticultural competitions and the other half to pay for a subscription to the *British American Cultivator* or another appropriate journal.[68]

The Toronto Horticultural Society was founded at a 19 January 1844 meeting, for which "the attendance of the Gentry" had been requested.[69] Toronto had witnessed a first attempt to create a horticultural society a decade earlier, lasting only two years,[70] and the city's resurgent interest in horticulture and floriculture coincided with a wider pattern of improvement in North America. Tamara Plakins Thornton notes that when the Massachusetts Horticultural Society was formed in 1829, it was one of just five such societies in the United States; in 1837, it was one of ten; and by 1852, it was one of forty. "The very practice of horticulture," she asserts, "was "evidence of an advanced civilization and the formation of a horticultural society an emblem of cultural refinement."[71] In his report of the January 1844 meeting, Edmundson echoed this sentiment, asserting that the founders of Toronto's horticultural society were "determined to be not one whit behind the citizens of the principal cities of the United States in efficiently sustaining [such] an institution." Likewise, a commentator on the Toronto Horticultural Society's July 1845 exhibition argued for "the influence of gardening on man's moral nature," believing "the progress of floriculture a sure indication of the progress of civilization and refinement."[72]

The Toronto Horticultural Society, which held its exhibitions in May, July, and September at either City Hall or the Government House, was more than just another auxiliary branch of the HDAS. Its first president and treasurer were William B. Jarvis and William Atkinson, respectively, who held the same offices in the HDAS.[73] Two years later, the mayor of Toronto, William H. Boulton, was elected president, while the city clerk, Charles Daly, served as the society's secretary.[74] Despite such prominent support, the society ceased to exist by 1848, and it is unclear if the Toronto Horticultural Society continued to be considered a branch of the HDAS at the time of its demise. Individuals revived the society briefly in 1849, and finally, it was reborn with some permanence in 1853, although no longer a branch of the HDAS, which had ceased to exist.[75]

## Commerce

It is not surprising that the HDAS took an interest in the commercial aspects of the province's agricultural economy in the 1840s, considering the society was led by executive members who lived in the city or at its edges and that it held its meetings and cattle shows in the rapidly developing commercial centre of Toronto. The society invited Benjamin Thorne to deliver a speech to the nearly two hundred people who attended its spring 1840 cattle show dinner at City Hall.[76] Thorne was a miller, exporter of flour and an importer of metal, groceries, and dry goods, around whose enterprises and influence developed the town of Thornhill. In 1830, he had helped the York elite establish the fledgling HDAS by becoming a director (although there is no indication of any longer connection with the society) and, by 1840, had become one the most successful businessmen in the province, rapidly becoming its largest flour exporter.[77] On this evening, Thorne presented his knowledge of the "commercial and agricultural relations" in the province. Four years later, he would deliver a speech on "Agriculture and Commerce" at a similar occasion.[78] Although he was at the peak of his business career between 1840 and 1844, Thorne would have short-lived success. With Britain's repeal of its corn laws in 1846 as it shifted to free trade, his circumstances illustrated the vagaries of commerce. Due to his being caught with too large an inventory of flour bought with his personal credit, the markets of 1848 brought his financial ruin. He took his own life.[79]

A Board of Trade for Toronto was established in 1841, and the HDAS invited its president, George P. Ridout, to speak to guests attending the society's spring 1841 cattle show dinner. Welcomed with a members' toast to "The Commercial Interests of Canada," Ridout delivered a "neat and appropriate speech observing that it is impossible to separate the agricultural from the commercial interests of the country, – they must go hand in hand and prosper together."[80] In the years ahead, executive members of the HDAS, the MDAS, and the NDAS took action to support the commercial interests of the province. Independently or using the position they held within their agricultural society, they helped draft and circulate several petitions in response to Britain's mid-decade shift to free trade and the elimination of preferential duties for Canadian produce. Britain's decision caused political and economic turbulence in British North American provinces. Early 1840s petitions, led by W.B. Jarvis, pleaded for the maintenance of preferential duties for Canadian produce and protection against the free importation into the province of American produce,[81] while an 1846 HDAS conference with the Toronto Board of Trade to discuss the repeal of preferential duties expressed united support for the opening of the St. Lawrence River to shipping from all nations. Meantime, in Frontenac County, a meeting was held to consider a petition to imperial authorities requesting the maintenance of preferential duties for Canada.[82]

What was the return on investment to Upper Canada for using public funds to support district agricultural societies? For most of the 1830s, the public funding did little more than permit district agricultural societies to host a cattle show once or twice per year, plus intermittent purchases of livestock and seed. By the 1840s, if commentators are believed, agricultural societies were using their regular annual funding to exert some influence on the quality of livestock and agricultural practices across the province, an effort also led by individual livestock importers and breeders affiliated or unaffiliated with an agricultural society. Whether the power of emulation had much success in rooting improvement as a widespread activity among farmers of a district is unclear. At the very least, a cycle of spring and fall cattle shows, as well as ploughing matches, had become part of the social fabric of Upper Canada, ranging from small cattle shows in some of the province's townships and counties to larger district shows hosted at more populous centres. As of 1846, Upper Canadians could also compete in an annual provincial agricultural exhibition. A culture of exhibitions remains a part of rural Ontario society. Local agricultural societies host community fall fairs, and local branches of the Ontario Plowmen's Association host annual county ploughing matches.[83]

Perhaps the province's main return on investment was not these exhibitions of varying sizes and quality, but the handful of district agricultural society executives who, in 1846, created a provincial agricultural association that would host the annual provincial exhibitions. While district agricultural societies were devolving into the role of coordinating county and township branch society activities, these executives used their prominence and experience gained leading their district agricultural societies to establish and operate a provincial organization. By the early 1850s, it would be an influential lobby group that would convince the government to make the promotion of agricultural improvement a more central activity of the state and convince it to rely on the established network of semi-public agricultural societies operating within the counties and townships of the province.

# 8 The Agricultural Association of Upper Canada, 1846–1852

J.B.W.'s September 1846 letter to the *British Colonist* announced ongoing efforts to establish a "Provincial Agricultural Society in humble imitation of the Royal Agricultural Society of England and of similar societies in the United States." To follow the example of the "first men of England" who were members of the RASE and similar associations, J.B.W. extended invitations "to the Bench, to the Bar, and to the members of the other professions." Only then did he extend an invitation "to the farmer, to the merchant, mechanic and citizen to lend his aid and assistance." It is likely that J.B.W. was William B. Jarvis, then vice-president of the HDAS, because he noted that the idea of creating a provincial association "had long been a favourite one" among some members of that institution; plus, J.B.W crowed about the HDAS having set "a noble example to its sister societies" with its donation to the new provincial organization "of *one hundred pounds*."[1]

Founded several weeks earlier by a handful of gentlemen improvers, "The Provincial Agricultural Association and Board of Agriculture for Canada West" had a title much grander than its initial capabilities and influence. This club of gentlemen improvers, established without any enabling legislation, had no claim to being a board of agriculture and quite a limited claim to province-wide support. Remarkably, after a precarious beginning, this association's annual exhibition would become a significant event in the province, attracting thousands from the Canadas and the United States. This chapter examines the association's first six years, during which a private club of gentlemen improvers initially struggled to attain the level of provincial authority it bestowed on itself at its founding but then evolved into a well-organized association that hosted increasingly larger and better-planned exhibitions aimed at highlighting the improvement and progress of the Upper Canadian economy as a whole: agriculture, commerce, and manufacturing – both domestic and industrial. It also analyses the association's rhetoric and actions, which increasingly emphasized the importance of expanding agricultural commerce to grow the province's economy, particularly the need for infrastructure to facilitate and expand trade with Britain and with the United States. In fact, Upper Canada's agricultural association would develop a close relationship with the NYSAS and encourage Americans to attend its annual exhibition, if not submit an entry for competition, to compare improvements between the colony and its neighbouring state. By the end of the 1840s, the agricultural association successfully lobbied the provincial government for a public grant to support its hosting of the annual agricultural exhibition, while it also lobbied for the creation of a board of agriculture, a subject explored in chapter 9. But first, here, the founding of the AAUC and its hosting of an annual provincial exhibition provides a lens to view the evolving meaning of improvement in Upper Canada by mid-century.

As noted in chapter 4, William Edmundson worked diligently in 1844 to secure support for his planned province-wide agricultural association, which included the reformation of Upper Canada's agricultural societies to better connect township, county, and district levels into a coordinated provincial system.[2] Although he claimed in March 1844 that a convention was soon to be held at Cobourg or Hamilton to create a provincial association, no such meeting occurred.[3] His expectations of early 1845 also went unfulfilled, having reported that the HDAS was about to draft a petition requesting the government establish a provincial agricultural association. It did not do so.[4] He hoped provincial legislators would enact his reformation of district agricultural society leadership, including a "Grand Provincial Agricultural Society" in the new agricultural societies bill it was debating. It also did not do so.[5] Regardless, the ever-determined Edmundson continued to press his plans within the HDAS, and in November, the society formed a committee to draft the form of a "Provincial Agricultural Society" and petition the legislature for its funding. The committee was to also draft a plan to establish "a Board of Agriculture in the Province of Canada" to be presented to the provincial legislature and petition the government for the Council of King's College in Toronto, opened in Toronto in 1843, to "found and endow a Professorship of Agriculture." Edmundson was to publish a report of these proceedings in his *British American Cultivator* as an invitation to other district agricultural societies to send delegates to the HDAS annual meeting in February when the society's committee would report on the success of its efforts.[6]

What this committee accomplished or did not accomplish is not clear, for there is no record of a February 1846 HDAS meeting. But what can be said is that support for Edmundson's highly structured plans had reached its end. At the society's 13 May meeting, held at its spring cattle show, President Thomson put forth a resolution expressing the society's opinion that "the cause of agriculture would be greatly promoted through the agency of a Provincial Agricultural Society." His motion received unanimous support and the society struck a new committee to invite the province's district agricultural societies to send delegates to a convention scheduled at the Toronto courthouse from 15–17 July to support the proposed organization.[7]

No report of the convention provides any clear indication of how many individuals attended from what number of districts, but it does seem rather odd that the committee invited agriculturists from across the province to attend an agricultural convention during the busiest time of the farm year. Literally, many would have been on their farms trying to make hay while the sun shone, instead of travelling to Toronto to organize a new association. In fact, the "convention" was no more than a one-day meeting, and even Edmundson would later admit: "Owing to the busy season the meeting was not numerously attended."[8]

Those who did attend drafted a constitution for "The Provincial Agricultural Association and Board of Agriculture for Canada West." The title is quite remarkable, for without any sponsoring legislation, the latter half of the name was completely void of authority. How could a board of agriculture be established by private means? Moreover, the eight resolutions agreed upon represented a near-complete disregard of the detailed plans Edmundson had championed and refined for this project. It proposed no connection to township, county, or district agricultural societies. Instead, this was to be a top-down privately sponsored gentlemen's club, unconnected to the existing district or township societies beyond the authority that its august name presumed. In fact, the only connection to existing agricultural societies was the proposal that two delegates from each of the province's district agricultural societies meet annually to elect officers to serve as the new association's board of directors. But that connection was not critical. Should there be no such delegates, the president and secretary of each district society would be considered *ex officio* members. The association would exist whether societies from district to district offered their support – or whether the government provided legislative or financial support.

Two classes of membership were proposed: regular and life. Attracting those who could pay life membership fees (£2 10s or more compared to the regular membership of 5s per year) would provide the association immediate prominence and the necessary funding for the association to accomplish its singular goal. Whereas Edmundson had proposed in the autumn of 1843 that a provincial board should host an annual provincial exhibition in a different district from year to year, as *one* of its many functions, a final resolution of the July meeting indicated that the *sole* function of the proposed new organization would be the hosting of "Annual Fairs or Exhibitions" to advance "the general interests of the country, and especially the agricultural and the manufacturing."

Those in attendance agreed that if the proposed association was to be "truly national in its character and all its bearings … the next meeting should not be held in Toronto." Thus, they selected the Hamilton courthouse as the venue for a 17 August meeting to endorse the draft constitution for the proposed association. But that meeting was no better attended or representative of the districts of the province. Only eighteen men attended, representing just seven of Upper Canada's twenty districts. One individual each from the districts of Johnstown and Newcastle and two from the Colborne District were the only representatives from the eastern half of the province. In the Midland District, no Kingston newspaper advertised or reported on the July meeting, and no delegates from that district travelled to Hamilton. One attendee travelled from the Huron District in the west, while two individuals each represented the Home, Wellington, and Brock Districts. The largest group of eight individuals represented the host district of Gore, of which Hamilton was the administrative centre. No delegates from the neighbouring district of Niagara attended even though the *Niagara Chronicle* had published the resolutions of the July meeting along with the request for delegates to attend.[9]

At Hamilton, Thomson was selected chair of the meeting and Edmundson its secretary. However, some delegates questioned whether they should proceed with discussion due to the poor attendance and lack of province-wide representation. Eventually, those in attendance agreed with the opinion of Sheriff Wilson S. Conger of the Colborne District, who argued that he and his colleague "had come one hundred and fifty miles to attend this Association, and he hoped the meeting would not break up without effecting its object." Apparently, there was also "a good deal of discussion" about whether a provincial association could even be successful in Upper Canada. What turned opinion in favour of the proposed constitution and the utility of the proposed association was a general appreciation of the positive example provided by the New York State Fair, held annually since 1841 in different cities of the state.[10]

The draft constitution provided a slightly better outline of the association's purpose than had the resolutions of July. The stated aims were "the improvement of Farm Stock and Produce; the improvement of Tillage, Agricultural Implements, &c.; and the encouragement of Domestic Manufactures, of Useful Inventions, and generally, of every branch of Rural and Domestic Economy."[11] Following approval of the draft constitution to establish the "Provincial Agricultural Association and Board of Agriculture for Canada West," attendees elected a temporary set of officers. Thomson was chosen president, and John Wetenhall of the Gore District Agricultural Society and Henry Ruttan, sheriff of the Newcastle District, were elected vice-presidents, with Edmundson chosen to serve as both secretary and treasurer. Thomson accepted his post rather reluctantly, for he declared in his acceptance speech that he had hoped to meet the Honourable Adam Fergusson, who was not in attendance, and propose that he serve as the association's first president.[12] Fergusson, the legislative councillor, continued to be one of Upper Canada's pre-eminent agricultural improvers on his Woodhill estate.[13]

The location of the proposed first provincial agricultural exhibition also provoked a "long discussion" at the August meeting. The representative of the Johnstown District Agricultural Society offered £20 and a portion of the society's public grant regardless of where the exhibition was held, but this did little to settle the question. Chairman Thomson informed the meeting that the HDAS would offer £100 in support of this exhibition if it was held in Toronto but only half that amount if another location were chosen. This seems to have trumped the debate and a committee of management was established to host the association's first exhibition in Toronto in October.[14]

It was in response to this meeting's decisions that J.B.W. boasted of the HDAS's exemplary £100 donation in his September 1846 letter to the *British Colonist* as a challenge to other district societies to fund even half that amount, for he warned there "must be no jealousy or backwardness upon the part of the inhabitants of the country." Failure in the first year would be "deplored," he argued.[15] Elsewhere, in advertising the upcoming exhibition, most newspaper editors looked forward to a successful event in Toronto and hoped that the society would be well supported by farmers from across the province.[16] George Brown noted in his *Globe*: "there cannot be any better mode of promoting the prosperity of the country," while the *Christian Guardian* told its readers that the "importance of such an exhibition ... can only be fully appreciated by those who have witnessed the impulse given to [agriculture, horticulture, and domestic manufactures] by similar exhibitions in other countries." Charles Lindsey, always sceptical of the tory elite's activities, expressed only guarded enthusiasm in his *Examiner*: "We are not about to shower unmeasured laudations upon a thing yet scarcely in existence."[17] Such guarded optimism proved wise.

A last-minute scramble launched Upper Canada's first "Grand Provincial Exhibition of Agricultural, Manufacturing, and Horticultural Products, The Fine Arts, &c." Edmundson assured readers of his *British American Cultivator* that preparations were being made "on a grand scale," although he confessed that the organization of the exhibition "may not be as complete as would have been the case if more time had been given the Committee of Management."[18] In effect, the society's £100 donation had purchased the HDAS's hosting of Upper Canada's first provincial exhibition with no proof that the society possessed the organizational skills or the support from inside and outside the district to make the event a success. The HDAS had wisely decided to not host its annual fall cattle show. It had been just two years since the society's aborted grand exhibition for the district, and in a repeat scenario, it would abandon the advertised location for the 1846 exhibition at the Caer Howell Grounds to the north-west of the city just prior to the event for unknown reasons.[19]

Instead, the exhibition was held on 21 and 22 October, on the grounds of Government House plus those of Upper Canada College at the corner of King and Graves (now Simcoe) Streets. The Provincial Agricultural Association offered some £300 in cash prizes and £100 in book awards across eighteen categories of competition, such as Durham, Hereford, Devon, and "other Improved breeds," as well as specific breeds of sheep, namely, Leicester, South Downs, and Merinos or Saxons. Given the association's leadership, it is not surprising that a provincial exhibition emphasized similar improvements as the HDAS: livestock improvement and the importance of supplying butchers with quality animals to feed a growing province while adding a need to supply quality wool to increase textile production. The association also advertised competitions for horses and pigs, plus seed crops, root crops, and horticultural produce. Makers of agricultural implements were encouraged to compete, and the category of Domestic Manufactures was expanded to cover a wide range of production. Competition was provided for woollen and flaxen goods, dairy products and sugar, cabinet ware, iron and hollow-ware, potteries and the like, and bookbinding and printing. A ploughing match was scheduled for a nearby field, and organizers promised a display "of ingenuity in every department of skill and science," as well as an exhibition of "collections of paintings, whether the works of the old masters, or of living artists, – statuary, &c., and any other works of art."[20]

At the end of the exhibition's first day, between two hundred and three hundred people assembled in the Government House for dinner.[21] According to Scobie of the *British Colonist*, this first exhibition "realized all that its most zealous advocates could have expected" despite "bad roads and not the most encouraging weather." The Honourable Adam Fergusson addressed those at the dinner and delivered a lecture on agriculture from the verandah of the Government House the following day.[22]

The final event of this inaugural provincial exhibition was the first annual meeting of the agricultural association to elect officers for the ensuing year. Although President Thomson continued to believe Fergusson was the better individual to lead the association because of his experience with the prestigious Highland and Agricultural Society of Scotland,[23] he was elected to a second term, with Fergusson elected as senior vice-president. Henry Ruttan was elected second vice-president, and Edmundson was asked to retain his position as secretary and treasurer.[24] The members in attendance then approved several fundamental alterations to the nature of the association.

Jarvis proposed a name change from the "Provincial Agricultural Association and Board of Agriculture for Canada West" to "The Agricultural Association of Upper Canada," dropping any aspirations to be a board of agriculture. The change to "Upper Canada" from the official "Canada West" indicates how Jarvis, like other gentlemen of the province, remained rooted in their sense of provincial nationalism.[25] Jarvis bundled his motion with another requesting removal of the fourth clause of the association's constitution. As agreed at the August meeting in Hamilton, this clause had called for the association to be "governed by Delegates sent by the several Districts, who shall meet annually for the Election of Officers, and the transaction of business of the Association; and in case no such Delegates are appointed, then the Presidents and Secretaries of such Societies to be *ex-officio* Delegates."[26] Members agreed to its removal. Then, Jarvis proposed two more changes, agreed to by those in attendance. His first recommendation was: "[t]hat the President, Vice-Presidents and the Directors ... have the power to *nominate* a Committee from among the members of the Association, to assist in the management of the Association, which committee, during their continuance in office, shall have full power to speak and vote at all meetings of the Board, in the same manner *as if they had been elected Directors from any District of the Province* [emphasis added]." The second clarified the leadership structure, providing for a president and two vice-presidents and limiting the number of directors to forty, two from each district. The original proviso remained that, if no director was selected in a district, the president and the secretary of that district agricultural society would be made *ex officio* directors.[27] In an age in which others in the province were fighting for responsible government, the AAUC members in attendance agreed that, in appointing directors and forming a committee of management, the association's executive represented each district of the province, whether or not it was supported by the membership of the province's government-funded agricultural societies to which the AAUC remained legislatively unconnected.

Yet, constitutional changes did nothing to secure long-term funding. How might the AAUC raise funds for an 1847 exhibition? Eight district agricultural societies had made donations to the new association, as had the Canada Company, but all were one-time gifts. The only funding strategy that the AAUC could devise was petitioning district societies across the province for financial aid from their annual public grants and to petition each branch of the legislature for an annual public grant.[28] Three hundred thirty individuals had become regular members; however, those who paid for a life membership provide a true understanding of initial support for the AAUC. At its post-exhibition annual meeting, twenty-four gentlemen paid £2 10s for that status – ten times the rate for regular membership. In return, life members received a membership badge to allow them free entry into all future exhibitions, plus a silver medal struck for the new association.[29] But life membership was more than just a convenience paid for by wealthy gentlemen. The AAUC employed the weight of their names and public offices when it petitioned the Canadian legislature for an act of incorporation in 1847. The petition was successful, and the life members and other gentlemen who had taken an active role in the formation of the AAUC were listed in the preamble to the act. MPPs, government officials, lawyers, millers, merchants, and businessmen were among the thirty gentlemen named. Only six were not original life members of the AAUC, and the list belied the fact that the AAUC's greatest support came from those in and around Toronto. Seven of the thirty individuals were either executives or directors of the HDAS, while sixteen lived in Toronto, and another ten resided in the Home District.[30]

Following the 1846 provincial exhibition, Edmundson continued his calls for a better form of provincial agricultural leadership. The December issue of the *British American Cultivator* featured a three-page lead editorial on "Canadian Agricultural Societies," followed by another of similar length in his February 1847 issue. In the first, Edmundson returned to his recommendations of years past that the township, district, and provincial agricultural societies should work in concert to promote agricultural improvement, and he announced that a "Provincial

Images 8.1. and 8.2. Obverse and reverse of medal issued by the Agricultural Association of Upper Canada, ca. 1847 (Private collection.)

Board of Agriculture" would likely hold a first meeting before the end of winter. This board, he recommended, should promote the creation of an agricultural college, an experimental farm, an agricultural and mechanical museum, and it should consider the publication of its transactions.[31] While AAUC president Thomson did send a letter in late February 1847 to the president of each district agricultural society to notify them that the association intended to petition the government concerning "the subject of agriculture improvement and encouragement," he slyly suggested that the government's inclination to pay attention to its petition would "in all probability, be regulated in a great measure by the support which District societies may incline to give to the General Association." *Canada Farmer* editor R. Brewer seized on this comment and rightly questioned Thomson's request for donations, because the province's agricultural societies were left "profoundly ignorant" as to the purpose for which such donations would be applied. Brewer, who had just begun publishing his *Canada Farmer* in Toronto during the previous month, continued his column by expressing regret "that such an unbusinesslike [*sic*] document, couched in such indefinite language, should have emanated from [the AAUC]."[32] By July, Edmundson could identify donations from just four district societies for a combined total of £145. Most societies did not respond, while others made clear that they "positively refused to render any aid."[33] Belying the desperate need for funding, Edmundson calculated in a September 1847 column that if just one person in each township of the province became a life member of the association and five as regular members, £1,500 could be added to its treasury.[34]

Reports of October 1847 illustrate that while the AAUC could not find regular financial support for its organization, it could secure financial support from across the province for its hosting of a provincial exhibition. Twelve district agricultural societies and one county society donated funds to support the AAUC's 1847 exhibition at Hamilton, as did newly arrived Governor General Lord Elgin, who would attend the exhibition with Countess Elgin. Held on 6 and 7 October, the exhibition featured some 1,700 entries for competition. But it was a disorganized washout. Torrential rain on both days compounded problems generated by poor planning and mismanagement. Held at a low-lying location more than a mile outside of Hamilton, the site and the roads approaching it were "deep with mud," while many of the finer domestic manufactures entered for competition were destroyed by the heavy rains. At the exhibition's close, the AAUC was unable to account for over £288 of the £670 that the association had on hand at its opening. Secretary and Treasurer William Edmundson would receive most of the blame for this situation, although he was reelected to his executive positions at the annual meeting that closed the exhibition.[35]

Attendees to that meeting also elected the Honourable Adam Fergusson as the AAUC's president for the year.[36] Under his leadership, the association would attempt to re-establish financial stability, if not the association's credibility. At its February meeting, the AAUC formed a committee to investigate Edmundson's accounts and prepared a petition to the government that Vice-President John Wetenhall, MPP for Halton, would introduce in the legislature. It requested a sizeable annual public grant to the AAUC of £1,000, plus a one-time grant of £500 to cover its losses from the 1847 exhibition. The petition produced no results.[37]

The AAUC issued another financial appeal to the province's district agricultural societies of the province but not for debt recovery. Fergusson made it clear that any donations offered would fund prize money for the 1848 provincial exhibition. He also declared that "the Secretaries of the District Societies, shall be considered by virtue of their office, Assistant Secretaries to the Provincial Association" to better streamline connections with district agricultural society members who were the most likely individuals to submit entries to the AAUC's provincial exhibition. Farmers who intended to compete at the 1848 provincial exhibition in Cobourg would have to register their entries with the secretary of their respective district agricultural society and pay them the entrance fee. In turn, each district secretary would forward to the AAUC secretary all registrations and payments received.[38]

Fergusson's appeal produced results. To host its third annual exhibition at Cobourg, the AAUC was able to draw upon some £415 in donations from various agricultural societies, plus donations from municipalities, the Canada Company, and private individuals.[39] Held on 3–6 October 1848, the Cobourg exhibition expanded by two days, but a massive rainstorm on the day prior to the event delayed the arrival of many entries, meaning organizers had to extend the deadline for admission to midday of the second day. This, in turn, delayed the completion of the judging until late on the third day, defeating the purpose of extending the exhibition in order that judging could be completed early and spectators could view the results of their work on the final days of the event. Furthermore, several quality entries from western locations did not make it on time and were excluded from competition, and some otherwise quality entries were damaged by the high winds and rain during transit to the showground. Nevertheless, reports suggest that the exhibition was well attended and showcased a wide range of items, from livestock to "the produce of the loom, the garden and the foundry, the ladies' parlour and the artists' atelier."[40] Notably, the prize list also included the first competition for "Indian Prizes," as a means to encourage Indigenous peoples to join Upper Canada settlers in adopting the ideology of improvement. The AAUC awarded thirteen prizes to ten Indigenous competitors for items ranging from a bark canoe, paddles, tobacco, cradle, pipes (peace and war), moccasins, and fruit baskets. Thirteen "Additional Indian Prizes" were also awarded for items ranging from a war club to door mats to window sash. This category was offered again at the 1850 exhibition in Niagara and at the 1852 exhibition in Toronto.[41]

In several ways, the Cobourg exhibition was a turning point for the AAUC. First, it had enlisted the Bank of Upper Canada to serve as its treasurer in order that the fiasco of Hamilton was not repeated. The exhibition turned a small profit of some £74, although it was not nearly sufficient to settle its Hamilton-incurred debts. Second, due to another rained-out exhibition, the AAUC decided to move the event from the first week of October to the third week of September, beginning with its 1849 exhibition scheduled for Kingston. The timing would improve chances of good weather; plus, it was a better time to display freshly harvested crops and produce. In turn, each factor would improve competition and attendance, and increased gate receipts were essential for the AAUC's continued existence in the absence of government funding.[42]

The AAUC inched closer to clearing its indebtedness in February 1849 when a meeting of directors established another committee to investigate the outstanding Hamilton debts with the authority to pay them when able. Elsewhere, the AAUC's renewed lobbying to secure funding proved successful. First, it instructed its past-president Fergusson and First Vice-President Wetenhall – the former a legislative councillor and the latter an MPP who were headed to Montreal for the annual session of legislature – "to urge upon the Government" the need for a grant of money to discharge its debt. Second, AAUC president Henry Ruttan issued another urgent appeal to the province's district agricultural societies for donations from their public funds because the association needed at least £1,200 to carry out its aims for the coming year. At its next meeting of directors in May 1849, it also established a committee to travel to each district of the province to advocate directly for "the general interests of the Association" and to collect any funds offered.[43]

By the time the AAUC's petition was read in the legislature in February 1849, two petitions from John Clark, president of the NDAS, had already been presented, requesting amendments to the district agricultural societies legislation. His petitions generated the establishment of a select committee of five members, including Wetenhall, to consider their contents and deliver a report. The next day, two other members joined the committee, and it was granted additional authority to report by bill.[44] In mid-April, the committee delivered its report, and the legislature provided its approval, permitting Welland MPP Duncan McFarland to introduce a bill amending the district agricultural societies legislation. Unfortunately, the bill did not reach a second reading. Introduced and first read on 18 April 1849, its second reading was scheduled for 23 April, but it did not make the agenda for that day. Then, on the evening of 25 April, an angry mob rushed the Montreal parliament building in response to Governor General Lord Elgin's granting of royal assent to the Baldwin–Lafontaine Rebellion Losses Act that offered compensation to those who had lost property in the uprisings of 1837 and 1838. The incident started a fire, which completely destroyed the legislative chambers, offices, and library. The legislature continued its session by meeting in temporary quarters at Bonsecours Market.[45] As a consequence of these tumultuous events, the fact that many bills and records had been lost in the fire, plus the unsettled atmosphere that followed this shocking incident, the order for second reading of the agricultural societies legislation was discharged on 3 May.[46] Numerous other important bills had to be recovered, and the attack itself generated many new urgent issues to deliberate. Yet, all was not lost for the AAUC. In its report, the select committee had expressed its "anxious desire" for the association's success and had recommended unanimously that the legislature provide funding to the association as well as its counterpart in Lower Canada, established in 1847.[47] As a result, the government issued the AAUC a grant of £250, plus a one-time grant of £350 to discharge its debts.[48]

In January 1850, the AAUC provided opportunities in both Hamilton and Toronto for those who had outstanding claims from the 1847 Hamilton exhibition to collect monies due to them. It also settled legal bills for the defence of two suits of claim brought against the AAUC in consequence of the 1847 exhibition. Doing so left the association with a deficit of some £13. Not only were these debts the result of William Edmundson's incompetent administration of the 1847 exhibition, but Edmundson had also caused more recent financial and legal issues for the AAUC to solve. He had purchased books in the United States and charged them to the AAUC. As a result, the American booksellers contacted Sheriff Jarvis demanding he secure the monies they were owed. Although the committee believed the AAUC did not have a legal obligation to do so, they recommended the bill be paid.[49]

By then, William Edmundson was no longer associated with the AAUC and no longer in Upper Canada. He had ceased publication of his *British American Cultivator* in December 1847, and eight months later, he and William McDougall launched the *Agriculturist and Canadian Journal*. The partnership did not last long, and Edmundson began his own journal, the *Farmer and Mechanic*, which presented mostly material copied from other periodicals. His business failed in August 1849, and he made a hasty departure to the United States under a cloud of controversy. Edmundson's life in the United States would be short, for he died in 1852 on a farm near Nauvoo, Illinois, at approximately thirty-seven years of age. As Edmundson's biographer notes, the lack of any obituary for him in the agricultural press of the day is evidence of how far from grace Edmundson had fallen among Upper Canadian improvers.[50] Although his Upper Canadian career ended in shame, Edmundson played two significant roles in the promotion of agricultural improvement and in the process of state formation. First, he founded the province's first agricultural journal of any duration and had been a tireless promoter of agricultural improvement, including as a member of the HDAS and founding executive of the AAUC. Second, he was responsible, in part, for encouraging the well-known English agriculturalist and member of the RASE, George Buckland, to settle in Upper Canada. This gentleman would provide the AAUC with both intellectual heft and substantial practical support for the rest of his life.

George Buckland had first toured Upper Canada and parts of the United States in the early 1840s, at which time he met Edmundson. They continued to correspond upon Buckland's return to England. Then, in 1847, Robert Baldwin and Adam Fergusson (both legislators and life members of the AAUC) encouraged Buckland to return to Upper Canada, holding out the promise that he would receive a professorship in agriculture at a reformed King's College in Toronto. He took up their invitation, but upon his arrival later that same year, he found the university reform issue mired in a seemingly endless political conflict.[51] Meantime, Buckland joined Edmundson as co-editor at the *British American Cultivator* from October 1847 until the business closed at the end of December. The AAUC invited him to deliver the agricultural address at the 1848 provincial exhibition at Cobourg, and at the general meeting that followed, members elected Buckland to replace Edmundson as the association's secretary. With Buckland and Fergusson, the AAUC had two improving gentlemen of considerable reputation within the improving circles of the British Isles and British North America holding executive offices.[52] Subsequently, Buckland resumed his career as an agricultural editor. After William McDougall stopped publishing his *Agriculturist and Canadian Journal* in December 1848, he and Buckland launched the *Canadian Agriculturist* the following month. It may not have been the official journal of the AAUC, but Secretary Buckland ensured the activities and interests of the association were promoted prominently within its pages.

With most of its financial burdens put to rest, the AAUC agreed to make its 1849 exhibition at Kingston its grandest event yet. In May of that year, AAUC president Henry Ruttan, announced that he had invited "one of the most eminent agriculturists in Great Britain," Professor James F.W. Johnston, to address the Kingston exhibition after learning that he would be delivering the annual address to the New York State Fair at Syracuse in the week prior to the Kingston exhibition.[53] Indeed, Professor Johnston was one of the most recognizable names in agricultural chemistry in the 1840s, a subject he taught at the University of Durham. He was a member of the RASE, a foreign member of the Royal Swedish Academy of Agriculture, and a chemist to the Highland and Agricultural Society of Scotland. His 1844 *Catechism of Agricultural Chemistry and Geology* had been published in numerous editions.[54] By September, Buckland reported happily that Professor Johnston had accepted the association's invitation and would deliver an evening lecture. A student of Johnston's, Professor John P. Norton, Professor of Agricultural Chemistry at Yale University, also indicated his intention to travel to Kingston, prompting Buckland to predict that the "presence of such eminent individuals in the walk of agricultural science … will be an additional means of attracting a great number to the Exhibition."[55]

The 1849 provincial exhibition was held 18–21 September on a ten-acre government-owned site at the south-west end of Kingston during weather that was "cool and pleasant." A ploughing match was held elsewhere on the fourth day as was a "Grand Provincial Regatta," which was open to all competitors. Some 12,000–13,000 people attended,[56] of which a large number were American exhibitors and visitors, including the secretary of the NYSAS as part of a three-person delegation from that organization. With such great numbers of Americans, however, organizers were disappointed at the small crowd that listened to Professor Johnston's address on the second evening as well as the poor attendance at the dinner on the third evening of the show. The reason for both disappointments was simple: many American visitors could not stay because they needed to catch the evening steamboat back to their side of Lake Ontario. A second reason as to why a third of the dinner tables were left empty was that Governor General Lord Elgin had been expected as a guest but did not attend. The "no, nos" provided by some dinner guests as an initial response to the toast to the governor general at the dinner provide some sense of Elgin's disinterest in attending public events with the continuing political disquietude of the Canadas in 1849.[57]

During the AAUC's meeting that followed the Kingston exhibition, Marks and Thomson proposed a series of regulations to provide clear instructions as to the planning and financial management of future annual exhibitions. Shortly thereafter, the AAUC's new president, John Wetenhall, wrote an open letter in support of such revisions to the management processes, because he did not believe any of the AAUC's first four exhibitions had "worked well, or that we can reasonably hope for the introduction of an *uniform* system of management, until our present system has been entirely changed." He welcomed any changes that would prevent "a repetition of blunders, which have already nearly destroyed the Association." During the Kingston exhibition, Wetenhall discussed the organizational structures of the RASE with Professor Johnston, and by late November, he had drafted up a set of twenty proposed amendments to the constitution, based on the resolutions put forward by Marks and Thomson. For his assistance, the association granted Professor Johnston an honorary life membership. The amendments were approved at the association's annual meeting in February 1850 in the form of twenty-three bylaws to be attached to the AAUC's constitution.[58]

The matter of securing funding and support from the province's agricultural societies remained. To further this goal, the AAUC offered badges for free admission to members of each agricultural society within the host county if they provided all their funds for that year to the AAUC, including their government grant. The executives of other county and agricultural societies would receive the same benefit should a county society provide no less than £25 and a township society no less than £10. Reflecting the expansion of the ideology of improvement in the 1840s to include commerce, manufacturing, canals, and railways, as well as chemistry and other sciences, the AAUC extended similar benefits to office holders of Mechanics' Institutes, established for the education of tradesmen and other workers, and any other scientific society willing to donate no less than £25.[59] By the time of the 1850 exhibition, the plan produced modest success: three townships and six counties donated funds, although three of the county donations and one of the township donations did not meet the prescribed level to qualify for the offer.[60]

While the AAUC matured with the adoption of these bylaws, the association lost President Wetenhall's valuable leadership and the potential of his political power and influence. In September 1849, his elevation from first vice-president to president appeared fortuitous for the AAUC, because MPP Wetenhall had been promoted into the Baldwin–Lafontaine cabinet. He was to replace Malcolm Cameron, MPP for Kent, as assistant commissioner of public works following Cameron's sudden resignation. Under parliamentary rules of the day, Wetenhall had to seek re-election in his riding to take up his cabinet appointment. In the campaign leading up to a March 1850 by-election, Wetenhall faced Caleb Hopkins, who had lost the Reform nomination to Wetenhall in 1844. Hopkins, eager to challenge Wetenhall again, found immediate support from Cameron and the Clear Grits, a new political faction he had joined after resigning his cabinet post in the Reform administration. Wetenhall's mental health was already deteriorating by early 1850 (both the by-election campaign and his mental health likely prevented him from attending the AAUC's February annual meeting in Toronto), yet this did not stop Cameron, Hopkins, and their associates from launching an unrelenting series of attacks on Wetenhall over the increasing conservatism of the Baldwin–Lafontaine ministry. As a result, Hopkins won an unexpected and resounding by-election victory on 11 March, while the mentally shattered Wetenhall had to be admitted to the Toronto asylum. He lived only until 20 June, dying just a few months shy of his forty-third birthday of an infection caused by wounds inflicted by another patient.[61]

First Vice-President J.B. Marks took up the task of leading the society through the rest of the year. The annual exhibition to be held in September at Niagara required planning, and the AAUC was lobbying for the creation of a board of agriculture for Upper Canada as well as the establishment of a chair of agriculture at the University of Toronto along with an experimental farm. Lobbying for the latter cause gained urgency in 1850, for the province had shuttered the Anglican King's College and recently passed legislation establishing a non-denominational University of Toronto, which began organizing its curriculum and hiring professors to teach it.

Here, the AAUC entered a brief, but significant phase of its existence and purpose, that of a lobby not just for its own survival or its exhibitions but also for state support of agricultural improvement in general. Those 1850 lobbying efforts resulted in the Baldwin–Lafontaine government establishing a board of agriculture (discussed fully in chapter 9), thereby also legislating the AAUC and its annual provincial exhibitions as state agents for agricultural improvement. By way of the new legislation, which the AAUC had drafted, the Board of Agriculture would henceforth act as the board of directors for the AAUC, along

Image 8.3. Sketch of the Provincial Agricultural Exhibition at Niagara, 18–20 September 1850. *London Illustrated News*, 14 December 1850 (Private collection).

with presidents of the county agricultural societies across the province. The board's secretary – a board appointee – would serve as *ex officio* secretary of the AAUC. Lobbying the government would now be performed by the Board of Agriculture, leaving the AAUC to perform the singular task of organizing and hosting an annual provincial agricultural exhibition. Thus, the "Provincial Agricultural Association and Board of Agriculture for Canada West" that had been announced in the summer of 1846 became a reality. Yet, its central problem remained unsolved. Both pieces of legislation that governed the AAUC, its act of incorporation and the Board of Agriculture legislation, failed to secure it permanent funding or provide it access to the provincial funding offered to county and township agricultural societies.[62] Instead, the government provided the AAUC with a one-time grant of £600 to assist preparations for the provincial exhibition at Niagara in September. This was a significant sum, being £100 more than the local committee for the 1849 exhibition at Kingston had recommended the AAUC request for the 1850 event and double the donation provided by the host town of Niagara.[63]

Blessed with good weather on 18–20 September, the provincial agricultural exhibition was attended by as many as thirty thousand visitors, including many from New York State.[64] These were by far the event's largest crowds to date, even though the exhibition had been reduced to three days. The chosen fourteen-acre site between the steamboat landing on the Niagara River and the Town of Niagara proved, perhaps, to be the correct choice of location rather than Niagara Falls as some members of the AAUC had wished. The association invited Professor Henry Croft to deliver the annual lecture at the courthouse in Niagara. Croft had been the chair of chemistry and experimental philosophy at King's College since 1842 and had recently been appointed vice-chancellor of the new University of Toronto. In the absence of a chair of agriculture who could deliver the lecture, the AAUC had requested that Croft "not to dive too deep into the ocean of Science, but to give a plain popular description of the principal elementary substances, which enter into the composition of the soil which the farmer tills, and the plants which he raises." A large audience attended this "able and interesting lecture," which included "a series of illustrative experiments." In thanks, the AAUC made Croft an honorary life member.[65]

The AAUC elected J.B. Marks to continue his service as president for what was effectively the association's final months as the pre-eminent provincial agricultural organization. Although the association concluded the year with a balance of some £227,[66] at its meeting in Toronto in June 1851 it prepared a petition to the government "praying that the Government would recommend to Parliament such a grant of money to the Board of Agriculture, as would enable the Board to sustain the Provincial Association in a state of efficiency, and also to carry out the several important objects contemplated in its establishment." It suggested a grant of £1,000 was required to accomplish its yearly activities. AAUC directors delivered this message to Inspector General Francis Hincks, who was the government's *ex officio* member of the new Board of Agriculture. The provincial government complied. At the end of the three-day meeting, the AAUC transferred its directorship to the Board of Agriculture.[67]

This was not the only important change happening within the AAUC. Increasingly, the exhibitions and speeches of its presidents demonstrated agriculture improvement's growing embrace of commerce and the technology to facilitate agricultural improvement and trade. The Kingston and Niagara exhibitions, and their proximity to the United States, were practical expressions of the AAUC's increasing rhetoric focused on the intersection of agriculture and commerce in the wake of British free trade and the instability of the British North American commercial trade during the end of the 1840s. The instability caused financial panic, fueled the anger of the mob that attacked the parliament buildings in April 1849, and caused Montreal businessmen to issue their Annexation Manifesto later that year calling for annexation of the Canadas by the United States. When addressing the economic future of the Canadas, the AAUC's rhetoric and actions highlighted Upper Canada's dual heritage continuing under the new necessities of securing trade relationships with Britain and the United States. While Anglophilia remained a driving force within the association, with Buckland as its expert improver, Professor Johnston's advice on the RASE's efficiency of operation, and an honorary membership extended to His Royal Highness, Prince Albert,[68] the practical examples provided by agricultural improvers, breeders, manufacturers, and the agricultural press in the United States drew closer Upper Canadian and American improvers at mid-century. Upper Canadian agricultural journals reported regularly on the activities of the NYSAS, and Upper Canadians certainly attended the New York State Fair, particularly when it was hosted at a location close to the border. By the end of the 1840s, proximity to the United States and the timing of the New York State Fair became an important factor in the AAUC's decisions where and when to host the provincial exhibition. Increased gate receipts from American visitors was just one part of this calculus. The AAUC desired as much to display the progress and potential of Upper Canada's agriculture and manufactures, as it wished to attract the attention of the NYSAS and have American livestock breeders and manufacturers exhibit to Upper Canadians as a spur to local ingenuity and entrepreneurship.

American manufacturers from Rochester, Albany, and even Boston had brought examples of agricultural equipment to display at the Cobourg exhibition in 1848. Although AAUC regulations did not permit American manufactures to compete for prizes,[69] this would be remedied in advance of the following year's exhibition at Kingston. Organizers established the AAUC's first "Foreign Department" prize category as "a means of increasing the Exhibition by considerable additions both from Lower Canada and the United States." In fact, Americans interested in competing or merely showcasing their equipment nearly overwhelmed the Kingston customs officials in their efforts to admit the American-manufactured machinery in time for the exhibition. In the end, fourteen prizes, from stallions to agricultural implements were awarded to competitors from New York state.[70] At its February 1850 directors meeting, the AAUC instructed Secretary Buckland to approach a contact in Buffalo, New York, to arrange for American judges to assist with the upcoming exhibition at Niagara, which they did.[71] At that exhibition, twenty-six prizes and commendations were awarded in the foreign category. At the conclusion of the exhibition, the AAUC also offered life memberships to two Americans, one to Silas M. Burrows of Medina, Ohio, for a generous donation to the association and another to E. Coulson Williams of Rochester for bringing several tents and marquees that he had manufactured, plus a large British flag he had manufactured and donated to the AAUC after having flown at the showgrounds throughout the exhibition. The *Canadian Agriculturist's* report of the exhibit trusted "that his handsome present will remain for many a long day, a memento of the mutual friendship and good will which should ever characterise [*sic*] a people who can proudly boast of possessing a common origin and language, and now forming the two greatest nations upon earth."[72]

To deflate the ongoing clamour in the Canadas that a more secure economic future rested in annexation to the United States, Governor General Lord Elgin engaged the AAUC to assist with his essay competition aimed at promoting the economic strength and opportunities for future growth provided by the Great Lakes–St. Lawrence trade route to Britain. Elgin offered £50 to the individual who could write "the best treatise on the bearing of the St. Lawrence and Welland Canals on the interests of Canada, as an agricultural colony." He requested the AAUC appoint two of its members to a three-member jury. President Ruttan and past-president Thomson were the association's choice and were joined by John Young, chosen by Elgin. He was a Montreal merchant who was an avid free trader and one of the few Montreal merchants who refused to sign the infamous Annexation Manifesto. By February 1850, ten essays had been received, with the winner being that submitted by Thomas C. Keefer, a civil engineer from Montreal whose father, George, had been the first president of the Welland Canal Company. According to one source, the 111-page pamphlet served its purpose, claiming that Keefer's submission was "pervaded by a spirit of ardent and noble patriotism ... No man can read it with attention without being impressed with the vast underdeveloped treasures and capabilities of our country, or without admiring the fore-sightedness of those who have projected the canal and other improvements in our internal navigation."[73]

In his December 1850 report to executives of county agricultural societies across the province, AAUC president Marks made clear why the association chose Brockville to host the 1851 provincial exhibition. Steamboats operating between Kingston and Montreal made Brockville easily accessible. Plus, it was located across the St. Lawrence River from Ogdensburg, New York, where railroads that stretched out across the state provided easy access to the exhibition for residents of Lower Canada's Eastern Townships as well as Americans from New York City, Boston, and other eastern states.[74] The local committee planning the Brockville exhibition echoed similar sentiments in its public address of February 1851 by anticipating a successful and well attended exhibition "in an age when Telegraphs, Railroads, Steam Navigation, and the various appliances of modern invention, have happily supplied the place of the one-handled Plough, and the shoulder-bruising pole-propelled Durham Boats of days now happily bygone in the history of Canada."[75]

The AAUC extended invitations to improvers and manufacturers from Lower Canada and the United States to compete at Brockville, and when the exhibition's organizers learned it would conflict with the New York State Fair at Rochester, the AAUC shifted its dates for the Brockville exhibition from the third to the fourth week of September out of respect for the larger, more senior exhibition in the neighbouring state. The AAUC hoped that those competing and visiting Rochester would continue to Brockville to engage the competition and exhibition there. Rescheduling also recognized and accommodated the number of Upper Canadians who visited the New York event.[76]

Not everyone was happy with the choice of Brockville. As close as that location was to the United States and Lower Canada, it was not a convenient location for many Upper Canadians living in the increasingly settled townships of the province's south-west. During the AAUC's executive meeting that followed the exhibition at Niagara, T.C. Dixon of Hamilton and James Dougall of Essex County had recommended that the 1851 exhibition be held at London. But members chose Brockville's proposal instead. Buckland's front-page editorial of November 1850 suggests the association needed to heal some rifts created by this decision. He responded to some complaints expressed by a recent agricultural society meeting in Guelph, at which members wondered why they should use their finances to support the AAUC, knowing that the provincial exhibition would likely never be held in their town. Buckland countered this "narrow and fallacious objection" by looking to the near future when "a grand trunk railway" would soon cross through the western part of the province, thus opening a much wider choice of location for the provincial exhibition. "Railway or no railway," he declared, "London, at least, must shortly have the Show," adding that suitable infrastructure might also soon make Guelph similarly accessible. To continue the success of the AAUC, he concluded, "We require a *long pull, and a strong pull, and a pull altogether*." Marks's December 1850 letter to the executives of county agricultural societies, extolling the location of Brockville, was no doubt intended to serve a similar purpose of placating those from the western part of the province.[77] London would have to wait until 1854 to host the provincial exhibition, following its return to Toronto and then, Hamilton.

The few reports of the exhibition, held on 24–26 September, indicate that the local committee had more than £2,000 in funding to organize the event, including the government's grant of £1,000.[78] While the categories of competition continued to be diverse, a general impression is that some competitions, such as livestock, were not as strong as the previous two exhibitions. Also, attendance was lower, estimated at ten thousand visitors.[79] One remarkable feature of the Brockville exhibition, however, was its impressive Floral Hall, hosted under a large tent supplied by a Rochester manufacturer. Creating "the greatest point of attraction" on the exhibition grounds, the display represented the growing interest in improved horticulture and floriculture. The tent also displayed "Ladies' Department" entries "to bring prominently forward the handy work of the Country Lass, as well as the Town Maiden." Increasingly, women were being encouraged to adopt the ideology of improvement in their spheres of activity. Women did not need to be a member of the AAUC to compete, and if they were the wife of a member, they and their children under the age of fourteen were admitted for free. Women's work had been encouraged at the initial provincial exhibition in 1846 under the category: "Ladies Department – Useful and Ornamental." Two years later, female judges were appointed for the Cobourg exhibition to evaluate entries to this category. By 1850, the AAUC had actively requested "the ladies of families [across the province] to devote at their pleasant fire sides, some part of their winter evenings, in preparing both articles of taste and usefulness," and a report advertising the 1852 exhibition in Toronto would define women's place at the competition this way: "Ladies ... are members *ex-officio* and have full rights to exhibit of their treasures."[80]

The Johnstown District Agricultural Society (centred in Brockville) had announced in spring 1851 that it would award a gold medal of £10 in value "for the best Essay upon Agriculture and its advantages as a pursuit" as part of the AAUC exhibition. Judged by directors of the Johnstown District Agricultural Society, individual competitors read their essays before members of the association at an evening event during the Brockville exhibition. The four prize-winning essays, including William Hutton's gold-medal-winning entry, later appeared as front-page content of various issues of the *Canadian Agriculturist*. This was due in large measure to a November 1851 agreement between the Board of Agriculture and William McDougall that his journal would be enlarged to include the transactions of the board while offering reduced subscription fees to agricultural society members. The agreement stipulated that the board secretary (George Buckland) would continue to edit the journal.[81]

Although these essays repeated many of the same tropes found in agricultural treatises published across the previous century, praising the practice of agriculture as a purer pursuit, unquestionably separate from commercial activities, this was no longer the message promoted by presidents of the AAUC. The Brockville exhibition featured no agriculturist or professor of international status. Instead, the association's President

J.B. Marks delivered a wide-ranging lecture that saw a different future for agriculture in the provincial economy than, say, what the former AAUC president Henry Ruttan had pronounced in his long-winded address to a crowd attending the Kingston exhibition in September 1849. Ruttan, the son of a United Empire Loyalist family, presented an agrarian, mixed-farming future for the province, one that saw farmers spending more attention on raising dairy cows and domestic manufacturers. But this was not for export. Instead, he desired Upper Canadians look to the example of the earlier settlers who lived more frugally and had to be more self-sufficient. The simple agrarian life, expanded only by a cow or two was, in his opinion, the way by which Upper Canada would get out from under its massive public debts.[82] In contrast, Marks, the British-born gentleman some fifteen years older than Ruttan, focused his speech mostly on the interconnected subjects of agriculture, manufactures, and commerce. In an age of increasing free trade, and communication improvements shaped by steamboats and railways, Marks expressed his optimism as to the bounties to be reaped by "bringing forth the latent resources of the country" and supporting merchant efforts to increase trade to England via Montreal, including ocean-going steamships, and new railways connecting the upper province with Montreal and Quebec. Trade with the United States held promise, but export duties for Upper Canadian wheat shipped through New York only highlighted the need for improvement of the St. Lawrence River trade to England. Reciprocity with the United States might also provide opportunities, Marks posited, but fear of market domination by American production of the same articles that Upper Canadians produced made that a possibility to consider only with considerable caution.[83]

The AAUC's leadership would further embrace the world of commerce with the elevation of its first vice-president to the role of president following the Brockville exhibition. Thomas Clark Street of Niagara was one of the largest landholders in the province, as well as one of its most prominent businessmen. The son of Samuel, the United Empire Loyalist miller who had been granted a life membership by the NDAS for his financial support at that society's founding, Thomas was also a lawyer and would soon be elected the MPP for the riding of Welland. Following the death of his father in 1844, Street had inherited over 15,500 acres of land scattered throughout south-western and central Upper Canada. Thus, when Street was elected to lead the association for 1851–2, he brought to the office the very embodiment of an increasing connection between agriculture and commerce and further integration of agricultural improvement into the activities of the Canadian state and the politics of the day.[84]

In fact, the executive of the AAUC elected at the Brockville exhibition had made a distinctive commercial shift. Joining Street as the first of two vice presidents was William Matthie of Brockville, whose profession was wholesale and retail trade within several partnerships he had formed over the previous decade. He was also a promoter of early attempts to use steamboats to connect Brockville with Montreal and on the triangular route of the St. Lawrence River–Rideau Canal–Ottawa River.[85] C.P. (Charles Platt) Treadwell, the other vice president, owned the large seigneury at L'Orignal on the Ottawa River that his father had secured in the 1790s, along with several thousand other acres in the area. A former sheriff of the Ottawa District and president of the Agricultural Society of the United Counties of Prescott and Russell, Treadwell was devoted to improvement, both agricultural and infrastructural. In 1856, he would author *Arguments in Favor of the Ottawa and Georgian Bay Ship Canal*, as a means of expanding trade to the upper Great Lakes and more western territories.[86] Also representing the ties of agriculture and commerce were the AAUC's secretary, George Buckland (via the editorials of his *Canadian Agriculturist*); James Fleming, the AAUC's official seedsman;[87] and the association's treasurer, the Bank of Upper Canada. This is not to say the gentleman farmer had disappeared entirely from the AAUC's executive offices. R.L. Denison served as AAUC treasurer while Thomson, Fergusson, Ruttan, and Marks continued to serve as ex-presidents.[88]

After Brockville, the AAUC's agricultural exhibition would return to the major commercial city of Toronto, whose thirty-two thousand residents[89] generated a sizeable market for farm produce and livestock on top of what was delivered to the city's wharves for export. In the six years since the AAUC hosted its initial provincial exhibition in Toronto, the city's population had nearly doubled. Both the AAUC and the City of Toronto wanted to create a spectacle, featuring "Canadian Industry and Enterprise" of quality and quantity not seen before in one location in the province. The local committee, created in March, was chaired by Toronto Mayor John G. Bowes, a merchant and railway promoter. Joining him as committee members were stalwarts of agricultural improvement in the Toronto area, including William McDougall and Professor Croft. Several were members of the County of York Agricultural Society. The government provided its annual grant of £1,000, while York County donated £100, and the City of Toronto offered £200. Thus, by June, the AAUC felt confident in its finances to abolish the usual entry fees charged to exhibitors. By September, the association was delighted to learn that the City of Toronto had increased its donation by an additional £600 and had authorized Mayor Bowes to declare the Thursday afternoon of the exhibition a public holiday "to afford all classes of the citizens an opportunity of visiting the Provincial Fair." With these developments, the local committee announced free admission to students from all schools, if supervised by their teachers. Highlighting the theme of "Agriculture and Commerce," Toronto butchers offered prizes for the best and second-best fat ox or steer, while the contractors of the new Ontario, Simcoe, and Huron Railway (the first steam-powered railway in Upper Canada) being constructed north from Toronto to Collingwood, donated some £12 to the AAUC.[90]

Railway connections lay in Toronto's future. For the exhibition, special rates had been coordinated with steamboat operators, as well as the city's hotels. All the preparations and financial support were rewarded by the reported thirty thousand paying visitors and school children with free entry to the seventeen-acre exhibition site on the western edge of the city on 21–24 September. The first day featured torrential rain, but the weather on the second and third days was beautiful. Following the layout of the Brockville exhibition, the centre of the site was occupied by the Floral Hall and an attached hall displaying Fine Arts and Ladies Work. Flanking this central building was an Agricultural Hall on its left and a Mechanic's Hall to its right. To the rear of these buildings were the Horse Ring and the displays of livestock. Dozens of the latest implements and agricultural tools were displayed by their manufacturers throughout the site. To further unite country and city, agriculture and commerce, on Thursday – the city holiday – a "grand procession" of livestock, along with thirty carriages, the Toronto Brass Band, and numerous men mounted on horses paraded through the downtown streets to the exhibition grounds. The display, according to the *Canadian Family Herald*, had never been seen before in Toronto "and was well calculated to impress strangers with an idea of the wealth and capabilities of the province."[91] The exhibition in Toronto succeeded in highlighting agriculture and agricultural improvement as the foundation of the province's economy. And to build on what the thousands of school children had witnessed at the exhibition, Egerton Ryerson, superintendent of schools for Canada West, indicated in his speech at the event a desire that agriculture be soon a regular subject of study in the province's common schools.[92]

In his presidential address, Street reflected on the state of the colony and its agricultural potential, pronouncing it to be one of the "finest agricultural portions of the world." However, much more could be done. In the second half of the century, more labour-saving farm tools

Image 8.4. Sketch of the Provincial Agricultural Exhibition at Toronto, 21–24 September 1852. John Allanson (1812/13–1853) (Toronto Public Library, Pictures-R-5343.)

and machinery needed to be manufactured in the Canadas to reduce reliance on American industries. Farmers should also better capitalize on new opportunities, for new markets were opening for Upper Canadian produce with the dawning railway era in British North America. Eastern seaboard cities of the United States, once distant, were closer because of railways and more miles of track were being constructed in North America every year. But much of the market was for beef, not wheat; thus, farmers should consider raising cattle. He also expressed concern that Upper Canada was too reliant on imports. Cheese, for example, could be made at home with the raising of more dairy cattle. In conclusion, Street expressed his belief that the government had "done all that could be expected from it, to promote the cause of agriculture." It was now the time for Upper Canadians to "become as skillful in the development of [the province's] resources, as many of them have become in the cultivation of wheat" in order to fulfil the "high destiny" of the province.[93]

A variety of commentators recognized the significance of the 1852 exhibition. It was not just another event in a growing metropolis and province, it was a showcase for the improvement of agriculture as well as the encouragement of manufactures and commerce. Hugh Scobie suggested to readers of his *British Colonist* that this was a moment in the province's history to reflect on the abundance now produced after little more than half-century of settlement as well as what might be produced in another half-century.[94] Elsewhere, Susanna Moodie, an Upper Canadian author who had recently published her *Roughing it the Bush, or, Life in Canada*, devoted an entire chapter to the subject of the "Agricultural Show" at Toronto in her subsequent book, *Life in the Clearings versus the Bush*. An English author, Moodie had settled in Upper Canada in 1832 following her marriage to half-pay officer J.W. Dunbar Moodie. Her chapter indicates that she and her husband had travelled from their residence in Belleville to Toronto to see the 1852 provincial exhibition. From her perspective, "no other subject was thought of or talked about" in the city during the week. "Every district of the Upper Province had contributed its portion of labour, talent, and ingenuity, to furnish forth the show," she noted. "The products of the soil, the anvil, and the loom, met the eye at every turn." In her boosterish book, she claimed that soon Canada would have no need to "be beholden to any foreign country for articles of comfort and convenience." The establishment of agricultural societies was the central reason for "the rapid advance of the province," she argued. These organizations had "stirred up a spirit of emulation in a large class of people, who were very supine in their method of cultivating their lands; who, instead of improving them, and making them produce not only the largest quantity of grain, but that of the best quality, were quite contented if they reaped enough from their slovenly farming to supply the wants of their family, of a very inferior sort." As to the AAUC, she lauded the fact that "[a]ll the leading men in the province are members of this truly honorable institution" and that the provincial exhibition created "a laudable struggle among the tillers of the soil as to which will send the best specimens of good husbandry to contend at the provincial shows, where very large sums of money are expended in providing handsome premiums for the victors." The agricultural improvements on display at Toronto in 1852, along with the implements that showcased the rising "genius of the mechanic," and the exhibits of "fancy work and fine arts" highlighted the advancing province. From personal experience, she claimed that her husband's second prize for wheat at the 1851 show in Brockville had instilled her with as much pride "as if it had been the same sum bestowed upon a prize poem."[95]

The 1852 exhibition and associated events at Toronto provide a measure of the AAUC's maturation following its founding in

1846. Its exertions across that period represented many of the changes occurring within the province's economy, politics, and society. While the RASE may have provided the inspiration and ideal to its Georgian gentleman-farmer founders, such as William B. Jarvis, increasingly the NYSAS provided the practical examples and reasons for ever-closer connections with the United States. Its annual state fair was a model for the AAUC to attain, as were America's advancing agricultural output, agricultural press, and industrial production of farming tools and machinery. Moreover, it offered new markets for Upper Canada's agricultural produce and nascent manufactures as Britain ended its preferential duties for the British North American colonies and shifted to a free-trade system. The AAUC's exhibitions and the addresses by its presidents demonstrate how the improving ideal of mixed farming was promoted to improve *and* diversify Upper Canadian agricultural production to fulfil new markets at home and abroad. Manufacturing and infrastructure claimed a greater place in the AAUC's exhibitions and rhetoric, too, for the labour savings provided by farm machinery inventions promised improvements in agricultural output. And for that increased output, steamboats, steamships, and railways held the promise of better transporting Upper Canadian goods to both local and distant markets. By 1854, a reciprocity agreement would be secured with the United States, facilitating the trade anticipated by AAUC presidents of the early 1850s. Shortly thereafter, the railways that had been proposed at mid-century would connect Upper Canada to Lower Canada and beyond to the Atlantic coast and the United States by bridges crossing the Niagara and Detroit Rivers.

In a brief, but significant, moment of its evolution, the AAUC helped shape the Canadian state, defining more precisely the role that responsible governments would play in leading agricultural improvement as part of their mandate. The AAUC successfully lobbied the provincial government to fund its hosting of annual provincial exhibitions, restructure the governance of the province's existing agricultural societies, and further integrate agricultural improvement as a function of the colonial state by establishing a board of agriculture for Upper Canada. Doing so deviated some distance from the "humble imitation" of the apolitical RASE that J.B.W. hoped to emulate in 1846. Far from that apolitical ideal, the AAUC had to dive into the turbulent politics of provincial governments transitioning to a responsible form of governance to secure its survival and the advancement of its agricultural improvement agenda. This chapter has identified several the association's successes that the AAUC's lobbying achieved, aided in large measure by a growing agricultural press and its skilful editors. A final chapter explores the

AAUC's success in securing a board of agriculture for Upper Canada, and, in turn, the board's work in advancing agricultural education at the new provincial university, plus its more fundamental task of further rooting the promotion of agricultural improvement as a government function by way of a bureau of agriculture, led by a cabinet minister who represented the entire Province of Canada, not just its upper half.

# 9 A Board and a Bureau of Agriculture, 1850–1852

The Honourable Malcolm Cameron addressed an over-capacity crowd at Toronto's St. Lawrence Hall on the final evening of the 1852 Provincial Agricultural Exhibition. Having accepted the new position of minister of agriculture earlier in the year, Cameron provided an overview of an agricultural bill soon to be debated in the upcoming session of parliament in Quebec that would establish a bureau of agriculture for the Province of Canada. The bureau would provide ministerial leadership and coordination of all agricultural societies established in both Upper Canada and Lower Canada, including a board of agriculture appointed in each part of the colony. Before he took the stage to speak, Cameron was keenly aware of the mixed reactions to the news that the Reform ministry of Francis Hincks and Augustin-Norbert Morin had created a new cabinet position and a new bureau of agriculture. For agricultural improvers, this was the direct state governance of agricultural improvement that realized long-sought dreams and lobbying efforts; however, opposition politicians of the day and the newspaper editors who supported them viewed the expansion of the Hincks–Morin ministry as nothing more than a political job with Cameron, the skilled master of political graft, as its sole benefactor. Newspaper reports of Cameron's speech would reflect these divided opinions. The *Toronto Examiner* claimed that "the audience appreciated the interest which the Government manifested in the agricultural prosperity of the county," while the *Globe* reported that people listened to Cameron, "but they were cold and incredulous, and all his petty, little bits of flattery to the farmers fell on unsympathetic ears."[1]

How had the plan for a Bureau of Agriculture, with a minister attached, emerged so quickly in 1852? To answer that question, this chapter ties together developments analysed in chapters 4 and 8 by returning to the February 1850 moment when the AAUC agreed "to draft an amended Bill for the Provincial Association; also an amended Bill for the Regulation of Agricultural Societies; and to prepare an address to both branches of the Legislature on the importance of establishing a Chair of Agriculture in the University of Toronto; – a Board of Agriculture and an illustrative or experimental farm." By May, William McDougall and George Buckland had drafted the bill and given it to the AAUC's former president E.W. Thomson to deliver to the ministry, along with the association's petitions and a request that the bill be introduced to the legislature immediately.[2] This degree of lobbying represented a remarkable change in character and action for the AAUC. Why all the new initiatives, and why did it believe the provincial government would accept and promote its draft bill as a government measure?

This chapter examines the AAUC's successful lobbying for a board of agriculture for Upper Canada and that board's initial character and activities, particularly its own lobbying for a chair of agriculture and an experimental farm at the University of Toronto. It also examines why, in 1852, the provincial government appointed the minister of agriculture to lead a new bureau of agriculture and what that meant in the short term to agricultural improvers and their long-held desire to see the promotion of agricultural improvement become a permanent function of the colonial state. Notably, each of these new measures came about during a period of often-nasty political battles as elected ministries of the United Province of Canada tried to govern successfully and effectively under a responsible government model while continually juggling a myriad of interests divided broadly between Upper Canada and Lower Canada, English and French populations, and Protestant and Roman Catholic faiths. And they did so while attempting to hold together a government drawn from each half of the "united" province with elected members representing a spectrum of reform and conservative views loosely affiliated into political groups and factions. Meanwhile, as discussed in the previous chapter, the economic footing of the Canadas was in flux, following Britain's shift to free trade and prior to a reciprocity agreement with the United States in 1854. Despite all these divisive and transformative forces, a basic unity of ideology at mid-century was that agriculture remained at the heart of the Upper Canadian economy and its improvement in conjunction with the expansion of commerce and infrastructure for trade would lead the province's economic development. To some degree, rifts among the shifting political factions of the day were not insurmountable as far as all were working towards the common goal of agricultural improvement and economic development of the Canadas; however, one might define specifics within those broad terms. Divisive legislation of the era, such as bills to provide compensation for losses during the rebellion, establish a provincial university, improve the province's schools with standardized administration and curriculum, or resolve the long-standing question of the millions of acres of clergy reserves tend to mask the fact that such unity was possible. Each elected member of Canada's colonial parliament had some direct or associated connection to agricultural and commercial enterprise, whether he identified as agrarian, commercial, professional, or crassly political. Undoubtedly, constituents in any member's riding certainly did.

To properly understand the motivations for creating a board of agriculture, quickly followed by a bureau of agriculture, we need to examine the mid-century colonial government, particularly the Upper Canadian half that transitioned from Robert Baldwin–led Reformers to Francis Hincks–led Reformers. Few historians have noted that during a brief window of time in 1852–3, it was the Hincks–Morin ministry that produced the greatest increase in state involvement in agricultural improvement in more than twenty years by making the only significant addition to the departmental structure of government established by Lord Sydenham in 1841.[3] Rather than simply adopt the standard characterization of the Francis Hincks and Augustin-Norbert Morin Reform ministry as the commercial, industrial, and professional ministry focused largely on railway schemes,[4] we need to focus on characteristics of Hincks and his leadership that positioned him as a bridge between the agricultural and the commercial interests who understood and promoted the collection and dissemination of statistical information as a means to attract new settlers and to expand the Canadian economy and investment opportunities. In doing so, we also need to challenge a blanket characterization of the emergent "Clear Grit" faction as radical. Lobbying for a board of agriculture, drafting legislation for the ministry in power, and doing so in collaboration with conservative gentleman farmers can hardly be considered the actions of agrarian radicals, determined to bring about American republican-style democracy and a smaller government to Upper Canada. Their successful lobbying, which led to the government establishing a bureau of agriculture with a Clear Grit appointed as its initial minister, challenges us to focus on the ideology of improvement shared by the "agrarian" and the "development-minded" Reformers who represented the electorate at mid-century.

The draft bill that McDougall and Buckland prepared in the spring of 1850 set forth the central duties of a proposed board of agriculture. Buckland's explanation captured the mid-century tone of agricultural and commercial interests: "the main duties of such a board would be to collect and disseminate information, and generally to watch over the agricultural interests of the country; similar to what is done by a chamber of commerce, or board of trade, for commercial objects."[5] Although the purpose of the bill was clear, the politics of the lobbying was not. The AAUC was asking the provincial government to provide additional support for agricultural improvement by way of new legislation and additional funding from the provincial treasury. It was a significant request of the Baldwin–Lafontaine ministry, which was facing pressure from the Clear Grit faction that McDougall had recently helped form in meetings with dissatisfied Baldwin Reformers at his Toronto office.

During 1849, the Clear Grits coalesced around two interrelated factors concerning the political and economic maturation of the Canadas. First, some dissatisfied Reformers did not believe responsible government to be an end unto itself, challenging the views of liberals at the heart of the Baldwin–Lafontaine ministry. Instead, they saw it as delivering the political tools necessary to launch further constitutional changes. Second, they were upset that the government dismissed outright the option of annexation to the United States as was called for in the Annexation Manifesto. Although not annexationists, Clear Grits saw the possibility in the same light as constitutional reform. Republican democracy and the institutions of the American republic should not be dismissed outright. Instead, they needed to be debated thoroughly to determine the best future for the province. As Michael Cross explains, Clear Grits were frustrated by the Baldwin moderates who were willing to foster gradual change while preserving "the best of the gentry past" and willing to encourage increased participation in government but not rapid change and not democracy.[6]

Undoubtedly, some reformers and conservatives did fear the energy and ideas of the Clear Grits' blend of a youthful core that included McDougall and David Christie (in their late twenties in 1849) with old reformers from the rebellion era, such as Peter Perry (thirty years their senior). As the Clear Grit platform developed through conventions that smacked of republicanism, it included planks that included substantial parliamentary and legal reforms, universal suffrage, free trade, and the sale of clergy reserves to fund education. It was, as Cross points out, an attempt "to yoke the old reform tradition of pre-rebellion days with modern concerns informed by the Chartist movement that had roiled English public life for over a decade." There were other attempts at unity within this movement, too. Cross notes: "Traditional agrarians were seeing common cause with liberal professionals," although he argues that such commonalities "posed more dangers for agrarians than they imagined for modern liberalism was on a path to financial and industrial capitalism, a path that had no lane for farm folk."[7]

No, the "farm folk" were not being sidelined. Reformers such as Christie and McDougall had joined with conservative agrarian gentlemen in the AAUC beginning in 1846, and by 1849–50, the association was asserting its power as a lobby of agricultural improvers, skilfully chopping out their own "lane" through the political brambles of the day to ensure increased and permanent government support of agricultural improvement. To them, the expansion of commerce, advanced infrastructure, greater circulation of information, and increased settlement relied on that support. For the AAUC, the first destination down its "lane" had been the AAUC's annual agricultural exhibition, which increasingly highlighted the hand-in-hand improvement of agriculture and commerce. The next destination was the creation of a board of agriculture, and hopefully, the appointment of a government minister for agriculture would not be a very far journey thereafter.

As veteran tories gave way to reformers in the Upper Canadian elections of the 1840s, reformers assumed leadership roles in the province's agricultural societies. David Christie was an agricultural improver and stockbreeder at "The Plains," his estate in Dumfries Township. A young Christie had emigrated with his family from Scotland in 1833, had been a founding member of the AAUC in 1846, and had helped organize the provincial exhibition in Hamilton the following year. In fact, it may have been Christie's words that generated the name for the political faction he helped form. "We want only men who are *clear grit*,"[8] he apparently said of those whom the movement wished to attract. The veteran Reformer Peter Perry of Whitby Township had been recently elected in a by-election in York East, following his split with Reformers, for he refused to denounce the possibility of a future annexation to the United States. A member of the HDAS and the AAUC, he had also worked with Christie and others to organize the Hamilton exhibition. Perry's health would soon fail, and he would die in 1851, but in 1849, he was a model to the younger Clear Grits for refusing to let Reform Party leaders silence his unpopular political views. Another veteran Reformer, now in his forties, was Malcolm Cameron, MPP for Kent. He was the former assistant commissioner for public works in Baldwin's cabinet. A politically unpredictable individual with an independent streak, his foremost interests were his own in lumber, flour milling, and railway construction. Cameron, ever the slippery fish, had supported Reformers and then withdrawn that support on several occasions. He resigned his cabinet post in the Baldwin–Lafontaine government in December 1849 due to several reasons that likely included pressing business interests and dissatisfaction with both the ministry and his public salary. Baldwin was happy to see him go because Cameron's actions generated even more battles for an unstable ministry to fight. Shortly after his resignation, Cameron began to associate with McDougall, Christie, Perry, and other Clear Grits.[9]

Cameron was a life member of the AAUC, but as noted earlier, his political actions in early 1850 inflicted significant damage on that organization. His and Caleb Hopkins's hounding of the mentally unstable John Wetenhall is recognized as having driven the Reform candidate and AAUC president to a mental breakdown and, shortly thereafter, his untimely death. Thus, we return to the muddy politics of the AAUC's lobbying efforts to establish a board of agriculture and its use of the *Canadian Agriculturist* to pronounce the need for such a board. While it was doing so, McDougall was hosting meetings at his office that spawned the vindictive Clear Grit attacks on Wetenhall, the ministry appointee in the Baldwin–Lafontaine government who might have helped secure what the AAUC desired.

Despite all these interwoven political machinations, the AAUC received initial support for its comprehensive agricultural bill when AAUC past-president (and former conservative MHA) E.W. Thomson presented the association's petition and bill to Baldwin and his inspector general, Hincks, shortly after the session of parliament opened at Toronto in mid-May. Each indicated their approval and promised to present the bill in the legislature. However, the AAUC's draft bill remained "entirely overlooked" by late July. The ministry had not even read its petition in the legislature, and the session of parliament was nearing its end.[10] Other, more immediate priorities filled the legislative agenda. All the while, Clear Grits launched numerous attacks on Baldwin's leadership and ministry during the session, weakening not only the Baldwin–Lafontaine reform alliance but Baldwin's mental and physical health as well.[11]

To salvage the measure, McDougall approached others in the government for support, securing a promise from Provincial Secretary James Leslie that he would introduce the bill into the Legislative Council. With the clock ticking, drastic action was required. The Honourable Adam Fergusson recommended dividing the bill in the hopes that the greater aim of establishing a board of agriculture could be secured before the end of the session. Reorganizing the governance of existing agricultural societies in Upper Canada could be left for a future session. On 2 August, Leslie introduced "An Act to establish a Board of Agriculture for Upper Canada" in the Legislative Council on behalf of the government. The council made a few important amendments, passed the bill, and sent it down to the Legislative Assembly for its approval. That chamber added one small detail and passed the bill with equal speed. Taking just eight days from introduction to approval, the bill received royal assent on the last day of the session.[12]

The AAUC's lobbying succeeded. At its request, the government established a board of agriculture of Upper Canada. However, that board was a gelding, not the stallion the AAUC had proposed. Without the larger portion of the AAUC's draft bill, expunged for the sake of expediency, the act created a board of agriculture possessing no direct authority over the county and township agricultural societies; they continued to be operated and funded under separate legislative control. The draft bill had envisioned a board of agriculture directly connected to the colonial government. Its chair was to be appointed by the governor from among members of the Legislative Assembly or Legislative Council, and the board's operations were to be funded by the public treasury, including a full-time secretary who was to have a government-provided office and draw a public salary.[13] Legislators amended this clause so that the seven members of the board would elect annually an unpaid chair from among their ranks. Board members would be elected from among the province's agriculturists, with a process set for board renewal by replacing two directors each year. In effect, this board election process represented the only operational connection between the Board of Agriculture and the county and township agricultural societies of Upper Canada. As for a connection to government, the act provided an *ex officio* board position for the inspector general. The University of Toronto's proposed professor of agriculture would also be an *ex officio* member when that appointment was filled. But the legislation provided no additional public funding. The board and the AAUC would have to draw on donations and the £1,000 annual grant that the government provided to the latter association on an ad hoc basis. Board members could seek reimbursement of travel expenses to attend meetings at the board offices but only "out of any funds at the disposal of the Board," not the public treasury.

Once the inaugural board was elected, the AAUC ceased to be the premier agricultural association of Upper Canada, and the hosting of the annual provincial agricultural exhibition would be its sole responsibility. In future, the AAUC's board of directors would be composed of members of the Board of Agriculture and the presidents of the county agricultural societies. The Board of Agriculture alone would serve as the council of the AAUC, provided with powers to act as directors in the periods between annual directors' meetings. The AAUC's connection to the board would be through the secretary, who would also serve as *ex officio* secretary of the AAUC.[14]

This half-measure act did not satisfy its original promotors, particularly McDougall, who wrote a spite-filled, front-page editorial for his September 1850 *Canadian Agriculturist*. He argued that, during the recent session of parliament, the government had approved unlimited public funds "for objects of little use to the people of Canada ... yet *one* thousand pounds a-year to be expended in collecting and diffusing information relating to that branch of productive industry on which *all* our wealth depends – expended by a popularly organized Board making a yearly return to the Legislature of their labors and expenditures could not be thought of, or voted, for want of time!!" Following some miserable Clear Grit rhetoric about parliament being filled with "place-hunting, time serving and ... *ignorant* lawyers," McDougall (a lawyer) presented his buried lede. The government planned to enact the other, more substantial portion of the AAUC's draft bill during the next session of parliament and was likely to grant funds to the Board of Agriculture at that time.[15] It is perhaps not coincidental that McDougall wrote this bitter editorial about the same time as he launched his semi-weekly newspaper devoted to advancing the Clear Grit platform. The first issue of *The North American* appeared on 11 September 1850.

Much of the act's preamble, which set forth the board's purpose, contained rhetoric mostly unchanged since the birth of the province. However, added to the long-stated claim that "the improvement of Agriculture is an object of the first importance to the people of this Province" was the indication that the board would promote such improvement by collecting and disseminating "statistical and other useful information concerning the agricultural interests and resources of the country." Although the government had rejected the AAUC's plan for the chair of the Board of Agriculture to be a member of parliament and a member of the Board of Registration and Statistics, the appointment of the inspector general as an *ex officio* member of the board was significant. That cabinet role included service as one of three members of the Board of Registration and Statistics, which had been established in 1847 by the Census and Statistics Act as an agency to oversee the execution of the census and collection of statistics in general. As for the Board of Agriculture, its legislation required the board to submit an annual report to the legislature, inclusive of important extracts of proceedings of the AAUC and "interesting and useful" information from the annual reports of county and township agricultural societies. In return, the government pledged to publish such reports as the legislature deemed appropriate. Effectively, the government was enlisting the willing members of this new, semi-public Board of Agriculture as a voluntary bureaucracy to accomplish these and other tasks mandated by the act in the absence of any government minister of agriculture or paid bureaucrats.[16]

By way of this act, the government also supported the AAUC's proposal for the Board of Agriculture to plan and manage "an experimental or illustrative farm in connection with the Chair of Agriculture in the University of Toronto, or in connection with the Normal School, or otherwise, as they may design best." The act directed the board to submit plans to the legislature "as soon as practicable," and more broadly, it instructed the board "to make any recommendations they may think expedient for extending agricultural education throughout the Province." This task was urgent, for the recently established University of Toronto had yet to fill the chair of agriculture provided by its 1849 founding legislation, and there was some indication that the university might decide to leave the appointment unfilled.[17]

Months would pass before the Board of Agriculture could meet. First, county agricultural societies across Upper Canada had to host their annual meetings, at which time the election process for board members would begin. Such meetings were not scheduled until the early months of 1851. As legislated, directors of each county agricultural society were to draw up a list of seven potential candidates for directors who resided "in the vicinity of, or at convenient distances from the City of Toronto." To expedite the process, AAUC president J.B. Marks sent a December 1850 reminder to executives of each county society, acknowledging the somewhat unreasonable expectation that everyone would know of qualified individuals from the Toronto area. Thus, he forwarded names of several gentlemen, being careful to remind county executives that they were free to suggest any other individuals they believed to be best qualified. The seven individuals who received the highest number of votes across the province would become directors of the inaugural Board of Agriculture of Upper Canada.[18]

In mid-June 1851, the government gazetted individuals selected during the election process, allowing for a first meeting of the Board of Agriculture in Toronto in early July. Voting had elected directors from across the province and not just the immediate Toronto area. Past AAUC presidents E.W. Thomson, the Honourable Adam Fergusson, and Henry Ruttan had been elected. David Christie along with John Harland, who farmed "The Poplars" in Guelph Township and served as secretary of the Wellington County Agricultural Society,[19] were two other directors elected who had been involved with the AAUC since its formation. Current AAUC president Marks and R.L. Denison rounded out the list of directors. George Buckland was appointed Secretary of the board; however, the board deferred selecting a chairman for the year, because both Thomson and Fergusson were unable to attend the inaugural meeting. Instead, directors requested Marks serve as interim chair until the board reconvened in November. At that meeting, the board elected E.W. Thomson, the gentleman improver who had served as the first president of the AAUC, as its first chairman.[20]

At its July 1851 meeting, the board considered two important issues. First, the government had not yet introduced a bill into the current session of parliament containing the omitted portions of the AAUC's draft bill. In this, directors faced a familiar situation; the session neared its close and the Baldwin–Lafontaine ministry continued to fall apart. When the session had opened in late May, Baldwin was both mentally and physically unwell, and as the session progressed, repeated attacks from the now politically strengthened Clear Grits only served to make things worse. On 30 June he resigned his office, effectively retiring from politics. By the end of the session, LaFontaine would resign, too. That the bill was even placed on the government's legislative agenda was either a testament to the influence of the AAUC and the Board of Agriculture as a lobby or the sign of a disintegrating government that considered the bill as a means for placating the Clear Grit faction. Perhaps it was some of each.[21]

Three days after Baldwin's resignation, James Hervey Price, MPP for York South, introduced into the Legislative Assembly a bill "to provide for the better organization of Agricultural Societies in Upper Canada." That afternoon, the Board of Agriculture performed its role as a volunteer bureaucracy when it met at the parliament buildings in Toronto to discuss the bill with legislators. It recommended several modifications. Details of this legislation, which received final reading and royal assent on 27 August 1851,[22] were analysed in chapter 4, but a few words relative to the Board of Agriculture are in order here. With passage of the substantial portion of the AAUC's 1850 draft bill, the Board of Agriculture now possessed a clear, direct relationship with each county and township agricultural society of the province. This semi-public board was now the state's supervisor of the province's agricultural societies, drawing their activities more fully into the operations of the colonial state. The act mandated a new system of financial reporting and provided a method to circulate knowledge of agricultural improvements, statistics, and other information throughout the province. Officers and directors of each county society were to be ready to answer any general or individual queries the Board of Agriculture sent to them, "touching on the interests or condition of Agriculture in their County."

Treasurers of county agricultural societies were now required to submit to the board secretary yearly financial accounts and annual reports of their activities by 1 April and forward reports of the township societies within their boundaries. In turn, the government would print these county and township reports as appendices to the annual report that the Board of Agriculture was required to produce. While the county and township agricultural societies could rely on their annual provincial grants to fund the production and delivery of these reports, if need be, the board received no such funding.[23]

During its July 1851 meeting, the Board of Agriculture also investigated options for an experimental or model farm. Directors visited a site selected by the University of Toronto, at the north end of the University Park, to the west of the Taddle Creek ravine. The Reverend John McCaul, the university's president, joined the board on this tour, along with Henry Croft, chair of chemistry and experimental philosophy, and William Nicol, professor of *materia medica*, pharmacy, and botany. All agreed that the chosen site was "quite favourable" because the soil varied in type from south to north, and most of the land had been cleared of trees. When the board met again in November, it ordered at least three acres of the experimental farm to be ploughed as soon as possible in preparation for planting plots of hemp, flax, and sugar beet.[24]

Agricultural improvers had promoted the need for an experimental farm for Upper Canada during much of the previous decade. Repeating opinions of improvers in Britain and the United States, they viewed it as an indispensable proponent of agricultural improvement and education. Rhetoric promoting such a farm had increased sharply in 1847 with news that the recently formed Lower Canada Agricultural Society planned to establish an agricultural college modelled on that in Cirencester, England, complete with an agricultural museum and library. R. Brewer had crystallized the implications of this announcement for readers of his *Canada Farmer*: "Now they really threaten to go ahead of Canada West; to outstrip us in the race of improved and enlarged production, and to pluck the laurel from our brow, while with a hearty huzza they proclaim the peaceful conquest." Overdramatic, perhaps, for there was an initiative in Upper Canada that would reap long-term benefits. A month after Brewer's comments, William Edmundson published an April 1847 editorial, "Agricultural College in Toronto," in which he ensured readers that they would soon know George Buckland, arriving shortly from England, as the new professor of agriculture at the University of Toronto. He included a passage from a letter Buckland had written to him that emphasized the need for agricultural education and his intentions to revisit the agricultural

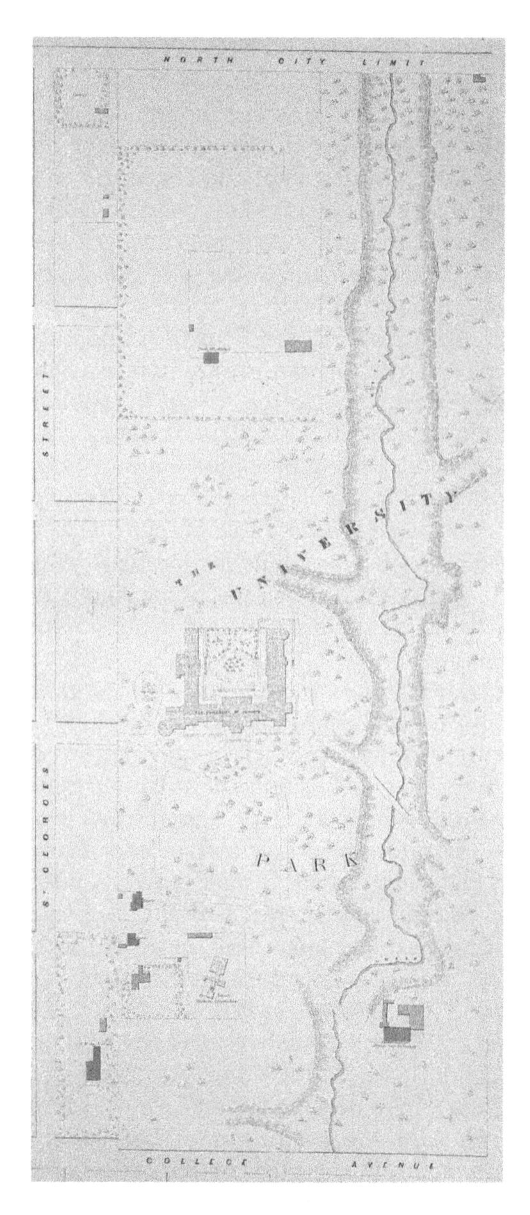


Maps 9.1. (above) and 9.2. (opposite page) Maps of University Park, Toronto, 1858. The map detail identifies the experimental farm, a house (labelled "Profesr Buckland"), and its other outbuildings. Detail from Plate V *Atlas of Toronto,* Surveyed & Compiled by W.S. & H.C. Boulton Toronto. Lithographed & Published by Jno Ellis 8, King Street, 1858 (Toronto Public Library, 912.71354 B594 1858.)

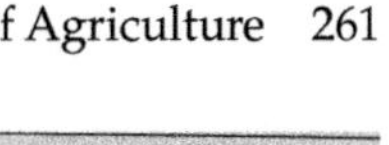


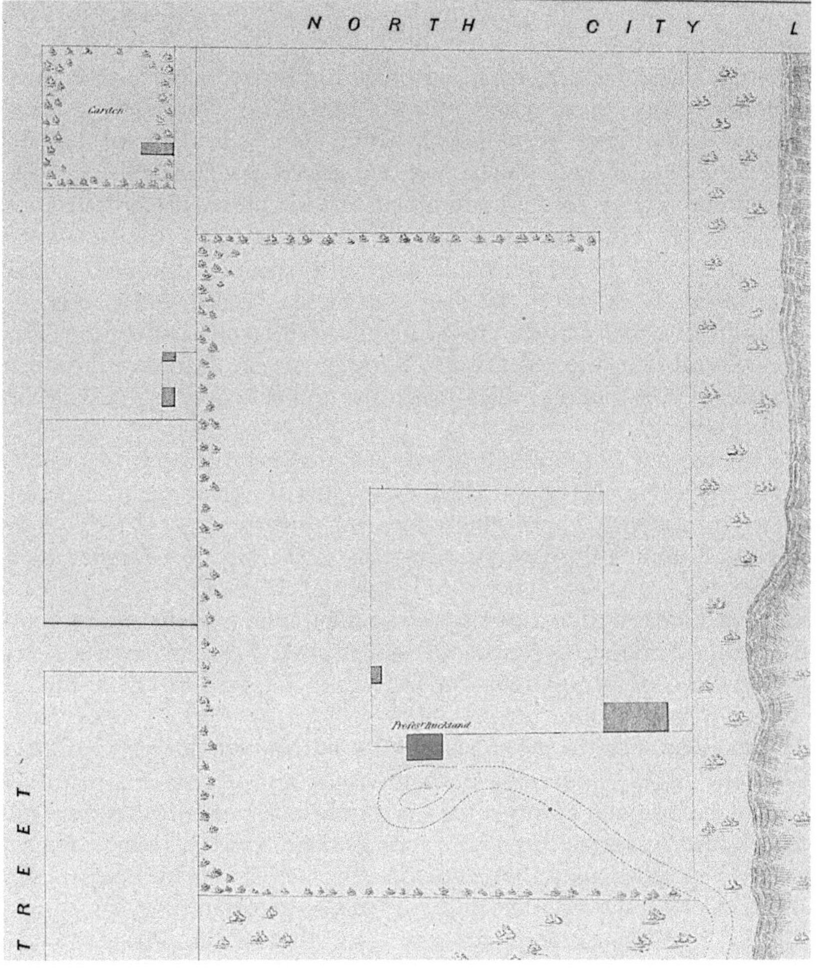
N O R T H C I T Y L
Garden
Profesr Buckland
T R E E T

college at Cirencester for information and ideas prior to his departure for Upper Canada.[25]

When Buckland arrived in Toronto, Edmundson paid even more attention to the subject of agricultural education and model farms, publishing another letter from Buckland that stated his plans for a model farm and his determination to see one established.[26] But the creation of a university at Toronto continued to be a political Gordian knot. In early 1848, Edmundson learned "from very good authority" that a chair of agriculture would not be part of the proposed university bill. In response, Buckland wrote a series of three editorials for the *Agriculturist and Canadian Journal*, providing information about institutions for agricultural education in Europe. Simultaneously, Edmundson wrote additional editorials on the subject and published letters of support from readers.[27]

After lengthy, acrimonious debate the provincial government finally agreed upon legislation in 1849 to establish a non-denominational University of Toronto. Shortly thereafter, the government created a commission to define the chairs for the new university. In February 1850, McDougall expressed concern to readers of the *Canadian Agriculturist* that the chair of agriculture may not be filled immediately, and the journal featured numerous articles on agricultural education in the ensuing months to convince the government to fill the chair of agriculture as soon as possible.[28]

Thus, we return to the starting point of this chapter when, in May 1850, the AAUC petitioned parliament for a professor of agriculture and an experimental farm at the same time as it presented its draft bill for a board of agriculture and a reorganization of the system of agricultural societies across the province. By November, McDougall and Buckland happily reported that the University of Toronto Senate had decided to fill the chair of agriculture and that its plan included some fifty acres for "an experimental or illustrative farm to be under control of the Board of Agriculture."[29] Professor Nicol introduced a motion to accomplish both things at a 16 November meeting of the university senate, and agreement was reached at its meeting of 11 January 1851. The Senate set the chair of agriculture's salary at £250 per year and began accepting applications. By the deadline of 1 August, three individuals had applied: a "Mr. Fyfe," Henry Youle Hind, and George Buckland.[30] Fyfe was likely David Fife, of Ontonabee Township, near Peterborough. In 1842, a friend in Glasgow had sent him a sample of wheat taken from a ship that had arrived from the Baltic port of Danzig (now Gdańsk, Poland). By 1849, this variety of spring wheat that Fife and his neighbours had planted over successive years became known as "Red Fife."

Its use spread rapidly across Upper Canada because farmers found it well suited to the environment. It threshed well and it produced excellent flour.[31] Henry Youle Hind had arrived in Toronto from England during the winter of 1846–7. In the following year, he was hired as the assistant master of the Normal School where, among other subjects, he lectured on agricultural chemistry. To promote the subject more broadly, several of his lectures were bound and published, a second edition of which was to be released in 1851. Hind would withdraw his application from consideration, a decision coincidental with the laying of the cornerstone of the new Normal School in Toronto that summer. Hind had developed a relationship with Upper Canada's chief superintendent of education, Egerton Ryerson, who employed him in efforts to adopt agricultural education into the provincial school curriculum.[32]

The Senate committee considered the remaining two applicants, Fife and Buckland. In November, it reported that Buckland's testimonials were "much superior both in number and in value." (The AAUC had highly recommended Secretary Buckland, "as a fit and proper person, in all respects.") At a late January 1852 meeting, the Senate appointed Buckland as the University of Toronto's chair of agriculture. For his part, Buckland had played coy throughout the lengthy selection process, noting in the February 1851 *Canadian Agriculturist*, "a sense of delicacy will only permit us to say, that we fully appreciate the worth of their good opinion and wishes." Even after the year it took the university to fill the position, Buckland announced his employment by modestly reprinting the government notice of his appointment from the *Canada Gazette* to the back page of his February 1852 *Canadian Agriculturist*.[33] Nevertheless, his coyness in print did not stop him from planning the experimental or model farm, the other aspect of agricultural instruction at the University of Toronto.

In December 1850, Buckland's lead editorial, "Professorship of Agriculture in the University of Toronto," had related his belief that the entire 180 acres of the University Park were to be under the professor's superintendence, with some 70 to 80 acres cleared for the purposes of experimental and practical farming. This latter portion, he argued, should not be a *model* farm, for in a country as extensive as Canada, there were so many variations in local farming conditions that one model could not possibly suit all. Instead, the province needed an *experimental* farm in which experiments of practical utility could be conducted on a small scale. Repeating the RASE motto "Practice with Science," he claimed the combined benefit of a chair of agriculture in control of an experimental farm would be "[t]he social *status* of our farmers will become elevated, by associating the Science and practice of their pursuits, with a liberal course of academical learning; while existing systems of farm practice will be necessarily improved, by imparting to the young, sound, practical knowledge, and the results of carefully conducted experiments."[34]

When appointed professor of agriculture in 1852, Buckland acquired a much smaller parcel of land for the experimental farm, and his authority over it was less than he had anticipated. He had joined the Board of Agriculture on its visit to the experimental farm's location during the previous July. In the month that followed, a committee of Senate decided what portion of the university grounds would be used for an experimental farm (and the botanical garden).[35] The farm would occupy no more than the fifty-acre parcel of land examined and approved by the Board of Agriculture and the university would offer it on a ten-year rent-free lease for use in conjunction with the chair of agriculture. At the expiry of the lease, the farm would have to be relocated, for the terms of the lease stated that, in ten years, the board must remove any wooden structures it had erected on the farm, while the university would purchase any stone buildings that it might have constructed.[36] During the summer of 1852, the Board of Agriculture and Buckland (who remained its secretary) improved some thirty acres of land on the University Park from a "state of nature," uneven and full of roots, stumps, and stones. After considerable effort levelling the land, some sixteen acres were sown with wheat in autumn. The good first crop it provided helped recoup the costs of preparing the land. In the following spring, some improvements were made on the remaining twenty acres.[37]

Highlighting the Board of Agriculture's increasing bureaucratic role, the government provided it a room in the Toronto parliament buildings by May 1852. This permitted space for its growing library. Professor Nicol had donated a complete set of the transactions of the original British Board of Agriculture, William McDougall had donated a three-volume set of *British Husbandry*, while Frederick Widder, commissioner of the Canada Company, had donated seven books on flax cultivation. Adam Fergusson donated books, too. To augment these donations, the board had established a committee by the end of 1851 to start a library with a budget of £50. Priority purchases were identified as both the American and English stud and herd books, for the board decided that an important service it could provide would be the compilation of similar breed records of Canadian livestock. The room at the parliament buildings would also provide space for a planned museum, the first items for which were samples of grains obtained from those displayed at the Great Exhibition of the Works of Industry of All Nations, held in London in 1851. Selected seeds from these samples were to be planted for experimental purposes.[38]

To further advance agricultural chemistry the board purchased one hundred copies of the new edition of Hind's *Lectures on Agricultural Chemistry* in November 1851 to distribute to county agricultural societies across Upper Canada. Sometime prior to May 1852 it also appointed Professor Croft as its consulting chemist. The board also ordered the printing and free distribution of five hundred copies of that month's *Canadian Agriculturist* issue containing C.P. Treadwell's essay that explored the agricultural, social, and industrial state of the Canadas. The board established prizes, too, from first to fourth, to be awarded for the best agricultural report submitted to it by a county society, as encouragement to fulfil the obligations set by the new legislation. To publish its annual transactions and prize essays, the board enlisted the *Canadian Agriculturist* as publisher. In response, the journal enlarged its masthead to the *Canadian Agriculturist and Transactions of the Board of Agriculture of Upper Canada*.[39]

The 1850 Board of Agriculture act required the inspector general to serve as an *ex officio* member. It was Francis Hincks who held this appointment in the Baldwin–LaFontaine ministry. This was a good fit for the board, for Hincks represented a new generation of politician who understood both the ideology and practice of agricultural improvement, plus the central role of commerce within the largely agricultural economy of the Canadas. It soon provided additional benefits, because Hincks assumed leadership of the Upper Canadian half of the Reform government in June 1851 following Baldwin's resignation from politics. In doing so, he faced two immediate pressures in his attempt to keep the Reform alliance afloat. First, he had to mollify the increasingly vocal Clear Grit faction. By August, he secured a confidential agreement with McDougall and Christie, supported by the rest of the Grit leadership. The *North American* would become a ministerial organ and Hincks would add two veteran Clear Grits to his cabinet. One was Malcolm Cameron, who would join as president of the council. On their side of the bargain, the Grits had to set aside the radical parts of their platform. It would prove to be an awkward marriage in many ways. The Clear Grits would not influence the ministry to the extent that they had hoped, and Hincks, although quieting radicalism in Upper Canada, would not eliminate Clear Grittism as a political force in Upper Canada.[40] Yet, it would be this political deal that spawned Cameron's appointment as the minister of agriculture to oversee a new bureau of agriculture for the entire Province of Canada.

With the 1851 retirements of Baldwin and Lafontaine, Hincks and Augustin-Norbert Morin led a reconstructed ministry that put forward its mandate to the electorate that December. After the results were

Image 9.1. The Honourable Malcolm Cameron, MPP (1808–1876), 1854. Lithograph, Theophile Hamel (Toronto Public Library, T 31101.)

tabulated, Morin brought with him a solid group of liberals who moved French liberalism closer to the centre of the political spectrum, and this group would form the main power behind the government. In Upper Canada, election results saw the former Baldwin moderate Reformers lose ground under Hincks. His ministry needed Clear Grit support, and they produced half of the Upper Canadian portion of the ministry. Significantly, the conservatives in opposition had also increased in strength. Gone were most old Tories, and a new force of liberal-conservatives staked its ground on the centre-right of the political spectrum.[41] These election results, plus the reformulated ministry, would all have significant ramifications on agricultural leadership as the Hincks–Morin ministry prepared to meet its first parliament in 1852.

Francis Hincks is a complex and often misunderstood character in Upper Canadian history. His *Dictionary of Canadian Biography* entry categorizes him as a banker, journalist, politician, and colonial administrator, but he is also portrayed as a politician, and a railway promoter, builder, and profiteer.[42] By the time Francis left for Upper Canada in 1832, there is little doubt that he was fully versed in agricultural improvement by way of his father's interests and likely his own education. When he was a child in Cork, Ireland, his father, Thomas Dix Hincks, was a member of the Royal Irish Academy and a salaried officer of the Royal Cork Institution, where he lectured in chemistry and natural philosophy until 1813. During this period, he also edited the *Munster Agricultural Magazine*. Although he left these positions to move on to other opportunities in Fermoy and then Belfast, Thomas became a member of most scientific societies of Ireland.[43] As a newspaperman in Montreal in 1844, Francis Hincks made clear to readers of his *Pilot* that he understood the economic interests of Canada and the need to draw together into one the worlds of agriculture, manufacturing, and commerce. As a politician, he promoted capitalism and the emergent middle class. His acceptance of the office of inspector general in the Baldwin–Lafontaine ministry as of 1848 placed him on the Board of Registration and Statistics and, later, the Board of Agriculture. Both roles contributed to his and others' vision of the province's future. Statistical and associated information about agricultural practices and economy in the Canadas needed to be shared domestically and abroad with prospective settlers. Increased knowledge of improved forms of agriculture would increase agricultural output, thus advancing commercial opportunities and the need for infrastructure to ship produce to markets. A greater number of settlers flowing into the province would, at once, increase the scale of the colonial economy, increase the ranks of the middle class, and drive the need for massive infrastructure investments, particularly railways. Thus, by the time he became co-premier of the Canadas, Hincks was not only trying to placate the Clear Grits, but he was also shaping a unity of improvement with the liberal professionals his Reformers represented, in association with the Board of Agriculture and the Clear Grit agrarians, regardless of what vituperative rhetoric might be published in a newspaper of a particular political faction or shouted across the legislature floor. The ideology of improvement remained sufficiently adaptable to accommodate such unity.[44]

One of Hincks's biographers, George A. Davison, has situated the Hincks–Morin ministry in the context of mid-century provincial development, arguing that it was Hincks who "transformed the theory of responsible government into a practical system for the new governing power, the businessman politician." When the Hincks–Morin ministry met parliament in Quebec in August 1852, the government benches were certainly filled with the "businessmen politician[s]" focused on provincial development.[45] As a means to hold together the wide ranges of interest with his Reform Party, Hincks employed the "politics of interest" by offering patronage at an individual level to draw businessmen politicians into his government, while at a broader level, he formulated policies that advocated economic progress and provincial development, particularly the construction of railways. By doing so, Davison argues, Hincks was able to transform the old marriage of economic development and politics that had existed under the control of the Family Compact by replacing the old tory control with business interests that he drew from the periphery of the political world into the centre of his Reform Party. It created a core that spanned the liberal-conservative centre of the political spectrum and paved the way for the dominance of liberal-conservatives in the decades to come. That Hincks employed the promise of a bureau of agriculture as a means to keep the Clear Grits within the party fold and that he appointed a businessman-politician from their ranks to be the minister of agriculture demonstrated his political acumen as much as it did his ministry's understanding of agricultural improvement's role in fulfilling government policies focused on commercial growth and infrastructure development.[46] That said, the process of doing so generated considerable controversy. Malcolm Cameron continued to be a difficult individual to keep happy, and Hincks's solution to yet another Cameron-generated problem would be decried as proof of the infant ministry's corruption.

The Clear Grit Cameron had been brought into the Hincks–Morin ministry before it faced the electorate in December 1851, and he believed he would assume the cabinet position of postmaster general in a new government. Although he won the riding of Huron easily – campaigning as an independent candidate – Cameron did not receive the appointment. Instead, he was appointed president of the council, a position some commentators suggested Cameron had recently declared to be useless. In one of its first tasks as the ministerial organ, the *North American* denied Cameron had ever made such a statement. Instead, it claimed that Cameron had first heard the news of his ministry appointment through the press and, not surprisingly, had denied having accepted the office.[47] Eager to pacify the Clear Grits and balance the numbers he brought into the ministry with the Lower Canadian appointments, Hincks remained determined to find a place for Cameron. By 23 January, the *North American* was able to break the news of a solution: Cameron had agreed to accept the position of president of the council; plus, he would take responsibility as a minister of a new bureau of agriculture that would be attached to his presidency.[48]

For the next month, opponents of Cameron, the Clear Grits, and the Hincks–Morin ministry tried to manufacture scandal from this deal, while McDougall employed his *North American* to fend off the attacks.

In his announcement of the deal, McDougall reasoned why a bureau of agriculture with Cameron at its head would benefit the province. First, he pronounced ideology promoted by agricultural improvers for decades: "Agriculture is emphatically our *leading* interest ... All other branches of industry spring out of this, are adjuncts to it, or depend directly upon it." The fact that the provincial government was only now providing ministerial responsibility for agriculture derived from the unfortunate situation in which "'what's everybody's business is nobody's.'" Singing a familiar Clear Grit refrain, he claimed the dominance of lawyers in previous parliaments had much to do with this state of affairs and was pleased to point out how new elections had reduced their number significantly. Not only did McDougall claim Cameron to be "peculiarly fitted" to lead the new department, but he also proclaimed: "The machinery for originating, conducting, and perfecting improvements, and promoting individual and national success in that department of industry in which five-sixths of our population are directly engaged, is now complete." Most of all, McDougall believed, people across the Canadas would support this bureau, because it required no additional public expense; Cameron would draw only his £800 salary as president of the council.[49]

Others did not see it that way. Hugh Scobie concluded in his *British Colonist* that the entire deal was nothing more than a political job. The question of a new bureau should have been submitted to parliament for its consideration, not announced with reference to a particular individual. He had a valid point. How could someone be appointed to a position for which there was not yet establishing legislation for parliament to debate and approve? Scobie was certain the appointment would force Cameron to seek re-election and he doubted the electors of Huron would care to support Cameron a second time around. George Brown came out swinging in his *Globe*, claiming this to be nothing more than a "bare-faced attempt to throw a gauze veil over Mr. Cameron's acceptance of the 'useless office.'"[50]

It is not surprising that Brown became such a vociferous opponent in the legislature and in the pages of his widely circulated and increasingly influential *Globe*. Hincks had failed to pacify Brown prior to the election, and he ran as an independent reformer. After winning his riding he became Hincks's most significant political foe. Furthermore, during the election, Cameron's most eager campaigning had been done in his former riding of Kent and not Huron, where he was a candidate. He was determined to ruin Brown's rookie attempts at winning that seat, but his gambit failed and Brown, now the MPP for Kent, was determined to see Cameron fail again, just as he was determined to attack him with editorials in his *Globe*.[51] In the specific case of the bureau bill headed to parliament, Brown wondered how Cameron and the Clear Grits – vocal critics of reform corruption – could possibly support the creation of a new office without the consent of parliament, particularly since neither the bureau nor the minister had any defined duties.[52]

As January ended, McDougall defended the charge that Cameron's appointment was a political job. First and foremost, he attacked the narrow-minded individuals who believed there was no role for a minister of agriculture, for he would supervise boards of agriculture and the networks of agricultural societies that existed throughout each half of the Canadas. Moreover, the new minister would have authority over the existing Board of Registration and Statistics and its responsibility for the census and patents, an important development in an advancing province. "No reliable agricultural statistics have ever been collected or promulgated. Our public lands have been notoriously mismanaged," McDougall noted. The Canadas required more settlers, and a bureau of agriculture would not only help coordinate and disseminate statistical information to Canadian farmers, but it would also be most useful to advertise the benefits of the Canadas to emigrants overseas who, at the moment, tended to head to the American West. This was not a new role without purpose. In fact, Cameron might have his hands full with multiple, significant responsibilities.[53] McDougall's *Canadian Agriculturist* voiced support, too, although with much less political grandstanding. It congratulated the farmers of Canada "that henceforth our Cabinet will have a MINISTER OF AGRICULTURE!" And it responded to the critics by pointing out that a minister of agriculture "would be a fitting representative" of agricultural societies, the Normal School, the chair of agriculture, and the experimental farm, "and would be the means of promoting the great interests of the country in many other ways, as yet untried or unknown."[54]

Parliament was months away from opening at its new location in Quebec, and the debate surrounding Cameron's appointment raged on. The *North American* announced on 3 February 1852 that Cameron had, in fact, accepted leadership of the Bureau of Agriculture as part of his role as president of the council. But to critics such as Scobie, this announcement solved nothing. He did not believe Cameron to be a suitable candidate for president of the council in the first instance, and he would not become more suited for it "by being an agricultural minister as well." Scobie opened another front of attack. Would not the creation of a bureau require public funding? While Cameron might not draw a salary, he argued, "the machinery necessary to work it" most certainly would incur costs. This affair, he concluded, was "bad in principle, and bad in details."[55] Others chimed in, too: either the post was a political job or it was not; either the Bureau of Agriculture was a most necessary cabinet position or it was not required now and not by this method. For its part, the *Examiner* pointed out in early March that various state legislatures were requesting that the US Congress establish an agricultural bureau. Therefore, progressive reformers could be gratified to know that, in this matter, the Canadas were ahead of the neighbouring republic.[56]

Critics were somewhat silenced when Cameron won the necessary spring 1852 by-election in his Huron riding.[57] While Cameron campaigned in Huron, the Board of Agriculture was busy at Toronto considering details of the legislation that would establish the bureau, for a bill was to be submitted to parliament when it opened in late August. At the board's first deliberation, which again highlighted its role as a volunteer bureaucracy, McDougall was brought in to offer his views and experience concerning the agricultural legislation. What leadership role the new minister of agriculture might play was a chief concern of the board.[58] When it met again in August, a recent letter written by Cameron from the "Agricultural Office" in Quebec was awaiting a reply. Although, legislatively, no such office yet existed, Cameron inquired about a few matters related to Upper Canadian agricultural societies and indicated his view that legislation to regulate them might provide the basis for "unity of action and one general system throughout the Province." Cameron clarified his meaning in a similar letter to Pierre-Édouard Leclèrc, president of the Lower Canada Agricultural Society, which he copied to the board at Toronto. He indicated that the system of Upper Canadian agricultural societies "has been found to work well" and that the Lower Canadian agricultural societies might be brought into conformity with that model, with leadership centralized through his office for equal and efficient communication across the entire colony. Cameron also relayed to Leclèrc his clearest statement of what purpose a bureau of agriculture might serve. It was "to condense and arrange for practical use all the statistics of Agriculture, to attend to the Agricultural interest in the Executive and Legislative bodies, and to aid, by every possible means, its full development."[59] The Upper Canadian board replied to Cameron with its agreement that "uniformity of action" among all the agricultural societies of Upper and Lower Canada was of "highest importance" and that the Bureau of Agriculture might be a useful agency by which "to form a friendly and more frequent communication" between each half of the province. It was an easy reply to offer, considering all the reorganization would occur in Lower Canada.[60]

Before a bill to establish a bureau of agriculture was introduced in the legislature, the AAUC held its annual exhibition in Toronto in late September 1852. A scheduled event was a set of evening lectures to be delivered at Toronto's St. Lawrence Hall, chaired by AAUC president Thomas C. Street. One speech would influence parliamentary debate over the Bureau of Agriculture bill in the weeks ahead. While the first evening witnessed a near-capacity crowd to hear speeches by George Buckland, Adam Fergusson, and Egerton Ryerson, the second evening attracted an overflow crowd to hear Cameron, who was to outline his government's plan for promoting agricultural improvement. As Street introduced Cameron, he had to silence a few disruptive individuals in the crowd. It was a brief disturbance that would become its own controversy as the bill to establish a bureau was debated in parliament. In its report of the evening, the *Globe* treated the disturbance as clear evidence of the farmers' disgust with both the government plans and of Cameron as minister of agriculture. It claimed that such groans were indicative of a general malaise expressed across the province and Cameron had not produced one good argument in favour of establishing a bureau of agriculture. Likewise, it argued, after eight months on the job, Cameron's policy was simply to let the agricultural societies and the university's experimental farm do his job for him. McDougall countered this claim in his *North American* arguing that he and others knew that the editor of the *Toronto Patriot* had hired a group of drunken "b'hoys" to make a disturbance.[61]

Once the crowd settled, Cameron outlined the benefits of agricultural societies as well as the government's new bills to reform agricultural leadership in both Upper and Lower Canada. He also described the roles that he expected to perform as the new minister of agriculture. When he was done, Street offered thanks, noting that, although he was politically opposed to the government's proposals (as a conservative MPP), "he should always hail good measures, from whatever source they might come, with approbation."[62]

In Quebec, the ministry introduced two bills in the legislature on the day of Cameron's Toronto speech. Attorney General William Buell Richards, MPP for Leeds, first rose to introduce "a bill to provide for the better organization of Agricultural Societies in Lower Canada," which incorporated Cameron's ideas expressed in July that the 1851 legislated system of governance for Upper Canadian agricultural societies be applied to Lower Canada. That bill would pass through parliament and into law with few amendments.[63] Immediately after Richards introduced the first bill, he stood again to introduce a second bill "to provide for the establishment of a Bureau of Agriculture, and to amend and consolidate the Laws relating to Agriculture."[64] It would face a more difficult path to approval.

Cameron returned immediately to Quebec to provide his opening statements on the latter bill when debate began on 5 October. True to character, he offered a short, straightforward overview of the need for the legislation. Cameron contextualized the bill by reminding parliament that during the previous administration, a bill to establish a board of agriculture had been submitted to parliament at the request of interested agriculturists, who had wished "to establish an intimate connection between the Board and the Government." Although the inspector general was appointed as an *ex officio* member of the board, in practice, it had proved difficult for Hincks to fulfil his duties. Considering this fact, plus an agreed-on need to marry the operations of the statistical and patent offices with agricultural leadership, the government proposed creating a bureau of agriculture with the president of the Executive Council at its head. Doing so, Cameron argued, would facilitate the "close communion" between the Board of Agriculture and the government that the former desired.[65]

Henry Smith Jr., the conservative MPP from Frontenac, stood first to respond to Cameron, astonished that the minister had taken just three minutes to introduce a bill aimed at creating an entirely new department of government. He claimed that farmers who heard Cameron's speech in Toronto were not impressed, and Smith attacked the bill for its untold costs in staffing the new bureau. He could only conclude that the whole matter was a political job. Street, also arrived from Toronto, responded to Cameron again, claiming in this forum that he did not care under what circumstances the government's bill had been brought forward, he was pleased that it would make agriculture "a distinct department of Government ... would bring the agriculturists immediately into contact with the Government ... [would give them] greater power than they now have through the means of a Board of Agriculture ... [and would give agriculturists] a voice in the administration of public affairs." After reiterating the Board of Agriculture's displeasure with specific portions of the bill (discussed later), Street turned to the matter of Cameron's recent speech at St. Lawrence Hall. He assured the legislature that after he had silenced the hecklers when Cameron first entered the hall, the minister had been "listened to with a degree of patience which entirely surprised him." In closing, Street challenged all members of the legislature "to produce a single tangible reason" why the bill should not pass, beyond the fact that it was a government bill.[66]

Other speakers on this first day of debate included Hincks who, like Cameron, reminded the legislature of the agriculturists' lobby, noting that the motives for the bill "did not originate with the Government, but with the agriculturists, and for years was pressed by them so forcibly on the attention of the Government that the Government did not like to resist." Furthermore, the new role had been proposed to Cameron; it had not originated with the member for Huron. In response, George Brown rose to claim again that this was a scandalous political job, pure and simple, whether Hincks tried to convince the legislature otherwise. He queried, as had his newspaper: Could anyone point to the need for such a bureau? Could anyone prove that Cameron was the best individual for the job? David Christie, MPP for Wentworth,[67] pointed out that the Board of Agriculture, of which he was a member, had approved of this bill, despite most of the board members being conservatives. He, too, had witnessed the disturbance at St. Lawrence Hall and confirmed that only two individuals had been the cause. Lower Canadian MPPs who spoke to the bill were mostly pleased it would provide improved and more coordinated leadership of agricultural improvement in their part of the province, although some added their concern as to the politics behind the bureau's creation. In an effort to close debate, Cameron sought to move the bill to a committee of the whole, but members still wished to debate the principles and merit of what the bill would implement. Debate was adjourned.[68]

When debate resumed on 22 October, Cameron moved for a second reading of his bill, and debate continued in a similar fashion: some were dead set against it because it was a political job; others refused to believe there was any need for the new department. Throughout, members raised stinging questions as to Cameron's suitability for the job. The legislature divided on second reading, and the bill was sent to a Committee of the Whole.[69] This process generated a few amendments.[70] Once again, the legislative clock ticked loudly. Due to a cholera outbreak in Quebec, the legislature had voted on 3 November to adjourn parliament on the 10th and not return until mid-February. Third reading of the bill on 6 November was attained on division, and the bill was sent to the Legislative Council. The upper chamber provided its swift approval of the bill and the Lower Canada agricultural societies bill on 9 November. Both received royal assent the following day, immediately prior to adjournment.[71]

The United Province of Canada now had a Bureau of Agriculture, headed by a minister of agriculture. Unlike other cabinet responsibilities that were duplicated, with one minister each for Upper and Lower Canada, there would be a single minister. The agricultural societies bill passed for Lower Canada would standardize the leadership of agricultural improvement to the Upper Canadian model so that parallel boards of agriculture and county and township agricultural societies would operate in each half of the colony. The act repealed all bills governing agricultural societies and the Board of Agriculture of Upper Canada in order that their leadership could be consolidated within the bureau by means of one act. The minister of agriculture would replace the inspector general on the Board of Registration and Statistics and would serve as its chairman. Control of any future census or collection of statistics was placed under his direction. He would be responsible for reviewing inventions submitted to his office and, where warranted, for issuing patents. Through his office statistical and agricultural information was to be collected and distributed as deemed fit, as part of the minister's annual report published as an appendix to the legislative journals or by other means. This shift of governance of the Board of Registration and Statistics from the inspector general to the minister of agriculture (and, in 1855, its merger into the Bureau of Agriculture) had lasting significance for, as Bruce Curtis notes, the creation of the bureau and its control over the statistical apparatus for the Canadas under the control of a Clear Grit minister not only "extended the central government's knowledge-generating capacity," but it also "shaped the Canadian statistical practice in the longer term by tying it to concerns with agricultural development."[72]

Not all agriculturists were satisfied. McDougall disagreed with the opinion of the Board of Agriculture of Upper Canada as to the extent of the minister's leadership and the degree of integration of this voluntary bureaucracy into the colonial government. As drafted, the bureau bill had made the minister of agriculture the *ex officio* president and, thus, chairman of both Upper and Lower Canadian Boards of Agriculture, with the authority to appoint a vice-president to each board. The Board of Agriculture of Upper Canada had expressed its displeasure with this plan during its review of the bill, and AAUC president Street had reiterated this fact during legislative debate. No definitive statement of the board's reasoning exists; however, it appears that directors did not wish the boards to become political agencies, which the proposed arrangement might produce. A minister as board president might make the boards bend to the political agenda of a specific ministry. Moreover, the existing system of board appointment, drawing leaders from county agricultural societies provided expertise in improvement. This was something that could not be guaranteed with the political appointment of a president and vice-president.[73]

McDougall thought otherwise. He maintained his original idea, which was incorporated into the AAUC's 1850 draft Board of Agriculture bill, which had called for the board chair to be a government appointment. Did not leaders of the agricultural societies in each half of the Canadas wish a close connection to the government by way of the legislation? He reasoned that "[t]hese Boards must depend wholly upon the Government of the day for the *funds* with which to carry on their operations, and therefore there would have been a great advantage in having one member at least appointed by, and in the confidence of the government." But legislators agreed with the Upper Canadian board and amended the bill to keep the minister of agriculture as an *ex officio* board member only, with each board alone responsible for electing its president and vice president from among elected directors.[74]

In practical terms, what was this new Bureau of Agriculture? The third clause of the draft bill had stated: "That said Minister of Agriculture shall not be entitled to any additional salary; but he may appoint a clerk or clerks with such salary or salaries respectively, as the Governor in Council shall order."[75] This had sparked intense criticism from ministry opponents, because it contradicted the government's claim that the new bureau would require no additional expenditure. Only Cameron was to draw a salary as the president of the Executive Council. To silence the critics, the clause was withdrawn. But the government had, in fact, hired a clerk and a messenger by the summer of 1852. All of this, before the legislation received royal assent.[76]

At its creation, the bureau was a semi-public, volunteer bureaucracy. The minister of agriculture relied entirely on two boards of agriculture and the local agricultural societies that reported to them. County agricultural societies remained the funded foundation of the network, with each continuing to receive an annual grant of no more than £250 with portions thereof to be distributed to the township societies operating within its boundaries. To receive their funds, county societies had to produce an annual report of finances and activity and deliver it to the Board of Agriculture by 1 April each year. In turn, it was the unfunded board's task to collect all reports and deliver them to the minister of agriculture along with its own annual report. It was also the board's unfunded task to coordinate the distribution of the annual public grant to county agricultural societies. Additionally, the Upper Canadian board was required to operate the experimental farm at the University of Toronto and establish new experimental or model farms elsewhere should it and the minister see fit; it was to establish at Toronto "an Agricultural Museum and an Agricultural and Horticultural Library"; it was to encourage the importation of new breeds of livestock and varieties of grains, seeds, or other agricultural items, along with improved implements and machines; it was to publish a monthly journal; and board members were to join with presidents and vice-presidents of the county agricultural societies to form the board of directors of the AAUC. All were unfunded tasks.

For Lower Canada, the legislation establishing the Bureau of Agriculture called for the Lower Canada Agricultural Society "to take immediate steps to wind up its affairs" so that a board of agriculture could be established in its stead. This board would have the same relationship with the government and possess the same mandate and responsibilities as the Upper Canadian board. Although no provincial agricultural association existed in Lower Canada, the bill stated that it was "expedient" for one to be established according to the model of the AAUC in Upper Canada. Under the act to reorganize Lower Canada's local agricultural societies, they would be reformed along a county basis, according to the Upper Canadian model.[77] Although there were certainly Lower Canadian agriculturists who sought increased leadership of agricultural improvement, the legislative record indicates it was the Upper Canadian lobby that brought about change for its own system, which was then applied to Lower Canada in order to standardize the system of leadership under a single bureau and minister.

The first year of the bureau's operation suggests Malcolm Cameron's critics were correct: he was not the best candidate for minister of agriculture. For example, he did not author a ministerial report. Instead, he merely published what his volunteer bureaucrats had prepared and submitted to him. Twenty-three reports submitted by Lower Canadian county agricultural societies, and the first annual report of the Board of Agriculture for Upper Canada, authored by its president, E.W. Thomson, formed Cameron's report.[78] Then, despite the Hincks–Morin ministry's substantial efforts to secure Cameron's and the Clear Grits' political support, Cameron left the bureau in August 1853 to become postmaster general, the cabinet position he had originally expected to obtain. Although in his new position for a year, someone compelled him to write a minister of agriculture report in August 1854, in the weeks that followed the end of his own parliamentary career, having lost his seat in the July elections. This election also brought about the end of the Hincks–Morin government.

In his report, Cameron worked through the 1852 statute that had established the bureau and addressed each responsibility of the minister. He reported several activities, such as collecting European periodicals for reference and republication in the Canadas, and sending William McDougall to the New York Exhibition of the Industry of All Nations in August 1853 to look for agricultural inventions not yet known in Canada, plus any other objects he believed should be brought to the minister's attention.[79] He also highlighted the production of a small pamphlet about the capabilities of the province written in German that was to be sent to Europe to attract prospective immigrants. Significantly, Cameron identified the vital role that statistical information would come to play within the ministry, employing agricultural statistics and other information to attract new immigrants. He used data from the 1851 Census to calculate how "by an improved system of husbandry we could increase the average yield of wheat one bushel per acre" and how that single improvement would represent a gain of £4.73 million to the provincial economy. He also appended his own questionnaire to be answered by individual farmers concerning specific aspects of agricultural improvement, as well as a table he had prepared by which the principal agricultural crops of a township could be tabulated for the purpose of calculating "the average profit per acre" and "the profit per cent. on the value of each farm." He hoped agricultural societies might use this table to chart and track local agricultural improvement.[80]

Cameron's replacement as Minister of Agriculture did little to advance the bureau. John Rolph, MPP for Norfolk, had been the other Clear Grit brought into the Hincks–Morin ministry at its formation and, in 1853, he was shuffled from his ineffectual work as commissioner of Crown lands to the president of the Executive Council and the minister of agriculture. But the sixty-year-old former rebel of 1837 was well past his political prime and, by most accounts, more interested in managing his Toronto School of Medicine than his political duties at Quebec.[81] He served until the 1854 elections. After much political jockeying in the election's aftermath, the old Tory Alan Napier McNab offered to join his Upper Canadian conservatives with Morin, who had returned his liberal majority in Lower Canada to create a liberal-conservative government. McNab assumed the role of president of the Executive of the Council and the associated position of minister of agriculture.[82] This proved significant, for the formulation of this ministry sustained the Bureau of Agriculture; there was no demand to shut it down as the "political job" it had once been called. Moreover, with McNab as minister of agriculture, leadership of the bureau crossed from Clear Grit and Reform hands and into those of the Conservatives. It was an important step to the bureau's legitimacy and longevity of purpose.

The lobby for a board of agriculture of Upper Canada was, in many ways, the culmination of a decades-long, multi-generational campaign for a provincial agricultural organization to better promote and coordinate agricultural improvement in Upper Canada. By the late 1840s, the target to accomplish this goal was the colonial state's acceptance that leadership of the province's agricultural societies was an important role of the state, worthy of direct leadership and permanent funding. While the Board of Agriculture, as founded, was not quite the achievement of that goal, the Hincks–Morin ministry's creation of a bureau of agriculture, in consultation with the new board, propelled the agriculturalists' lobby in ways they had only hoped. The ministry established the bureau on such a broad foundation that its leadership of agricultural societies formed just one important component. Under a single minister of agriculture, the Upper Canadian system was applied to a reorganization of Lower Canadian agricultural societies so that parallel systems of leadership and information exchange occurred in each half of the Canadas. Furthermore, the governance of the Board of Registration and Statistics by the minister of agriculture brought supervision of the census under his control. This was a significant measure that brought together practical, statistics-based measures that united the agriculturists' promotion of agricultural improvement with proponents of expanded commerce, infrastructure, and settlement. That this unfolded during a very brief window of time due to specific political issues requires reflection.

The appointment of a minister of agriculture to a newly created bureau of agriculture in order to satisfy one politician's ambitions and to secure the Clear Grit support for the Hincks–Morin ministry demonstrated that underneath much vituperative political rhetoric within the provincial newspapers and parliament was support across the political spectrum for increased, permanent government leadership of agricultural improvement because of its foundational role in most facets of the developing the Canadian economy. The brief window in which to act on this unity closed rapidly with the collapse of the Hincks–Morin coalition. Holding together so many political and business interests was its own full-time task. Political scandals did not help. By late 1852, William McDougall had begun to speak privately of his disillusionment with both the Hincks administration and politicians such as Cameron. Following a reorganization of the ministry in August 1853, McDougall cut ties with the ministry and his *North American* became an opposition voice. In the elections of 1854, Cameron and Brown fought head-to-head for the riding of Lambton with the former going down to defeat. Soon, Hincks would leave politics and the province and, in 1856, accept an appointment as governor of Barbados and the Windward Islands.[83] Yet, once the Bureau of Agriculture was established, it was immediately legitimized by the agriculturalists who had lobbied for state leadership of improvement. A minister of agriculture of any political affiliation could rely on them as a willing volunteer bureaucracy to demonstrate to the government and to the public the benefits of state promotion of agricultural improvement.

# Conclusion

Malcolm Cameron's ministerial report of 1854 repeated two timeless observations: "The improvement of Agriculture is a subject of national importance, susceptible of aid from wise legislation ... [and] ... Eminence in the art of husbandry is reached only by intermediate steps, as the place of destination is ultimately arrived at only by successive efforts to move in that direction."[1] "Intermediate steps" encapsulates well the efforts taken by generations of Upper Canadian agricultural improvers since 1791. That a minister of agriculture was reflecting on their journey in his ministerial report indicated they had reached an important destination. By legislating into existence a bureau of agriculture with authority to lead agricultural improvement across the entire colony, from the Detroit River to the Gaspé Peninsula, the Province of Canada provided government leadership of agricultural improvement across the vast territory that the Quebec Agricultural Society had attempted to encompass in 1791, joined by its branch societies at Montreal and Niagara. But in 1852, hundreds of thousands of settlers throughout the Canadas were improving their farms to some small or greater degree. Some attended exhibitions hosted by agricultural societies and, perhaps, read an agricultural periodical published in Upper Canada or the United States. Reflecting the "intermediate" nature of this progress was the generation of younger Victorian improvers joining with or replacing elder Georgian gentlemen improvers who had first established district agricultural societies in the 1830s. Some from that elder cohort, such as W.B. Jarvis and E.W. Thomson, continued to be active in agricultural societies until their deaths in the 1860s. Thomson would die at age seventy-one while walking from his farm to Toronto for an April 1865 executive meeting of the AAUC.[2] But in the 1850s, individuals such as the forty-six-year-old Clear-Grit Cameron and the forty-year-old Street, conservative MPP and president of the AAUC, were now among the leading improvers who championed the cause along with other businessmen and farmers. In doing so, these improvers found in agricultural improvement a commonality that rose above the highly fractured political lines of the day. They represented the improving ideology's adaptability to encompass an increasingly capitalist world in which commercial arguments were key drivers of improved agricultural output, agricultural-related manufacturing, and infrastructure construction for expanded and efficient agricultural trade. Among this new generation of improvers was forty-six-year-old George Buckland, who taught the next generation of improvers at the University of Toronto and its experimental farm.

Upper Canada's agricultural improvers were agents of colonial state formation. They employed their private roles and public offices to establish publicly funded agricultural societies across the province as semi-public agents of agricultural improvement. Their agricultural societies hosted cattle shows to display to colonists the results of improvements in the hopes that attendees would attempt to emulate such efforts on their own farms. At first, an agricultural society for each district was the workable approach to secure public funds to promote agricultural improvement, particularly when established without a requirement for subservience to the oligarchy at the provincial capital. Out of this 1830s' generation of Upper Canadian improvers emerged a set of agricultural society leaders who would establish the AAUC in 1846, at first a private club of improving gentlemen that by the end of the decade would produce a critical step in further embedding the encouragement of improvement as a role of the colonial state. It became an effective lobby, with members employing their private and public roles to draft, debate, and approve legislation to reorganize the governance of Upper Canada's publicly funded agricultural societies under the guidance of a board of agriculture whose offices were populated by AAUC members.

The mid-century creation of a bureau of agriculture for the Province of Canada was part of a broader "intermediate step" in the pursuit and promotion of agricultural improvement. By the 1850s, exhibition culture had established itself firmly within provincial society, with Upper Canadians and visitors from beyond attending township, county, and provincial agricultural exhibitions. These events provided an annual public measure for the "steps" that improvement had produced to the quality and range of Upper Canada's crops, livestock, and manufacturers. That culture remains a rural Ontario tradition today. Other "intermediate steps" required further steps to complete the intended journey. Upper Canada's Board of Agriculture expected the experimental farm at the University of Toronto would provide Professor

Buckland opportunities to test new seeds, plants, and agricultural implements to evaluate how they might adapt to the soil and climate of the province. For its part, the board promised to report to farmers across the province any successful experiments and provide them with such seeds or plants. Work on the property commenced in 1853, with the board providing £250 to purchase a team of horses and implements and construct a house, other buildings, and fences. But the following year, updated plans for University Park required space for additional university buildings and, possibly, new legislature buildings for the Canadas on the eastern (after 1860, Queen's Park) portion. As a result, space for the experimental farm was restricted to no more than twenty acres at the north-west corner of the grounds. A board request for £500 in public funds to cover operational costs for the farm went unfulfilled, and in 1855, one individual moaned that the location of the experimental farm had been shifted several times, with no current "certainty to its location." The experimental farm's woes reflected the University of Toronto's agricultural program itself. It would close in 1864. By then, Buckland had taught only a small number of students who received a diploma in agriculture. It would be another decade before the provincial government established the Ontario Agricultural College near Guelph.[3]

On the bureaucratic front, much remained to be done to create a full department of agriculture for the Province of Canada, complete with a bureaucracy not entirely reliant on the volunteer labour provided by the colony's semi-public Board of Agriculture and its agricultural societies. Two mid-century improvers played key roles. A year after winning a gold medal for his essay, William Hutton secured a position in the colonial government as a clerk thanks to his cousin Francis Hincks, then inspector general. In 1853, Hutton was appointed secretary of the Board of Registration and Statistics, and two years later, he submitted to superiors his ideas for bureaucratic improvement. In his opinion, neither the Bureau of Registration and Statistics nor the Bureau of Agriculture was performing as mandated by law. He proposed that inefficiencies and duplications could be solved by merging the two into a new bureau of agriculture, registration, and statistics, with him leading the larger common secretariat of five staff. The government agreed. Bruce Curtis summarizes the long-term significance of this decision: "under Hutton's plan, a dedicated Canadian statistical office would draw on the extensive information-gathering powers contained in the Census and the Bureau of Agriculture Acts, both to provide detailed information to policy makers about colonial conditions and to press for 'improvement,' especially in the domain of agriculture."[4] The change set the bureau on a path to become responsible for a wide range of matters critical to the developing colony: census, agricultural societies and improvements, annual statistical reports on manufacturing, mines, fisheries, registering vital statistics, granting patents and copyrights, immigration and colonization, and more. J.E. Hodgetts claims that what had the potential to become one of the most valuable agencies of government became a "potting house" for a number of "oddments" the government was uncertain about undertaking as a role of state.[5] With the Bureau of Agriculture taking control of the immigration agencies operating in Quebec, Montreal, Ottawa, Kingston, Toronto, and Hamilton, the Province of Canada possessed a greater ability to employ statistical information to attract and settle immigrants to the province.[6] The decentralized form of leadership that emerged within the bureau, and then full department, was rooted in the external agencies and agents over which the minister and civil servants presided. In the specific case of the boards of agriculture and the agricultural societies the bureau was to oversee, Hodgetts labelled it the "enabling administrative mode." Existing laws provided annual funds to boards, associations, and societies; therefore, the bureau required only minimal staff to audit financial reports submitted to them and disburse the annual public funds to those who met the minimal requirements.[7] Therefore, through to Confederation in 1867, the bureau continued to rely heavily on the volunteer bureaucracy provided by the boards of agriculture and the county and township agricultural societies under their mandate to promote agricultural improvement locally and to provide information about local conditions.

Following Confederation in 1867, Hutton's work had national importance, for under the terms of the British North America Act most tasks of the pre-Confederation bureau became federal responsibilities. Immigration was a shared federal and provincial authority, and a few provincial staff managed that file in Toronto. With most of the bureaucracy of the Province of Canada converted into the federal bureaucracy of the Dominion of Canada at Confederation, the British North America Act provided for the establishment of new provincial governments and bureaucracy. For the province of Ontario, one of its five government offices was to be headed by a commissioner of agriculture and public works, under which a single secretary would manage the arm's-length administration of agricultural matters for the first decades of Ontario's existence. The provincial commissioner's secretary could rely on the volunteer bureaucracy provided by Ontario's agricultural societies and the provincial agricultural association. Moreover, the semi-public nature of these organizations was sustained beyond Confederation by way of an 1868 Ontario statute that continued the colonial legislation of the early 1850s with minimal amendments beyond the obvious need to remove mention of the Lower Canada Board of Agriculture, now under the Province of Quebec's mandate. By 1876, Buckland, who had been lecturing at the new Ontario Agricultural College, became the bureaucrat responsible for coordinating the grants to Ontario's agricultural societies and associations. He was given the title of Assistant Commissioner of Agriculture in 1882, at which time the province established a Bureau of Industries which, in practice, served as an agriculture department until 1888, when Ontario established the foundations of a distinct Department of Agriculture. Also in 1882, the province published the Ontario Agriculture Commission's report. Appointed in the previous year, eighteen commissioners investigated and reported on seven broad aspects of the province's agriculture. Its very purpose was improvement: it investigated the province's agricultural practices, the quality and care of livestock, and it addressed the lack of useful and trusted statistics on the agricultural output. In their report, the commissioners recommended the establishment of a provincial bureau of agriculture but cautioned against any radical alterations of the system of semi-public societies and associations. Instead, it recommended relying on the advice of the agriculturists at the forefront of these organizations for their valuable input on provincial governance of agriculture, just as the commission had done to produce its report.[8] Improvement remained a guiding force, and in many ways, this comprehensive survey of Ontario's agricultural practices of the 1880s was not that far removed from the British Board of Agriculture's county surveys of the late 1700 and early 1800s. In fact, between 1874 and 1881, detailed descriptions and maps of forty Ontario counties were published in thirty-two atlases, sold by subscription. Not only did an atlas's maps of each county's township display the ordered improvement of the Ontario countryside into rectangular and square farm lots, fronted by straight concessions and sideroads, but its collection of short biographies also highlighted the improving works of prominent settlers. The wealthiest improvers of the county could pay the publisher for their biography to be inserted along with a bird's-eye sketch of their improved farm to display their idyllic mixed farm.[9] Were he still alive, Robert Gourlay might have smiled.

As these two examples highlight, the endpoint of this study provides a gateway for others to investigate and analyse fully the continued relationship of improvement to the functions of the Canadian nation at both the federal and provincial levels. Much remains to be studied regarding the Province of Canada's Bureau of Agriculture from its creation through to Confederation. R. MacGregor Dawson observed in

1955 that "[f]ew students, one suspects, appreciate how great has been the influence of permanent officials in the years before Confederation, nor do they have an adequate comprehension of the degree to which administrative decisions of those days, both by Ministers and officials, determined many of the present practices."[10] There have been few historical investigations into Hodgett's "potting house" seedlings in the years before and immediately after Confederation. How, exactly, was the bureau replicated as a federal ministry with similar responsibilities for the Dominion of Canada? How did the agricultural administrations established in Ontario and Quebec from the remnants of the colonial bureau compare? What, if any influence, did the Canadian bureau – both pre- and post-Confederation – have on the development of similar bureaucracies in the other two original provinces of the dominion, New Brunswick and Nova Scotia?

There is much work for historians of more recent eras to accomplish, too. Despite numerous upheavals and transformations of Ontario's agriculture and rural communities, particularly during the last decades of the twentieth century, an arc from this arm's-length pre-Confederation administration of agricultural societies reaches the present day with the continued operation of local agricultural societies, associated as members of the Ontario Association of Agricultural Societies, operating under the terms of the Agricultural and Horticultural Organizations Act, supervised by the Ontario Ministry of Agriculture, Food and Rural Affairs. Here in the twenty-first century, federal and provincial ministries continue to uphold Cameron's minister of agriculture statement (whether they explicitly understand that they do so): "The improvement of Agriculture is a subject of national importance, susceptible of aid from wise legislation."

This study has demonstrated that agricultural improvement and its expression via agricultural societies at the district, county, and township levels, plus a province-wide agricultural association, a provincial board, and a government bureau and minister, are proof that the ideology of improvement was a crucial factor in the formation of the Upper Canadian state. Agricultural improvement and the agricultural societies that improvers created to lead it were an element in many of the historical touchstones of Upper Canadian state formation: Simcoe's initial efforts to secure support for his leadership of the new province from among the settlers of the Niagara peninsula; the Thorpe imbroglio; high-handed efforts to expel Gourlay from the province; the first reform-dominated legislature of Upper Canada; battles between the Family Compact and their tory supporters and Mackenzie and his reform supporters; the 1837 rebellion at York; border raids in the

Niagara Peninsula; Sydenham's creation of district councils in his government reforms; the establishment of a provincial university after years of acrimonious debate; a Hincks–Morin government attempting to continue and strengthen the form of responsible government established by its Baldwin–Lafontaine predecessor in the face of intense criticism by the Clear Grits and the fracturing and transformation of the Upper Canadian Reform movement; and a minister of agriculture being the first expansion of cabinet roles following Sydenham administrative reforms of 1841.

The first generations of Upper Canadian improvers had desired an impossibility – a province-wide agricultural society that would lead improvement from the provincial capital, relying on patrons of the outlying districts to employ their influence to extend the society's authority. The 1850s' Board of Agriculture and Bureau of Agriculture were forms of what the early improvers had desired, although they developed through different means and under terms of responsible colonial governance that would have been anathema to most of those who promoted a provincial agricultural leadership during Upper Canada's first decades. For these reasons, Charles Fothergill's legislative legacy is both significant and lasting. Although he saw minimal successes stem from his bill prior to his death in 1840, it established a legislative foundation for the government to assume leadership of agricultural improvement. But it did so with a considerable measure of independence at the heart of that support. Perhaps this was the key to long-term success. The 1830 act to establish agricultural societies in Upper Canada required no communication among district agricultural societies across the province. That arrangement, illustrative of the rule of local oligarchies within each district of the province at the time, was continued, infusing county and township agricultural societies with similar independence from each other. Most importantly, that independence was replicated at the levels of the Board and Bureau of Agriculture, too. Legislation creating the Board of Agriculture for Upper Canada established an umbrella organization to oversee all agricultural societies operating in the province, but it did not require communication between each society. Independence also permitted the Bureau of Agriculture to represent both halves of the Province of Canada, with a board of agriculture established for Lower Canada, with no requirements to communicate with its equal in Upper Canada, even though the upper province's system of agricultural societies had been applied to the lower province, which had evolved under a parallel but separate governance. As this study has highlighted, it is difficult to identify specific local or provincial improvements introduced to the province's agricultural practices between 1791 and 1852, and certainly, many of the agricultural societies' efforts at introducing improvements did not immediately cause "two blades of grass to grow where only one grew before." Nevertheless, they were benefactors of the colony, for their determined efforts eventually expanded the role of the state through intermediate steps to provide bureaucratic governance of agricultural improvement where only occasional and inadequate leadership had been offered before.

# Notes

## Introduction

1. Officially, the colony's name had changed to "Canada West," the western portion of the United Province of Canada as of 1841 when Britain joined the colonies of Upper Canada and Lower Canada within a single colonial government. This study employs the colony's original name, "Upper Canada," throughout the 1792–1852 period for clarity and to reflect choices made by improvers and legislators who continued its use long after 1841.
2. *CdnAg*, January 1852. Turner, "Hutton, William"; Boyce, *Hutton of Hastings*, 178. Boyce's book contains images of the medal awarded to Hutton.
3. Google Books Ngram search for "two blades of grass." https://books.google.com/ngrams
4. Swift, *Gulliver's Travels*, 176.
5. Stubbs, *Jonathan Swift*, 11–12, 16, 81–2. Hawes, "*Gulliver's Travels*: Colonial Modernity Satirized," 1–31.
6. Second- and third-place essays were also published as front-page material in subsequent months. *CdnAg*, January, February, and March 1852.
7. Hofstadter, *The Age of Reform*, 1–45.
8. Errington, *The Lion, the Eagle and Upper Canada*, 5.
9. "improve." In Stevenson, ed., *Oxford Dictionary of English*, https://www.oxfordreference.com/view/10.1093/acref/9780199571123.001.0001/m_en_gb0403460. Drayton, *Nature's Government*, 51–2; Hancock, *Citizens of the World*, 282.
10. Gascoigne, *The Enlightenment and the Origins of European Australia*, 1–2.
11. Gascoigne, *Joseph Banks*, 31–2, 187.
12. Gascoigne, *The Enlightenment and the Origins of European Australia*, 1 and, in general, 1–16.
13. Spadafora, *The Idea of Progress*, xiii, 6, 80–1, 84.
14. Gascoigne, *The Enlightenment and the Origins of European Australia*, 4.

15 Gascoigne, *Joseph Banks*, 185.
16 Canny and Pagden, eds., *Colonial Identity in the Atlantic World*, 11; Mokyr, *The Enlightened Economy*, 185.
17 Heaman, *Civilization*; McNairn, *The Capacity to Judge*, 9. The colony's history is lacking a study such as John Gascoigne's *The Enlightenment and the Origins of European Australia*, which clearly sets forth the role Enlightenment thought, improvement, and progress played in the contemporary establishment of the New South Wales colony with the arrival of the first fleet in 1788.
18 Mokyr, *The Enlightened Economy*, 175, 188, 194 and, in general, 171–97.
19 Drayton, *Nature's Government*, 50, 57, 63.
20 Gascoigne, *Science*, 30.
21 Drayton, *Nature's Government*, 89, 102.
22 Canny and Pagden, eds., *Colonial Identity in the Atlantic World*, 7, 10–11.
23 Emily Pawley makes this important distinction in her study of improvers in New York state. Many improvers were not farmers; moreover, the definition of a farmer was in flux in the antebellum United States, she notes. *Nature of the Future*, 6.
24 Hancock, *Citizens of the World* 15–17, 20, 279, 281.
25 Thornton, *Cultivating Gentlemen*, 1, 20–21, 32, 202. Marti, "Agrarian Thought and Agricultural Progress," viii.
26 Laurence Gouriévidis provides a succinct assessment of improvement's effects on the Scottish Highlands in his attempt to understand the modern historical presentation by Scotland's museums of this painful and controversial aspect of Scottish history. Gouriévidis, *The Dynamics of Heritage*, xi-xii, 19–23.
27 On the broader point of colonists establishing settler communities throughout the Atlantic World and their relationship with Indigenous populations in working out their relationship to the new land and environment, see Eliot, "Introduction in the Atlantic World" in Canny and Pagden, eds. *Colonial Identity in the Atlantic World*, 7. For an environmental approach, see Virginia DeJohn Anderson *Creatures of Empire*. For an intellectual approach, see Heaman, *Civilization*.
28 Bushman, *The American Farmer in the Eighteenth Century*, 4–5.
29 Stoll, *Larding the Lean Earth*, 17, 20–21, 30, 167–69. In the Upper Canadian context, Sean Gouglas investigates local environmental factors at play in famers' crop decisions, along with what would best sell at market. In this, he argues, persistence on the land was a key to success See: Gouglas, "The Influences of Local Environmental Factors on Settlement and Agriculture in Saltfleet Township, Ontario, 1790–1890."

30 Pawley, *The Nature of the Future*, 16.
31 Pawley, *The Nature of the Future*.
32 Samson, *The Spirit of Industry and Improvement*, 59, and, in general, 54–79.
33 In 1983, the *Dictionary of Canadian Biography* published a short explanation of its index of identifiers, noting that, "Improvers" as a sub-group under "agriculture" included "land agents, gentlemen farmers, and colonizers." See "Index of Identifications," vol. 5, *Dictionary of Canadian Biography* (1983), 951. The *DCB*'s sub-set identifier "Agriculture – Improvers and developers" is located under the drop-down menu of "Occupations and Other Identifiers" at http://www.biographi.ca/en/search.php.
34 Zeller, *Land of Promise, Promised Land*, 2–3; *Inventing Canada*, 3.
35 Lawr explored the intersection of scientific agriculture and the idea of progressive farming with mid-nineteenth-century ideas of institutionalizing and promoting such knowledge by establishing agricultural colleges, while Nesmith focused on the philosophy of agriculture in Ontario, analysing agricultural improvers' ideas about the promotion and acquisition of agricultural knowledge in the years prior to the college's founding and during its first decades of operation. Lawr, "Agricultural Education in Nineteenth-Century Ontario," and Nesmith, "The Philosophy of Agriculture."
36 Derry, *Ontario's Cattle Kingdom*.
37 Province of Canada, Board of Agriculture, *Journal and Transactions of the Board of Agriculture for Upper Canada*, vol. 1, 1856, 4. *CdnAg*, February 1, 1862. "A History of the Agricultural and Arts Association," in Ontario, Sessional Papers, vol. 28, part 6, 1896, *Fiftieth Annual Report of the Agriculture and Arts Association of Ontario 1895*, Appendix D, 139–40; James, "The First Agricultural Societies" in Ontario, Department of Agriculture. *Annual Report of the Bureau of Industries for the Province of Ontario*, 1901, Appendix No. 26, 111–35; James, "The Pioneer Agricultural Society of Ontario," *Farming World*, September 1902, 211–12; James, "The First Agricultural Societies," *Queen's Quarterly* 10 (October 1902): 218–23; James, "History of Farming," in *Canada and its Provinces*, Shortt and Doughty eds., vol. 18 (1914), 556–69; Middleton and Landon, "Agriculture in Ontario", *The Province of Ontario: A History*, vol. 1 (1927), 459–79; Talman, "Agricultural Societies of Upper Canada," *Ontario Historical Society Papers and Records* 27 (1931): 545–52; Jones, *History of Agriculture in Ontario 1613–1880* (1946), 156–69, 328–45; Guillet, *The Pioneer Farmer and Backwoodsman*, vol. 2 (1963), 129–46; Ontario Association of Agricultural Societies, *The Story of Ontario Agricultural Fairs and Exhibitions 1792–1967* (1967); Reaman, *A History of Agriculture in Ontario*, vol. 1 (1970), 99–105;

Scott, *A Fair Share: A History of Agricultural Societies and Fairs in Ontario 1792–1992* (1992).

38 Nurse, *Cultivating Community*; Heaman, *The Inglorious Arts of Peace*, 31–6, 44–8; McNairn, *The Capacity to Judge*, 92–7, 114; Ferry, *Uniting in Measures of Common Good*. Elsewhere in British North America, substantial studies of agricultural societies have been conducted for Nova Scotia. See Wynn, "Exciting a Spirit of Emulation Among the 'Plodholes': Agricultural Reform in Pre-Confederation Nova Scotia"; and Samson's discussion of improvement in *The Spirit of Industry and Improvement*, 54–79.

39 J.M.S Careless, for example, characterizes the Clear Grit faction that coalesced in 1849 as "radicals" while dismissing one of its young leaders, William McDougall, as the "tall young lawyer turned journalist who had already published a little agricultural weekly for the countryside." Careless, *The Union of the Canadas*, 167. Michael Cross is similarly dismissive of the agrarian "farm folk" who composed and supported the Clear Grit faction. Cross, *A Biography of Robert Baldwin*, 284–95. In a similar vein, both William Ormsby and George A. Davison note in their histories of reform premier Francis Hincks that a "George Brown version" of history has dominated our understanding of Hincks and his Reform ministry with co-leader Augustin-Norbert Morin of Lower Canada. Brown was a vocal political foe of Hincks, and his version of Hincks's premiership emphasizes him as the man who opened the door for widespread corruption to take over Canadian politics. Ormsby, "Sir Francis Hincks," 192; Davison, "Francis Hincks and the Politics of Interest," 21.

40 James, "The First Agricultural Societies," 218; Hodgetts, *Pioneer Public Service*, 229 and Chapter 14.

41 Jones, 163, 170, 174; Fowke, *Canadian Agricultural Policy*, 105. Romney, "A Man Out of Place," 516–17. Guy Scott, writing for the Ontario Association of Agricultural Societies, reaffirms the "humble beginnings" of the societies of the 1792–1830 period, noting that they were "largely failures." Scott, 13, 27.

42 Ian Radforth and Allan Greer's *Colonial Leviathan* is a key example of this temporal framework with their effort to identify, define, and illuminate state formation in the Canadas of the mid-nineteenth century. At the time of their writing in the early 1990s, they cited Hodgetts's 1955 study as a singular example that broke a trend in Canadian history in which "[t]he state was assumed more than studied." And they identified numerous studies of education as evidence of recent scholarship breaking that trend. Greer and Radforth, eds., *Colonial Leviathan*, 4.

43 See Samson, *The Spirit of Industry and Improvement*, 17.

44 Curtis, *True Government by Choice Men?*, 4–6, 9–11, 15–18, 29. Curtis's book on the emergence of the Canadian census also informs the latter chapters of this study. *The Politics of Population*.
45 Heaman, *A Short History of the State in Canada*, 1.
46 Curtis, *True Government by Choice Men?*, 21–2.
47 Johnson, *In Duty Bound*, 7–9, 244. See his figure 2 and table 2.1 for an outline of the Upper Canadian state and a listing of its main offices. He, too, laments the lack of a study of the 1783–1841 period, such as Hodgetts conducted for the Province of Canada from 1841–67.
48 Dehli, "Creating A Dense and Intelligent Community," 112, 128.
49 Armstrong, *Handbook of Upper Canadian Chronology*, 272.
50 For a detailed economic history of Upper Canada, see McCalla, *Planting the Province*. See Margaret Derry's concise assessment of the value of livestock to wheat in *Ontario's Cattle Kingdom*, 5–7.
51 Armstrong, *Handbook of Upper Canadian Chronology*, 272.
52 Kelly, "Notes on a Type of Mixed Farming Practiced in Ontario During the Early Nineteenth Century"; Derry, *Ontario's Cattle Kingdom*, 6–7.
53 A St. Catharines' newspaper editor used this term to announce a cattle show of the Niagara District Agricultural Society. *FJWCI*, 16 November 1831.

### 1 Transatlantic Improvers and the Niagara Agricultural Society, 1791–1807

1 *QG*, 12 April 1792.
2 Although Lieutenant Governor John Graves Simçoe would rename Niagara to Newark in his 1792 proclamation establishing it as Upper Canada's first capital town, most local inhabitants continued to use Niagara. Two years after Simcoe's return to England in 1796, a provincial act restored the town's original name. A century-plus later, Niagara adopted its present name, Niagara-on-the-Lake, to better sort mail destined for that town versus nearby Niagara Falls. For clarity, this book will refer to the town as "Niagara" throughout.
3 In 1946, R.L. Jones corrected earlier historians by suggesting a founding date of "the autumn of 1791", citing an obscure 1932 report by R.P. Gorham as evidence. But later historians appear to have disregarded Jones and Gorham, instead continuing to claim a date of 1792 or 1793. See Jones, *History of Agriculture in Ontario*, 157 and Gorham, "The Development of Agricultural Administration in Upper Canada during the Period before Confederation." For suggested dates of the agricultural society's founding, see Kirby, *Annals of Niagara*, 134; James, "The Development of Agriculture in Ontario,"37; James, "The First Agricultural Societies," 123; James, "The Pioneer Agricultural Society of Ontario," 211–12; James, "The First

Agricultural Societies," 222–3; Carnochan, "Members of the Agricultural Society, 1792–1805," 17; Carnochan, *History of Niagara*, 266–9; Talman, "Agricultural Societies of Upper Canada," 545–6; Guillet, *The Pioneer Farmer and Backwoodsman*, vol. 2, 129; Ontario Association of Agricultural Societies, *The Story of Agricultural Fairs and Exhibitions 1792–1967*, 1–5; Reaman, *A History of Agriculture in* Ontario, vol.1, 22; Scott, *A Fair Share*, 17–18; Heaman, *Inglorious Arts*, 35.

4 In 1902, C.C. James reported his discovery of a notice among the papers of Upper Canada's Deputy Surveyor General, David W. Smith, that he had been appointed "Vice-President of Agricultural Society" on 27 October 1792. See James, "The Pioneer Agricultural Society of Ontario," 212, and TRL, David W. Smith Papers, B 15, "A memorandum of the dates of the Honourable D.W. Smith's appointments," n.d.

5 The first mention of Simcoe as patron is found in the *UCG*, 9 May 1793.

6 The Agricultural Society in Canada, *Papers and Letters on Agriculture.*

7 *UCG*, 13 June 1793; 4 July 1793; "From E. B. Littlehales to the Secretary of the Agricultural Society, 25 April 1793," Cruikshank, ed., *The Correspondence of Lieutenant Governor John Graves Simcoe*, v.1, 318; "The Petition of the Agricultural Society of Niagara" in Cruikshank, ed. "Petitions for Grants of Land in Upper Canada, Second Series, 1796–99," 102–3; "Minutes of the Executive Council [Lands]," in Cruikshank, ed. *The Correspondence of the Honourable Peter Russell*, v.1, 266. "The Niagara Agricultural Society," *Ontario Heritage Trust Online Plaque Guide*, https://www.heritagetrust.on.ca/en/index.php/plaques/niagara-agricultural-society

8 Bowler and Wilson, "Butler, John." Also see Wilson, *The Enterprises of Robert Hamilton*, 35–47; Joy Ormsby, "Building a Town," in *The Capital Years*, 15–16.

9 Wilson, "Hamilton, Robert."

10 In collaboration with Bruce A. Parker, "Warren, John."

11 Ouellette, "Crooks, James"; Merritt, "Early Inns and Taverns," in *The Capital Years*, 200.

12 Wilson, "Dickson, William."

13 Moogk, "McNabb, Colin"; Wilson, *Enterprises*, 149; Seibel, *The Niagara Portage Road*, 18.

14 Roland, "Kerr, Robert."

15 Ormsby, "Building a Town," *The Capital Years*, 15–16; Wilson, *Enterprises*, 38.

16 Cruikshank, *The Story of Butler's Rangers and the Settlement of Niagara*, 116–17.

17 Wilson, "Clench, Ralfe."

18 In collaboration with Bruce A. Parker, "Street, Samuel."

19 Graymont, "Thayendanegea."

20 Ormsby, "Building a Town" in *The Capital Years*, 16; "Weddings at Niagara, 1792–1832," 54.
21 Wilson, *Enterprises*, 15, 19–20, 60–3, 132.
22 Wilson, *Enterprises*, 26; Seibel, 17.
23 Wilson, *Enterprises*, 19, 30–1; Seibel, 15.
24 See McCalla, "The 'Loyalist' Economy of Upper Canada, 1784–1806," 279–304.
25 Wilson, *Enterprises*, 93; Wilson, "Dickson, William." In a sense, Thayendanegea was the largest landholder in the society, for he oversaw the Six Nation's lands along the Grand River, which had been given to them by the British government for their loyalty during the American Revolution. This territory totaled some 570 000 acres, large portions of which Thayendanegea, Dickson, and Street sought to exploit for profit after Simcoe's arrival to the province.
26 Robertson, *The History of Freemasonry in Canada*, vol. 1, 257.
27 Harland-Jacobs, *Builders of Empire*, 2.
28 MacDonald, "Colonel John Butler: Soldier, Loyalist, Freemason, Canada's Forgotten Patriot, 1725–1796," 1319–22; Power, "Religion and Community," in *The Capital Years*, 124–5. McClenachan, *History of … Free and Accepted Masons in New York*, 151–3.
29 Talman, *Historical Sketch to Commemorate the Sesqui-Centennial of Freemasonry in the Niagara District 1791–1942*, 3–4; Dunquemin, comp., *A Lodge of Friendship: The History of Niagara Lodge, No. 2 A.F & A.M. G.R.C., Niagara-on-the-Lake, Ontario, Canada*, 1792–1992, 6–19, 207–14.
30 Harland-Jacobs, 4, 22, 53–4, 116.
31 McNairn, 69–83.
32 For the prolonged issue of establishing a provincial lodge following the relocation of the capital from Niagara to York, see Robertson, *The History of Freemasonry in Canada*, vols. 1 and 2. Chapters 19–23 and 28.
33 Dunquemin, *History of Niagara Lodge, No. 2*, 218.
34 Ormsby, "Building a Town," in *The Capital Years*, 15–16. "Memorandum From J.G. Simcoe to Hon. Henry Dundas, 30 June 1791, *Simcoe Correspondence*, vol. 1, 27.
35 Mealing, 'The Enthusiasms of John Graves Simcoe,' 314.
36 Craig, *Upper Canada: The Formative Years 1784–1841*, 31–2; Nelles, "Loyalism and Local Power," 100–101; David Mills, *The Idea of Loyalty in Upper Canada 1784–1850*, 19. Errington, *The Lion, the Eagle and Upper Canada*, 29–30; Mealing, "Simcoe, John Graves"; Fryer and Dracott, *John Graves Simcoe, 1752–1806: A Biography*.
37 Burns, "Jarvis, William"; Harland-Jacobs, 54.
38 The most recent biography of Simcoe devotes two chapters to his life as an improving gentleman. Fryer and Dracott, *John Graves Simcoe*, 85–104. Also see Robertson, ed., *The Diary of Mrs. John Graves Simcoe*, 31–5; AO, F47-6-0-1,

Simcoe Family Fonds, Subject Files, Wolford Estate; F47-16-0-3, Simcoe Estate, Property and Financial Records; Firth and Fahey, "Scadding, Henry."

39 Andrea Wulf has explored the existing avenues of botanical exchange across the Atlantic Ocean in two studies: *The Brother Gardeners: Botany, Empire and the Birth of an Obsession* (2009) and *Founding Gardeners: The Revolutionary Generation, Nature, and the Shaping of the American Nation* (2012).

40 Lord Kames (Henry Home), *The Gentleman Farmer*, 392–406.

41 Great Britain, Parliament, House of Commons, *The Parliamentary History of England from the earliest period to the year 1803*, vol. 30, 1817, 949–53; Gascoigne, *Joseph Banks*, 191; Drayton, *Nature's Government*, 88–9.

42 Gascoigne, *Science in the Service of Empire*, 128. For a history of the Board of Agriculture see, Mitchison, "The Old Board of Agriculture (1793–1822)," 41–69.

43 Gascoigne, *Science in the Service of Empire*, 65–6.

44 Quoted in Hudson, *Patriotism with Profit*, 2–3, 96–7. In Kames view, farming was an exercise for both the body and mind, offering greater personal satisfaction and benefits to society than the traditional gentleman's pastime of hunting. Lord Kames, *Gentleman Farmer*, preface.

45 Drayton, *Nature's Government*, 102.

46 Averley, "English Scientific Societies of the Eighteenth and Early Nineteenth Centuries," 333. Also see Sarah Wilmot's enumeration of the growth of scientific and agricultural societies from the last half of the eighteenth century though to the mid-nineteenth century; "'The Business of Improvement': Agriculture and Scientific Culture in Britain, c.1700–c.1870," 8–10.

47 The Agricultural Society in Canada, *Papers and Letters*, 1.

48 Greer, "Davison, George."

49 Davison's comments to John Brickwood, London, 14 January 1791 are found on the flyleaf of a copy of the society's *Papers and Letters* held by Special Collections and Rare Books at the Toronto Reference Library.

50 *Letters and Pape[rs] on Agriculture*, 12, 38; White, "Speed the Plough: Agricultural Societies in PreConfederation New Brunswick," 21.

51 James, "The First Agricultural Societies," 111–26; True, "The Early Development of Agricultural Societies in the United States," 295–306. Of the twenty-three members who founded the Philadelphia Society for Promoting Agriculture in 1785, four had signed the Declaration of Independence, four had been members of the convention to draft the Constitution, seven had been officers in the Revolutionary Army, seven were Congressmen, and two were Senators. Benjamin Franklin was a resident member, and President George Washington, an honorary member. Baatz, *"Venerate the Plough,"* 4–6.

52 Wall, "George Washington: Country Gentleman," 5.

53 See the Agricultural Society in Canada, *Papers and Letters*, and Society for Promoting Agriculture in the Province of Nova Scotia, *Letters and Pape[rs]*.
54 For a report on the soil and climate of the territory that would become Upper Canada, see LAC, MG11, CO42, v.50, "Information communicated by John Collins Esquire Deputy Surveyor General of Lands concerning the raising of Hemp & Flax in the New Settlements," Quebec, 9 January 1787.
55 "From J.G. Simcoe to Sir Joseph Banks, 8 January 1791," Cruikshank, ed., *Simcoe Correspondence*, vol. 1, 17–19; Scadding ed., *Letter to Sir Joseph Banks*.
56 See Gascoigne's two studies of Banks: *Joseph Banks*, and *Science in the Service of Empire*.
57 "Simcoe to Banks, 8 January 1791," *Simcoe Correspondence*, vol. 1, 17–19. Sometime during that same year, Simcoe also jotted down a few suggestions in a list he titled "Objects that may be worth the attention of the new settlers in Upper Canada." First among the items was "Growing Hemp and Flax." LAC, MG11, CO42, v.316, "Objects that may be worth the attention of the new settlers in Upper Canada, 1791."
58 Gascoigne, *Science*, 30.
59 Gascoigne, *Joseph Banks*, 31–2; Gascoigne, *Science*, 16, 30; Drayton, *Nature's Government*, 94–106.
60 "J. G. Simcoe to Henry Dundas, London, 12 August 1791," *Simcoe Correspondence*, vol. 1, 49.
61 Simcoe had purchased these books at the London bookstore of Thomas Payne. "Canadian Letters," 54; "J. G. Simcoe to Henry Dundas, London, 12 August 1791," *Simcoe Correspondence*, vol. 1, 49; TPL, S 106 John Graves Simcoe and Wolford Papers, vol.1, "Books intended for Canada by Colonel Simcoe, August 1791," 369–70, and vol. 2, "Alex Davison, B'ot of Thomas Payne, Mar. 26 [1792]," 51–2. Concerning the status of Arthur Young, see Gazley, *The Life of Arthur Young*, vii, 306.
62 *UCG*, 9 May 1793; "From E. B. Littlehales to the Secretary of the Agricultural Society, 25 April 1793," *Simcoe Correspondence*, v.1, 318.
63 *QG*, 12 April 1792.
64 In his study of London booksellers and customers in America, James Raven argues that we might consider the Atlantic Ocean "a zone of circulation and exchange" and that imported books to North America from Europe were "lifelines of identity" and "direct material links to a present and past European culture." Raven, *London Booksellers and American Customers*, 7, 14.
65 Carnochan, *History of Niagara*, 267; Carnochan, "Niagara Library, 1800 to 1820," 1–30; James, "The Pioneer Agricultural Society of Ontario," 212. For my attempt at identifying the full titles and authors of this collection, see "Appendix 2: List of books donated to the Niagara Library by the Niagara Agricultural Society, 1805," in Fair, "Gentlemen, Farmers, and Gentlemen Half-Farmers," 318–19.
66 *QG*, 12 April 1792.

67 Bowler and Wilson, "Butler, John."

68 Nelles, "Loyalism," 102.

69 Wilson, *Enterprises*, 1. The eighteenth-century North American mansion was as much a performer as the elites who resided in them, argues Richard Bushman. Mansions stood out on the North American landscape and were "stage sets for dramas." Bushman, *The Refinement of America*, 132.

70 Wilson, *Enterprises*, 142.

71 Noel, *Patrons, Clients, Brokers*, 43; Wilson, *Enterprises*, 143; duc de La Rochefoucauld-Liancourt, *Travels*, 241.

72 Turner, "Addison, Robert." This list of the executive is recorded on the Ontario Heritage Trust plaque, which stands in Simcoe Park, Niagara-on-the-Lake. While it is plausible that Dickson and Addison were vice-presidents, the source of that information is unclear. See Ontario Association of Agricultural Societies, *Story of Ontario Agricultural Fairs and Exhibitions*, 1. Janet Carnochan's list of members suggests that Simcoe was the society's first president. However, this is unlikely, as he was the society's patron. Carnochan, "Members of the Agricultural Society, 1792–1805," 17; C.C. James identified David W. Smith as a Vice-President. James, "The First Agricultural Societies," 123. This is confirmed by "A memorandum of the dates of the Honourable D.W. Smith's appointments." in TRL, B 15, David W. Smith Papers.

73 *UCG*, 13 June 1793.

74 There is no official membership list, but one of twenty-seven individuals has been created from a variety of sources. Carnochan, "Members of the Agricultural Society, 1792–1805," 17; James, "The First Agricultural Societies," *Queen's Quarterly*, 222; James, "The First Agricultural Societies," 122–6; Clark, "Memoirs of Colonel John Clark of Port Dalhousie, C.W.," 158; Quaife, ed., *The John Askin Papers*, 353–5.

75 Ormsby, "Building a Town," in *The Capital Years*, 28–9, 33, 35; "Canadian Letters," 56.

76 Wilson, *Enterprises*, 178.

77 *UCG*, 9 May 1793; 13 June 1793; 4 July 1793.

78 "Canadian Letters," 54. The identity of the author of these anonymously written letters is found in William Colgate, "The Diary of John White," *Ontario History* 47 (1955), 159n22.

79 This information was recorded in 1860 by Colonel John Clark of Port Dalhousie as he reminisced about his life in Upper Canada. He claimed that he had a "perfect remembrance of the first Agricultural Society," as his father had been a member. Clark, "Memoirs," 158; Also see Canniff, *History of the Province of Ontario*, 590; James, "The Pioneer Agricultural Society of Ontario," 211; James, "The First Agricultural Societies," *Queen's Quarterly*, 222; James, "The First Agricultural Societies,"122–3.

80 Phillip McCann highlights the use of invented tradition in his investigation of the relationship between culture and state authority in Newfoundland. He cites Governor Sir John Harvey's 1843 establishment of an agricultural society as a patriotic organization whose rituals were meant to reaffirm ties to Britain and its hierarchical form of society, thus bringing harmony to a colony recently divided by political social conflicts of the late 1830s. McCann, "Culture, State Formation, and the Invention of Tradition: Newfoundland, 1832–1855," 86–103.
81 "Canadian Letters," 54, 56.
82 "Canadian Letters," 58, 42, 41.
83 Bushman, *Refinement of America*, xiii-xiv.
84 Thornton, *Cultivating Gentlemen*, 68.
85 Thornton, *Cultivating Gentlemen*, 32.
86 Marti, "Agrarian Thought and Agricultural Progress," viii.
87 *UCG*, 26 October 1796.
88 *UCG*, 15 March 1797.
89 *UCG*, 29 March, 5 April 1797. Tiffany had previously published a 21 December 1796 editorial that opened with "Agriculture is justly thought to be the most ancient art; and it is certainly [by] far the most useful." Drawing upon the ancient writings of Virgil, it echoed the words of Lord Kames about the benefits of agriculture permitting a farming population to be self-sufficient and healthy in both body and mind. It is not clear if Tiffany wrote this or copied it from elsewhere. It appeared years later in the *Kingston Gazette*. *UCG*, 21 December 1797; *KG*, 25 September 1810.
90 *UCG*, 15 November 1806.
91 LAC, MG23-GIII7, John Porteus Papers, "Robert Hamilton to John Porteous, 9 March 1794." See Carnochan, *History of Niagara*, 266–67.
92 Wilson, "Clench, Ralfe."
93 In 1810, Richard Cartwright cited the annual fair at Queenston in the context of other "public markets" in York and Kingston. See [Cartwright, Richard]. Letters of an American Loyalist in Upper Canada on a Pamphlet Published by John Mills Jackson, 83–4. In 1901, C.C. James, Ontario's Deputy Minister of Agriculture, wrote brief histories of agricultural societies in each of the British North American provinces. He opened a section on "Early Markets and Fairs" by noting "Just how closely our Agricultural Fairs or Exhibitions are related to the old fairs of the British Isles and Europe is uncertain." James, "The First Agricultural Societies," 131. Fairs and agricultural society "cattle shows" then became conflated in the historical record. See: Middleton and Landon, *The Province of Ontario*, vol. 1, 460; Talman, "Agricultural Societies of Upper Canada," 546; Jones, *History of Agriculture in Ontario*, 159; Scott, *Fair Share*, 7–12, 18.

In E.A. Heaman's cultural study of mid-nineteenth century Victorian exhibitions in Canada, the Queenston Fair is portrayed as an actual NAS event, with the institution sponsoring prizes for specific competitions of agricultural produce. Heaman, *Inglorious Arts of Peace*, 35. See similar confusion as to the early events at Niagara in Walden, *Becoming Modern*, 11–12.

94 *CC*, 8 November 1799.

95 *NH*, 17 October 1801. Also see notice for the Niagara Fair earlier that fall, *NH*, 12 September 1801.

96 Addison, *English Fairs and Markets*, 59.

97 See, for example, the 21 October 1792 minutes of the Executive Council and Lieutenant Governor Simcoe's Proclamation establishing an annual fair "in the Neighbourhood of the Town of Newark." Cruikshank, ed *Simcoe Correspondence*, vol. 4, 351–2. Also see Osborne. "Trading on a Frontier: The Function of Peddlars, Markets, and Fairs in Nineteenth Century Ontario," 73–7; McCalla, *Planting the Province*, pp. 26–7.

98 For a study of the agricultural economy in this period, and the primacy of wheat cultivation, see McCalla, "The 'Loyalist' Economy of Upper Canada."

99 Hamilton's letter of 22 November 1795 in which he outlines his plans has not survived, only Simcoe's reply. "From J. G. Simcoe to Robert Hamilton, 30 January 1796," *Simcoe Correspondence*, vol. 4, 187.

100 "From J. G. Simcoe to Robert Hamilton, 30 January 1796," and "From Robert Hamilton to J. G. Simcoe, 21 February 1796," *Simcoe Correspondence*, vol. 4, 187.

101 Simcoe had promised his daughter, Eliza, in 1795 to "be totally a Farmer, Planter & Gardener when I can obtain leave to return home," and he was delighted that she continued to grow a garden at Wolford Lodge (for which he had sent seeds from Upper Canada). AO, F47-4-4-3, Simcoe Family Fonds, Subject Files, Personal Correspondence, "J.G. Simcoe to Eliza Simcoe, Upper Canada, 1795"; F47-1-2-4 Official Correspondence, "J.G. Simcoe to Eliza, August 22, 1792, and "Elizabeth Simcoe to Eliza, [1792]." Also see farming-related records in F47-6-0-1, Wolford Estate; and F47-16-0-3, Simcoe Estate, Property and Financial Records; Fryer and Dracott, *John Graves Simcoe*, 213, 224–30.

102 "The Petition of the Agricultural Society of Niagara, Newark, 26 August 1797" in "Petitions for Grants of Land in Upper Canada, Second Series, 1796–99," 102; Cruikshank, ed., "Minutes of the Executive Council. [Lands], Newark, 28 August 1797," *Correspondence of the Honourable Peter Russell*, vol. 1, 266.

103 There had been one last newspaper notice of an upcoming NAS meeting on 18 March 1797 with a request that members return any books on loan at that time. Although it was signed "R. Hamilton, formerly president," the August 1797 petition is signed by Hamilton as society president. *UCG*, 8 March 1797.

104 Janet Carnochan noted that the NAS continued in some form until 1807. *History of Niagara*, 268. That information seems to have been lost upon later historians. See Jones, *History of Agriculture*, 158. Scott, *Fair Share*, 18; Heaman, *Inglorious Arts of Peace*, 35.

**2 Imperial Defence, Agricultural Improvement, and the Upper Canada Agricultural and Commercial Society, 1801–1815**

1 *UCG*, 15 November 1806.
2 Walton, "An End to All Order,"171–2. Also see "Note D. – Political State of Upper Canada in 1806," *Report on Canadian Archives for 1892*, 32–135. Firth, *Town of York, 1793–1815*, vol. 1, lxvii–lxviii; Craig, *Upper Canada*, 60–4; Errington, *The Lion, the Eagle*, 48–51; McKenna, *A Life of Propriety*, 66–9; Patterson, "Thorpe, Robert."
3 During the 1790s, the price of Russian hemp increased to £61 from £32 per ton, causing Britain to spend upwards of £1,000,000 to fulfill its annual requirements. Crosby, Jr., *America, Russia, Hemp and Napoleon*, 8, 84; Macdonald, "Hemp and Imperial Defence," 388; Graham, *British Policy and Canada 1774–1791*, 99–115; LAC, MG11, CO42, v.120, "Sir Joseph Banks to Lord Glenbervie," Soho Square, 30 July 1802.
4 Crosby, Jr., 84–5. The sizeable Royal Navy required a constant and considerable supply of hemp to manufacture the cordage needed to rig the fleet, especially during times of war when ships required repair and new ships were built to replace the lost. Annual hemp requirements also included the cordage that Britain manufactured for her colonial fleets, stored at various dockyards throughout her colonies. A Navy Board report from 1801 noted that its annual consumption of hemp was some thirteen thousand tons valued at several hundred thousand pounds sterling. See Morriss, *The Royal Dockyards during the Revolutionary and Napoleonic Wars*; and Morriss, "Quantities and Dimensions of Standing and Running Rigging Required by a 74-gun warship in 1794,"197–8.
5 TRL, B 15, David W. Smith Papers, Papers Relative to the Culture of Hemp in Upper Canada, "Duke of Portland to Lieutenant General Hunter," Whitehall, 18 December 1800.
6 Concerning Russian sources for hemp, see Morris, "Naval Cordage Procurement in Early Modern Europe," 86; and Crosby, Jr., chapter 3.
7 *UCG*, 16 May 1801. For more details about this society and the efforts to cultivate hemp in Upper Canada, see Fair, "A Most Favourable Soil and Climate," 41–61. Prior to my research, the crop's cultivation in Upper Canada had been little analysed. See Macdonald, "Hemp and Imperial Defence," 385–98. Fowke, *Canadian Agricultural Policy*, 73–7, and Jones, *History of Agriculture in Ontario*, 43–4.

8 Upper Canada, House of Assembly, *Journals*, 29, 30 May 1801; and Upper Canada, Legislative Council, *Journals*, 2 June 1801.
9 LAC, MG11, CO42, v.327, Upper Canada, *Executive Council Minutes*, 16, 18 June 1801. The Executive Council determined that a ton of Upper Canadian hemp sold in Britain could receive a price no less than £50.
10 The cost of this labour-intensive work was estimated at more than £4 per sixty pounds – the amount of hemp that could be processed manually by one individual in a single day. Galt, "A Statistical Account of Upper Canada," 6.
11 Upper Canada, House of Assembly, *Journals*, 3–6 July 1801; Legislative Council, *Journals*, 6 July 1801; and LAC, MG11, CO42, v. 327, "Hunter to Portland (No. 30)," York, 16 July 1801.
12 Smith would leave Upper Canada for England in 1802. TRL, B 15, David W. Smith Papers, Papers Relative to the Culture of Hemp in Upper Canada, "Peter Hunter to John McGill and David William Smith," York, 28 July 1801; "To the Public," York, 3 August 1801; "James Green to John McGill and D.W. Smith," 6 August 1801; and *UCG*, 1 August 1801; 22 August 1801.
13 The RSA offered a gold medal or 50 guineas to the farmer who grew the largest acreage of hemp (at least ten acres); a silver medal or 25 guineas to the farmer who grew the second-greatest acreage (at least five acres); and a silver medal to the farmer who grew the third-greatest acreage (at least one acre). RSA, *Transactions* 19 (1801): viii, 59–61; TRL: 1801. Society of arts. Account of certain premiums offered in 1801 by the Society instituted at London for the encouragement of Arts, Manufactures and Commerce, to encourage the cultivation of hemp in Upper and Lower Canada, Adelphi, London, April 9, 1801; and *UCG*, 22 August 1801.
14 No copy of the pamphlet appears to have survived. *UCG*, 29 August 1801; 17 September 1801.
15 LAC, MG11, CO42, v.327, "Lord Hobart to Hunter (No. 1)," Downing Street, 3 September 1801; "Lord Hobart to Hunter (No. 2)," Downing Street, 13 October 1801.
16 RSA, *Transactions*, 21(1803): 442–62. Also see *UCG*, 22 August 1801, 2; RSA, *Transactions*, 20 (1802): vii, 62–4; *UCG*, 22 January 1803, Supplement.
17 LAC, MG11, CO42, v.120, "Sir Joseph Banks to Lord Glenbervie," Soho Square, 30 July 1802.
18 LAC, RG5 A1, Upper Canada Sundries, "Sheffield to Hunter," 30 June 1803, 973–4; Young, *Annals of Agriculture* (1804), 390; Board of Agriculture (Great Britain), *Premiums offered by the Board of Agriculture, 1804*, 11 and *Premiums offered by the Board of Agriculture, 1808*, XIX.
19 Upper Canada, House of Assembly, *Journals*, 8–28 February 1804; 1–9 March 1804; Upper Canada, Legislative Council, *Journals*, 11–29 February

1804; 1–9 March 1804; *Statutes*, 1804, 44 Geo. 3, c. 9; *UCG*, 24 March 1804, 1–2; Upper Canada, House of Assembly, *Journals*, "Report of Select Committee on Provincial Public Accounts," 26 February 1806; LAC, RG1 E3, Upper Canada, State Submissions to Executive Council, v.69, "James Green to the Executive Council," 22 March 1804, 25–9; RG1 E1 Upper Canada State Book C, Executive Council Minutes, 22 March 1804; and *UCG*, 24 March 1804. During the 1803 session, the House of Assembly had presented another address to Hunter but drafted it too late for consideration before the prorogation of the legislature. Upper Canada, House of Assembly, *Journals*, 3, 22 February, 4, 5 March 1803; and Upper Canada, Legislative Council, *Journals*, 5 March 1803.

20 Upper Canada, House of Assembly, *Journals*, 4–27 February, 2 March 1805; Upper Canada, Legislative Council, *Journals*, 26–27 February, 2 March 1805; and Upper Canada, *Statutes*, 1805, 45 Geo. 3, c. 10.

21 Taylor, *Remarks on the Culture and Preparation of Hemp in Canada*, [1805] CIHM 53716 and 20892. The CIHM copies of Taylor's pamphlet do not contain the engravings. They can be found in LAC, MG11, CO42, v.129, "Taylor, *Remarks*," Adelphi, London, 19 August 1805, enclosed in "Stephen Cottrell to Edward Cooke," 2 September 1805.

22 In the months that followed, the Navy Board appointed Isaac W. Clarke, the deputy commissary at Montreal, as hemp agent in the Canadas. Clarke had won the RSA's gold medal for his sample of hemp sent to London in 1802. Lower Canada provided its own legislative supports for hemp cultivation. RSA, *Transactions* 22 (1804): 345–52. LAC, MG11, CO42, v.131 "Stephen Cottrell to Sir George Shee," Whitehall, April 15, 1808; "George Harrison to Sir George Shee," Treasury Chambers, 2 June 1806; MG11, CO42, v.130, "William Windham to Thomas Dunn," Downing Street, 5 June 1806; "Thomas Dunn to William Windham (No. 12)," Québec, 22 August 1806.

23 LAC, MG11, CO42, v.131, "J. King to George Shee," 10 April 1806; v.130, "Thomas Dunn to William Windham," Québec, 30 May 1806; v.130, "Windham to Dunn (No. 5)," 6 August 1806; v.130, "Windham to Dunn," Downing St., 5 June 1806; and v.130, "Windham to Francis Gore (No. 3)," Downing St., 5 June 1806.

24 Patterson, "Thorpe, Robert."

25 "Thorpe to Edward Cooke," 24 January 1806," in "Note D," *Report on Canadian Archives for 1892*, 39.

26 Patterson, "Thorpe, Robert."

27 "Thorpe to Edward Cooke," 24 January 1806," in "Note D," *Report on Canadian Archives for 1892*, 39.

28 *UCG*, 25 January 1806, Supplement; LAC, MG11, CO42, v.129, "Stephen Cottrell to Edward Cooke," Whitehall, 2 September 1805; v.340,

"Alexander Grant to Lord Castlereagh," York, 23 December 1805; LAC, RG5 A1, UCS, "Lord Castlereagh to Peter Hunter (No. 2)," Downing Street, 5 September 1805 and "William Stanton to John McGill," York, 17 January 1806; "William Stanton to John Bennett," York, 30 December 1805; "R. Hamilton to W. Stanton," Queenston, 22 January 1806; Fleming, *Upper Canadian Imprints, 1801–1841*, 13–14.

29 "Judge Thorpe to Lord Castlereagh, March 4, 1806," and "Proceedings of the Upper Canada Agricultural and Commercial Society," in "Note D," *Report on Canadian Archives for 1892*, 40, 41–3; *UCG*, 15 February 1806. For notices of meetings of this society see, *UCG*, 15 February 1806, 5 April 1806, 10 January 1807, 31 January 1807, 23 December 1807, 30 December 1807, 12 February 1808.

30 "Judge Thorpe to Lord Castlereagh, York, March 4, 1806" and "Judge Thorpe to Edward Cooke, March 5, 1806" in "Note D," *Report on Canadian Archives for 1892*, 40, 44.

31 LAC, MG11 CO42, v.342, "Thorpe to Castlereagh," York, 2 May 1806.

32 LAC, MG11 CO42, v.342, "Thorpe to Castlereagh," York, 2 May 1806.

33 [John Cameron], *The Upper Canada Almanac for the Year of Our Lord 1810*, 3; "Conversation between Judge Thorpe and Lieut. Governor Gore, October 31, 1806" and "Judge Thorpe to Sir George Shee, York, December 1, 1806," in "Note D," *Report on Canadian Archives for 1892*, 69–71, 57. For more about the ropewalk, see "A Most Favourable Soil and Climate."

34 "Judge Thorpe to Lord Castlereagh" and "Proceedings of the Upper Canada Agricultural and Commercial Society." in "Note D," *Report on Canadian Archives for 1892*, 40, 41–2.

35 "The Burlington Board of Agriculture," formed 1 May 1806, may have been another branch of the UCACS, but the sole document noting its existence offers no direct evidence of the link. Robertson, "The First Agricultural Society in the Limits of Wentworth–1806," 93–5; Jones, *History of Agriculture*, 158. LAC, MG 24 E1, William Hamilton Merritt Papers, vol.1, "Gentlemen Proposed for Vice-Presidents and Directors of the Niagara Branch of the Upper Canada Agricultural Society."

36 The UCACS membership selection process upheld the gentlemenly tradition of blackballing wherein candidates for admission were chosen with white and black beans as ballots. If a proposed candidate received one black bean in three, he was considered excluded. Those accepted for membership had to pay one dollar to join and a two-dollar annual membership fee "No. 9. – Proceedings of the Upper Canada Agricultural and Commercial Society," in "Note D," *Report on Canadian Archives for 1892*, 41–3; [Cameron], *The Upper Canada Almanac 1810*, 3.

37 Patterson, "Weekes, William."

38 Jones, "Wyatt, Charles Burton."

39 Mealing, "Small, John."
40 McKenna, *A Life of Propriety*, 61–2, 66–8.
41 Upon the death of William Weeks, Hamilton noted to the lieutenant governor, the result of the duel was "the death of a Member of the Society." It is not clear if he meant Thorpe's agricultural society. Cruikshank ed., "Records of Niagara in the Days of Commodore Grant and Lieut.-Governor Gore 1805–1811," 33, 37; LAC, MG 24 E1, William Hamilton Merritt Papers, v.1, "Gentlemen Proposed for Vice-Presidents and Directors of the Niagara Branch of the Upper Canada Agricultural Society."
42 *UCG*, 15 November 1806.
43 "Judge Thorpe to Sir George Shee, December 1, 1806" in "Note D," *Report on Canadian Archives for 1892*, 57.
44 "Judge Thorpe to Sir George Shee, n.d." in "Note D," *Report on Canadian Archives for 1892*, 90.
45 Walton concludes that what brought Thorpe's Upper Canadian career to a swift end was a fear of "a domino-like collapse of the entire established order." Walton, "An End to All Order," 171–2, 174.
46 LAC, RG1 E3, Upper Canada, State Submissions to Executive Council, v.88; RG1 E1, Upper Canada State Book D, Executive Council Minutes, 4 July 1807.
47 Carnochan, *History of Niagara*, 268, *UCG*, 13 June 1807.
48 *UCG*, 26 September 1807, 30 December 1807. Years later, William Warren Baldwin, a friend of Thorpe's while the latter was in the colony, noted in a letter to Wyatt that he had over £9 of the former society's money in his hands. All Wyatt had to do was request it and Baldwin would send it to him. J. McE. Murray, "A Recovered Letter: W.W. Baldwin to C.B. Wyatt, 6th April, 1813," *OHSPR* 35(1943): 54.
49 *UCG*, 30 December 1807, 6 January 1808.
50 [Cameron], *Upper Canada Almanac for 1810*, 3–9.
51 Members agreed that a new society should be formed with the object of the general promotion of agriculture, "and in particular the Cultivation of Hemp," but there is no evidence that the meeting they scheduled for this purpose was ever held. *UCG*, 30 December 1807, 12 February 1808.
52 Jackson, *A View of the Political Situation of the Province of Upper Canada*, 25–6, 28–9, Appendices 6, 19, 20; Fraser, "Jackson, John Mills."
53 For this final record of NAS activity, see the letter to the editor by "Falkland" in *KG*, 10 December 1810. "Falkland" may have been Richard Cartwright. See McNairn, *The Capacity to Judge*, 180–1.
54 [Richard Cartwright]. *Letters from an American Loyalist in Upper Canada to his friend in England*, [1810], 90–1. The author of these letters was once considered to be William Dummer Powell but George Rawlyk and Janice

Potter explain in their biography of Cartwright the proof that Cartwright wrote them. See Rawlyk and Potter, "Cartwright, Richard."

55 LAC, RG5 A1, Upper Canada Sundries, "J. Baby to Robert Barrie," York, July 1822, 29447–9; Robert Garcia, "Fort Amherstburg in the War of 1812," http://www.warof1812.ca/fortambg.htm.

56 Gourlay, *Statistical Account of Upper Canada*, 361–2.

**3 Robert Gourlay, the Upper Canada Agricultural Society, and Independents, 1815–1830**

1 "Strachan to Rev. James Brown," York, 1 December 1818 in Spragge, *The John Strachan letter book, 1812–1834*, 183–5. Also see Strachan's pride in his students expressed in "Strachan to Lieutenant Colonel James Harvey," York, 27 July 1818, 171.

2 Milani, *Robert Gourlay, Gadfly*.

3 Craig, *Upper Canada*, 100–2.

4 Robert Gourlay himself published an exhaustive account of his survey, meetings, and trial in three volumes. Gourlay, *General Introduction to Statistical Account of Upper Canada*; Gourlay, *Statistical Account of Upper Canada.*

5 Milani, 20, 71.

6 Milani, 12–14.

7 Milani, 11–12.

8 Milani, 15–60.

9 Milani, 50, 58–9; Gourlay, *Statistical Account of Upper Canada*. vol. 2, vi-viii. Also see S.R. Mealing's introduction to the Carleton Library series edition of Gourlay, *Statistical Account of Upper Canada*, 8–10.

10 Nelles, "Loyalism and Local Power: The District of Niagara 1792–1837," 99–114; Milani, 87–90.

11 Wise, "Conservatism and Political Development: The Canadian Case," 191.

12 Fraser, "'All the privileges which Englishmen possess': Order, Rights, and Constitutionalism in Upper Canada," xxx, xxxii, xxxvi.

13 Wise, "Gourlay, Robert Fleming."

14 Gourlay, *Statistical Account*, vol. 1, 274.

15 As Wise has succinctly argued, Gourlay's challenge to the government witnessed the reemergence of "his more accustomed guise as antiauthoritarian." Wise, "Gourlay, Robert Fleming."

16 Craig, Upper Canada, 92–3; Errington, *The Lion, the Eagle*, 109–10; Wise, "Gourlay, Robert Fleming."

17 Wise, "Gourlay, Robert Fleming."

18 Milani, Robert Gourley, Gadfly, 33.

19 Bowsfield, "Maitland, Sir Peregrine."

20 Jane Errington notes that Gourlay was "conspiring against constitutional authority, encouraging the tyranny of the people, and, at a very personal

level, threatening the continued prominence of many of the tory elite." Errington, *The Lion, the Eagle*, 108.

21 Upper Canada, *Statutes*, 1818, 58 Geo. 3, c. 9. This act was passed on 31 October 1818 and given royal assent on 27 November 1818.

22 Milani, *Robert Gourley, Gadfly*, 191, 198–215.

23 *UCG*, 10 December 1818; 17 December 1818

24 *UCG*, 17 December 1818. Lois Darroch Milani suggests that the legal scheming began on 10 November 1818. Both agricultural society meetings were held before Gourlay's arrest on 19 December 1818. Milani, *Robert Gourley, Gadfly*, 186. Despite having been officially constituted as "The Agricultural Society of Upper Canada," it was always referred to as the "Upper Canada Agricultural Society." See *UCG*, 17 December 1818.

25 While there is no record of those gentlemen attending the initial December meetings, it can be deduced that the list of the society's officers elected at the 20 January 1819 meeting indicates which individuals established this society. *UCG*, 21 January 1819. For a complete list of officers from 1819–1820 see Fair, "Gentlemen Farmers," Appendix 5, 323. In 1820, the only significant changes to the society were the election of James Baby as president, and the addition of John Beverley Robinson, the attorney general of the province, as a vice-president. A brother of Peter, John had studied at Strachan's school in Cornwall and had articled in law with D'Arcy Boulton Sr. *UCG*, 9 March 1820. Saunders, "Robinson, Sir John Beverley."

26 Gourlay, *Statistical Account of Upper Canada*, vol. 2, vi-viii.

27 Cameron, "Robinson, Peter."

28 Craig, "Wells, Joseph."

29 Pemberton, "Sherwood, Levius Peters."

30 *UCG*, 17 December 1818.

31 Mealing, "Powell, William Dummer"; Lownsbrough, *The Privileged Few*; Armstrong, "Crookshank, George"; Clarke, "Baby, James"; Craig, "Wells, Joseph."

32 It followed the structure and timing of the plan set by Thorpe's UCACS in 1806. *UCG*, 17 December 1818.

33 *UCG*, 18 February 1819.

34 Bowsfield, "Maitland, Sir Peregrine."

35 *UCG*, 18 February 1819.

36 *UCG*, 18 February 1819. Britain's Board of Agriculture had employed this same argument in the 1790s. That board – firmly ensconced in the urban quarters of London – was, at the very time of the creation of the York agricultural society, witnessing a waning of support, particularly the financial support offered by Parliament. In fact, the British Board of Agriculture would last only one year beyond the expiration of the UCAS in 1821. Mitchison, "The Old Board of Agriculture," 66–7.

37 *Report of the Loyal and Patriotic Society of Upper Canada*, 12–13. Sheppard, *Plunder, Profit, and Paroles*, 66–7.
38 Wise, "Conservatism," 196–7.
39 UCG, 29 April 1819; KC, 28 May 1819; Ennals, "Burnham, Zacheus"; Senior, "Boulton, George Strange"; Pioneer Life on the Bay of Quinte, 136–9.
40 The society elected to the position of vice-president, two businessmen and landholders of the district, Daniel Burritt Jr., owner of a sawmill at Burritt's Rapids, and John Kilborn, a merchant and lumberman from Elizabethtown Township. The treasurer and secretary were two unidentified individuals, Roderick Easton and James Hall. KC, 9 April; 28 May 1819; UCG 10 June 1819. McIlwraith, "Jones, Charles."
41 *KC*, 2 July; 17 September; 15 October 1819; UCG, 30 September; 4 November 1819; 25 January 1821.
42 *KC*, 29 January 1819.
43 Osborne and Swainson, *Kingston Building on the Past*, 47–54; Firth, *Town of York 1815–1834*, xxxiv, lxxxii; Bindon, "Kingston: A Social History 1785–1830," 84.
44 AO, F32 Macaulay Family Fonds, "John Strachan to John Macaulay, York, 11 March 1819."
45 Lower Canada, *Journals of the House of Assembly*, 7, 20, 21, 31 January, 3, 9, 10, 14, 18, 25, 30 March, 1 April 1818. Lower Canada, 1818, 58 Geo. 3, c. 6, and 1821, Geo. 4, c. 5. *UCG*, 7 January, 4 February 1819.
46 *UCG*, 21 January 1819. Burroughs, "Ramsay, George, 9th Earl of Dalhousie." *Reports of the Provincial Agricultural Society of Nova Scotia* (1821); *An Abstract of the Proceedings which occurred at the two meetings of the Provincial Agricultural Society during the Session of 1823.*
47 Young, *The Letters of Agricola on the Principles of Vegetation and Tillage*; Martell, "The Achievements of Agricola and the Agricultural Societies 1818–25,"; MacLean, "Young, John (1773–1837)." *KC*, 3 December 1819; 7 January; 21 April 1820. Also see the 8 March 1825 letter from A.J. Christie to John Nielson at Quebec noting that most subscribers for *The Letters of Agricola* "were from Upper Canada chiefly from the vicinity of Kingston." AO, MU 2104, Miscellaneous 1825 #2, "A.J. Christie to Messrs. Neilson and Corran, Montreal, 8 March 1825."
48 *UCG* 11 November 1819. Flindall, *J.M. Flindall: The Uncommon Man*, 128, 140–4.
49 *UCG*, 21 January; 17 June; 24 June 1819.
50 The branch society in the Newcastle District made no mention of a cattle show in its founding resolutions of March 1819 (although, admittedly, further information about this branch has not been found), while the Agricultural Society of the Johnstown District decided at its May 1819 meeting to host two cattle shows per year, on the first Monday of May

and October. Its first show was held at Brockville in on 4 October of that year. At its February 1819 meeting, the ASMD set similar dates for its cattle shows and would host its first exhibition on 18 October. *UCG*, 29 April, 10 June, 4 March 1819. The Williamstown Fair in eastern Ontario claims to have had its founding in 1812, but there has been a conflation of the market fair established by Letters Patent in 1812 and the cattle show of a later agricultural society held in the same town. See my comments on this general conflation in chapter one. Williamstown Fair, History, http://williamstownfair.ca/about-us/history/

51 *UCG*, 24 June 1819.

52 Smith, "Russell, Francis, fifth duke of Bedford (1765–1802)."

53 Hudson, *Patriotism with Profit*, 53; Powell, *History of the Smithfield Club from 1798–1900*; Addison, 85–6.

54 Ritvo, *The Animal Estate*, 52, 45–54.

55 *UCG*, 24 June 1819. There was no indication of how many competitors there were at this event. A list of prizewinners ranged in location from Etobicoke, Pickering, and York.

56 *UCG*, 24 June 1819. At its annual general meeting in March 1820, new officers were elected for the coming year. The province's inspector general, Jacques Duperron Baby, graduated from his vice-president role to president of the UCAS, while Upper Canada's attorney general, John Beverley Robinson, was added to the executive, filling the vice-president position Baby had vacated. Others retained the positions to which they had been originally elected. *UCG*, 9 March 1820.

57 *UCG*, 16 March, 18 May 1820.

58 Errington, "Markland, Thomas."

59 *KC*, 19 February 1819; *UCG*, 4 March 1819.

60 Thomas Shaw was also a member, but his occupation and place of residence could not be identified. *KC*, 19 February 1819; *UCG*, 4 March 1819.

61 At this time, the County of Lennox and Addington was still often considered two counties. Thus, Macpherson likely represented Lennox, with Fraser representing Addington. *KC*, 19 February 1819; *UCG*, 4 March 1819.

62 It is unclear whether Macaulay was a member of the society, like his business partner Alexander Pringle.

63 Kathryn Bindon notes that the *Kingston Chronicle* was "deeply involved in the problem of commercial improvement throughout the decade of the 1820s." Bindon, "Kingston: A Social History," 150. Milani, 143.

64 *KC*, 2 July 1819.

65 *KC*, 9 July 1819.

66 *KC*, 30 July 1819; 8 October 1819.

67 *KC*, 22 October and 29 October 1819.

68 At 9s per membership, 110 members might have signed up, although the number may be smaller if several gentlemen had offered donations in addition to their membership fee. *KC*, 5 May 1820.
69 *KC*, 7 January 1820; 11 February 1820.
70 *KC*, 7 April 1820; 5 May 1820.
71 *KC*, 5 May 1820; Fraser, "Washburn, Ebenezer"; Fraser, "Dorland, Thomas"; *KC*, 2 June 1820; 9 June 1820. For a list of subscribers to the Hasting County Agricultural Society see the Lennox and Addington County Archives, William Bell Papers, 286–9; Shipley, "Meyers, John Walden."
72 *KC*, 30 June 1820; 27 October 1820.
73 George H. Markland likely withdrew from the society's executive because a few weeks after the ASMD's annual general meeting in April 1820, he was elected to be the MHA for the Kingston riding. Then, in mid-July, the thirty-year-old Markland received his appointment to Upper Canada's Legislative Council, thanks largely to the influence of his former teacher, Strachan. Burns, "Markland, George Herchmer."
74 *KC*, 11 May 1821; 18 May 1821; 8 June 1821.
75 *KC*, 15 June 1821; 6 July 1821.
76 There is no record of county societies ever being formed in either Lennox or Prince Edward counties. *KC*, 2 June, 9 June 1820 (Hastings); 27 July 1821 (Addington).
77 *KC*, 18 October 1822; Burns, "Markland, George Herchmer."
78 *UCH*, 18 October 1825. Evidence of activity between 1822 and 1825 is found in *KC*, 11 October 1822; 18 October 1822; 6 June 1823; 20 June 1823; 17 October 1823; 12 July 1825.
79 *KC*, 19 February 1819; *UCG*, 4 March 1819. *KC*, 29 October 1819.
80 Errington, *The Lion, the Eagle*, 36, 40–1, 124–5.
81 True, *A History of Agricultural Education in the United States 1785–1925*, 12–13. Lord, "Elkanah Watson and New York's First County Fair," 442–7. For further information on Elkanah Watson, see Van Wagenen, "Elkanah Watson–A Man of Affairs," 404–12; Neely, *The Agricultural Fair*, 51–78; United States, Department of Agriculture, Bureau of Animal Industry, *Special Report of the History and Present Condition of the Sheep Industry of the United States*, 218–20. Watson, *History of the rise, progress, and existing condition of the western canals in the state of New York* (1820).
82 *KC*, 8 October 1819.
83 *KC*, 30 July, 8 October, 22 October, 29 October 1819.
84 *KC*, 2 July, 15 October 1819.
85 *UCG*, 14 October 1826. Stanton had lived in Kingston and had served as Treasurer of the FCAS prior to his appointment as the King's Printer and relocation to York. Neary, "Stanton, Robert."

86 See Mitchison, "The Old Board of Agriculture," 64–5, and Powell, *History of the Smithfield Club*, 4–5. Sarah Wilmot has calculated at least 136 agricultural societies were operating in Scotland as of 1834, and about 90 in England by 1835. The vast number of these likely had roots reaching back into the 1820s and earlier. Wilmot, 9.
87 White, "Speed the Plough," 21–2. Wynn, "Exciting a Spirit of Emulation," 10–11. Lower Canada, *Statutes*, 1825, 5 Geo. 4, c. 13.
88 Hedrick, *A History of Agriculture in the State of New York*, 122. Plakins Thornton, *Cultivating Gentlemen*, 85–105.
89 *UCG*, 20 October 1827.
90 Galt, *The Autobiography of John Galt*, vol. 1, 97–8, vol. 2, 37; Hall and Whistler, "Galt, John"; Galt, "A Statistical Account of Upper Canada"; Burroughs, "Ramsay, George, 9th Earl of Dalhousie.
91 *GG*, 18 August 1827; 23 August 1828; *CA*, 30 August 1827; Galt, *Autobiography*, vol. 2, 97; Barker, "Thomas Coke"; Hudson and Luckhurst, *The Royal Society of Arts 1754–1954*, 69, 78.
92 Raible, *Muddy York Mud: Scandal and Scurrility in Upper Canada*, 139–47. Also see Hall and Whistler, "Galt, John."
93 *GG*, 18 August 1827; *CA*, 30 August 1827. See Galt, *Autobiography*, vol. 2, Appendix No. 1. For a list of the gentlemen who attached their names to the resolutions see, *GG*, 23 August 1828.
94 Wilson, "Dickson, William."
95 Weaver, "Hamilton, George (1788–1836)"; Kelsay, "Tekarihogen (1794–1832)"; Hall and Whistler, "Galt, John."
96 Unfortunately, the *Gore Gazette*, the main source of information regarding this society, ceased publication in June 1829. *GG*, 9 August 1828; 23 August 1828; LAC, RG5 A1, Upper Canada Sundries, "John Galt, President of the Agricultural Society of Upper Canada to Sir John Colborne, 24 November 1828," 50501–2.
97 LAC, RG5 A1, Upper Canada Sundries, "Henry Lelièvre to Major Hillier, Perth, 15 March 1828" and society resolutions enclosed, 48377–81. Brown, *Lanark Legacy*, 58, 273.
98 AO, F1078, Riddell Family fonds; "Historical Sketch of Northumberland Agricultural Society, [January 1892]"; AO, F535, John Steele fonds, "Report of elections of officers of the County of Agricultural Society, 30 July 1829"; LAC, RG5 A1, Upper Canada Sundries, "John Steele to Z. Mudge, Colborne, 1 June 1830," 56560–7; "Charles Fothergill to Z. Mudge, Ontario Cottage near Port Hope, 18 June 1830," 56790–56793; *CA*, 8 June 1830; *UCH*, 17 June 1828; 24 June 1828; 12 August 1829; 14 November 1829; 25 November 1829; 2 June 1830; 20 October 1830; *KC* 16 October 1830.
99 Dalton, *Thomas Dalton and the "pretended" Bank*, 10, 15–17, 23, 40, 77–9, 185.
100 Ennals, "Spilsbury, Francis Brockell."
101 *UCH*, 12 August 1829; 14 November 1829; 25 November 1829.

**4 Agricultural Societies as State Formation, 1821–1851**

1 Upper Canada, House of Assembly, *Journals*, 8 and 12 January 1830.
2 Romney, "A Man Out of Place," 516–7. For further details on this enigmatic gentleman see Baillie, Jr., "Charles Fothergill 1782–1840," 376–96; Romney, "Fothergill, Charles"; Romney, "A Conservative Reformer in Upper Canada," 42–61. In her assessment of the legislation, Elsbeth Heaman calls Romney's assessment as "overly harsh", though categorizes Fothergill's motive as patronage to secure votes from his riding's farmers. Heaman, *Inglorious Arts*, 45.
3 Fraser, "'Like Eden in Her Summer Dress,'" chapters 3 and 4. Upper Canada, House of Assembly, *Journals*, 5 February–14 April 1821; TRL, *First Report of the Select Committee appointed to take into Consideration the Internal Resources of the Province of Upper Canada in its agriculture and exports and the practicability and means of enlarging them, also to consider the expediency of granting encouragement to domestic manufactures* (1821).
4 Fraser, "Nichol, Robert"; Upper Canada, House of Assembly, *Journals*, 5 February 1821.
5 Upper Canada, House of Assembly, *Journals*, 9, 10, 12, 13, and 14 April 1821.
6 Upper Canada, House of Assembly, *Journals*, 31 March 1821.
7 Upper Canada, House of Assembly, *Journals*, 21 November 1821.
8 Upper Canada, House of Assembly, *Journals*, 21, 26, 30 November 1821; *KC*, Extra, 24 November 1821; 30 November 1821; 14 December 1821; 21 December 1821.
9 Upper Canada, House of Assembly, *Journals*, 9, 14 January 1822; Legislative Council, *Journals*, 14, 15, 17 January 1822. Although Hagerman's committee offered no support for tobacco, several farmers in the Western District petitioned the House of Assembly in early December 1823 requesting that it issue an appeal to the Imperial Parliament for a reduction in the duty on tobacco exported from Upper Canada to the England market. Apparently, by 1822, enough tobacco was being grown in the Western District to consider exports. Following joint committee meetings of the House of Assembly and legislative councillors, six resolutions were agreed on, and in an early 1824, Lieutenant Governor Maitland forwarded an address to British officials requesting a lowering of the excise duties levied against Upper Canadian tobacco. Although the Imperial Parliament granted the request, Fred C. Hamil suggests that Upper Canadian tobacco made it no further than Lower Canada, resulting in the flooding of that colony's tobacco market with an increasingly inferior product. The tobacco that did make it to the British market in 1827 and 1829 was of inferior quality and sold at an unprofitable rate. Hamil, *The Valley of the Lower*

*Thames, 1640 to 1850*, 123–7; Upper Canada, House of Assembly, *Journals*, 31 March, 30 November 1821; 23, 26 December 1823; 3, 6, 14, 15, 17, 19 January 1824. Legislative Council, *Journals*, 30, 31 December 1823; 2, 6, 14, 15, 17, 19 January 1824; LAC, MG11, CO 42, v. 372, "To the King's Most Excellent Majesty ..., 17 January 1824," in "Maitland to Bathurst, No. 115, 5 July1824"; v. 374, "John Galt to Bathurst, Downing Street, 13 March 1824"; v. 383, "P. Maitland to W. Huskisson, M.P., No, 12, Miscellaneous, York, 4 March 1828"; v. 474, "Macaulay's General Report to Governor Sir George Arthur upon Canada, 2 March, 1841"; *KC*, 14 December 1821, 21 December 1821; *UCH*, 18 July 1826.

10 Upper Canada, House of Assembly, *Journals*, 27 February; 4, 6, 11, 14, 16, 18 March; 1 April 1816. *Statutes*, 1816, 56 Geo. 3, c. 35. *Journals*, 27 February; 3, 4, 5, 9 March; 1 April 1818. *Statutes*, 1818, 58 Geo. 3, c. 7.

11 Upper Canada, *Statutes*, 1822, 2 Geo. 4, c. 17. For more on this subject see, Fair, "The Kingston Royal Naval Dockyard and Canadian hemp supply, 1822–34," 69–82.

12 *KC*, 14 December 1821; 21 December 1821.

13 In examining cultivated land rates and wheat production, McCalla suggests that, certainly by the 1860s, many farmers used their properties to raise cattle. Kelly suggests that, by the late 1850s, the improvers' message of mixed farming was being practiced. McCalla, *Planting the Province*, Chapter 5, 67–91; Kelly, "Notes on a Type of Mixed Farming Practiced in Ontario During the Early Nineteenth Century," 205–19.

14 Howison, *Sketches of Upper Canada*, 136–8; Goldie, *Diary of a Journey through Upper Canada and some of the New England States – 1819*, 21; *KC*, 24 May 1822. See Macaulay's earlier statement of this problem in *KC*, 19 September 1819.

15 Bell, *Hints to Emigrants*, 161–2; Talbot, *Five Years Residence in the Canadas*, vol. 1, 157–8; Fothergill, *The York Almanac and Royal Calendar of Upper Canada for the Year 1825*, 53–4.

16 *CA*, 18 May 1824.

17 Armstrong and Stagg, "Mackenzie, William Lyon."

18 Armstrong and Stagg, "Mackenzie, William Lyon."

19 *CA*, 18 May 1824.

20 See Marti, "Early Agricultural Societies in New York," 322; Marti, "Agrarian Thought and Agricultural Progress," 131–41.

21 *CA*, 18 May 1824.

22 *CA*, 18 May 1824.

23 Hofstadter, *The Age of Reform*, 1–45; Abbot, "The Agricultural Press Views The Yeoman," 35; Blanton, "The Agrarian Myth in Eighteenth and Nineteenth-Century American Magazines," 9–10; Marti, "In Praise of Farming," 351–75; Thorton, *Cultivating Gentlemen*, 2–3, 56, 68–77.

24 *CA*, 18 May 1824.
25 Fred Armstrong and Ronald Stagg argue that Mackenzie was really, "in essence, an early Ontario nationalist." Armstrong and Stagg, "Mackenzie, William Lyon."
26 Romney, "Fothergill, Charles."
27 Aileen Dunham argued in her 1926 *Political Unrest in Upper Canada* that the reformers may have dominated the tories in the House of Assembly by a thirty-five to fifteen margin, but they "proved that a reform assembly could accomplish less in the way of constructive legislation than a tory assembly." Dunham, *Political Unrest in Upper Canada 1815–1836*, 118, 136–7. More recent historians conform to G.M. Craig's 1963 assessment that between 1828 and 1830 the reformers "were able to harass and complain, but unable to produce a positive result against a powerful executive … Their best efforts were rejected or simply ignored by the upper chamber." Craig, *Upper Canada*, 195–6. Also see Errington, *The Lion, the Eagle, and Upper Canada*, 187–8; Mills, *Idea of Loyalty*, 72; Johnson, *Becoming Prominent*, 137–8.
28 Upper Canada, House of Assembly, *Journals*, 12 January 1830. For the reported minutes of this debate, see the *CA*, 21 January 1830.
29 Wynn, "Exciting a Spirit of Emulation," 11; White, "Speed the Plough," 21; Vass, "The Agricultural Societies of Prince Edward Island," 32–3; Lower Canada, *Statutes*, 1829 9 Geo. IV, c. 48; Hedrick, 122.
30 *CA*, 21 January 1830.
31 Upper Canada, House of Assembly, *Journals*, 12 January 1830.
32 Like Mackenzie's *Colonial Advocate*, Dalton's newspaper soon devoted less space to agricultural matters. He moved his business to York in December 1832, and after March 1833, *The Patriot* dropped *Farmer's Monitor* from its masthead.
33 Upper Canada, House of Assembly, *Journals*, 12 January, 17, 22, 24 February, 3 March 1830; Legislative Council, *Journals*, 25, 26, 27 February, 3 March 1830; Upper Canada, *Statutes*, 1830, 11 Geo. 4 c. 10.
34 Upper Canada, *Statutes*, 1830, 11 Geo. 4 c. 10; LAC, RG5 A1, Upper Canada Sundries, "Fothergill to Mudge, Port Hope, 18 June 1830," 56790–3.
35 Wilson, "Colborne, John, Baron Seaton."
36 *CA*, 9 December 1830; *UCH*, 15 December 1830; *KC*, 25 December 1830; Also see AO, F 380 Percy C. Band collection, "Suggestions Transmitted by His Excellency the Lieutenant Governor to the President of the Midland District Agricultural Society, [1830]"; Upper Canada, House of Assembly, *Journal*, "First Report of the Select Committee on Roads & Bridges," 11th Parliament, 1st Session, 1831, 1Wm IV, Appendix, 173–4. The MDAS discussed the matter in mid-February, after the committee had risen. Its lengthy report on roads, submitted to Colborne can be found in LAC, RG5

A1, Upper Canada Sundries, "Extract from the Minute Book of the General Board of the Midland District Agricultural Society, Bath, 11 February 1831," 59728–51. Later, the society published that report along with its constitution for circulation to the public. *KC*, 5 March 1831. In 1833, "Agricola" submitted some of this information to the editor of the *Courier of Upper Canada* because, in his view, the pamphlet contained much "sound practical knowledge in the mode of constructing roads" that its contents deserved much wider notice. *CUC*, 6 March 1833.

37 Craig, *Upper Canada*, 160; McCalla, *Planting the Province*, 167–8.

38 *UCG*, 6 June; 22 August 1833; *TP*, 28 June; 6, 10, 13 December 1833; *CF*, 5 September 1833; Upper Canada, House of Assembly, *Journals*, 4, 6 December 1833. As early as December 1831, newly appointed member of the Legislative Council, James Crooks, had announced his intentions to introduce a bill in the council regarding agricultural societies. No such bill was introduced. Upper Canada, Legislative Council, *Journals*, 13 December 1831.

39 Upper Canada, House of Assembly, *Journals*, 19 November 1833; 12 December 1833; *TP*, 22 November, 10 December 1833; House of Assembly, *Journals*, 26 November 1833, 9 January 1834.

40 McCalla, *Planting the Province*, 168.

41 Upper Canada, House of Assembly, *Journals*, 18 February 1835; "Report of the Select Committee, Appointed to examine and Report, what laws have expired and are about to expire," 6 April 1835, *Appendix to the Journal of the House of Assembly of Upper Canada*, vol. 2, no. 104, 18; House of Assembly, *Journals*, 7, 8, 9, 14 April 1835; *Statutes*, 1835, 5 Wm. 4, c. 11.

42 Cross, "Duncombe, Charles."

43 Upper Canada, House of Assembly, *Journals*, 19, 22, January and 23 March 1836. The committee members were John Gilchrist (Northumberland), Elias Moore (Middlesex), Caleb Hopkins (Halton), Denis Woolverton (Lincoln, First Riding), Peter Shaver (Dundas), John McIntosh (York, Fourth Riding), Samuel Lount (Simcoe), Charles Waters (Prescott), and Alexander Chisholm (Glengarry). Woolverton was a director of the NDAS and Samuel Lount was a vice president of the HDAS. Two other members had been appointed – John Roblin (Prince Edward) and Peter Perry (Lennox and Addington) – but their names are not attached to the report.

44 Upper Canada, House of Assembly, *Journals*, 23 March 1836

45 Armstrong, *Handbook of Upper Canadian Chronology*, 272.

46 Upper Canada, House of Assembly, *Journal*, 23 March 1836; Evans' *Treatise* was divided into five parts, the first being a history of agriculture, the second focused on the science of agriculture, the third and fourth on the cultivation of various crops, and the last part on the raising of cattle.

Evans, *A Treatise on the Theory and Practice of Agriculture*; Robert, "Evans, William."

47 Upper Canada, House of Assembly, *Journals*, 23 March 1836. As the official organ of the NYSAS, *The Cultivator* "behaved as a lobby, almost as a political faction," argues Donald B. Marti. Marti, "Early Agricultural Societies in New York," 324–8. Hedrick, 120–1, 318–22. In 1835, the MDAS ordered twelve copies of *The Cultivator* to be purchased "in order to give the members of the Society information on the American system of Farming." *BW*, 19 May 1835.

48 Upper Canada, House of Assembly, *Journals*, 23 March and 13, 19 April 1836; Legislative Council, *Journals*, 14, 15, 16, 19 April 1836.

49 Craig, *Upper Canada*, 235–6; Dunham, *Political Unrest*, 170–2.

50 Upper Canada, House of Assembly, *Journals*, 9 November 1836.

51 Upper Canada, House of Assembly, *Journals*, 15, 18, 29 November 1836, 8, 9, 15 February 1837. The editor of the *Kingston Chronicle and Gazette* commended Marks for his "most creditable perseverance" in promoting this bill through its various stages of passage. *KCG*, 10 May 1837. Upper Canada, *Statutes*, 1837, 7 Wm. 4, c. 23. Preamble.

52 Upper Canada, *Statutes*, 1837, 7 Wm. 4, c. 23.

53 Noel, *Patron, Clients, Brokers*, 149.

54 The agricultural societies legislation was continued by "An Act to continue for a limited time period the certain Acts therein mentioned." Province of Canada, Legislative Assembly, *Journals*, 10, 11, 13, 15, 18 September 1841. Province of Canada, *Statutes*, 1841, 4&5 Vic, c. 23. For the legislation affecting Lower Canada see Province of Canada, Legislative Assembly, *Journals*, 21 January, 3, 17, 24, 27, 28 February, 5, 6, 7, March 1845. Province of Canada, *Statutes*, 1845, 8 Vic, c. 53. Province of Canada, *Journals*, 17, 20, 21, 23 February and 5, 6, 7, 25 March 1845.

55 LAC, RG 5, B 21, Upper Canada and Canada West: Civil Secretary, correspondence on Immigration, vol. 1, parts 1 and 2, 1840–44. *KCG*, 17 January 1846.

56 Careless, *Union of the Canadas*, 38. For a thorough study of the pre-Confederation administrative reforms see Hodgetts, *Pioneer Public Service*.

57 Despite Evans's success in publishing *A treatise on the theory and practice of agriculture* in 1835, followed by a *supplementary volume* the year after and *agricultural improvement by the education of those who are engaged as a profession* in 1837, an agricultural journal was a completely different sort of venture, and it failed after only two issues. See *SCJ*, 20 September 1838; Robert, "Evans, William."

58 *KCG*, 3 April 1841. This was a reprint of Evan's letter to the *Montreal Gazette* of 26 March 1841.

59 *CFM*, 16 August 1841. In the final issue of his paper, Garfield published a letter in support of his plan from Archibald MacDonald, President of the County of Russell Agricultural Society of the Ottawa District. *CFM*, 15 October 1841.

60 MacKenzie, "Edmundson, William Graham." During the last half of the 1830s, several unsuccessful attempts had been made to establish an agricultural periodical in Upper Canada. The following list of agricultural periodicals did not progress beyond a publication of their prospectus: *The Upper Canada Farmer* was to be published in February 1837 at the office of the *Cobourg Star*. See *CUC*, 24 December 1836. Subsequent failed attempts were made to establish this journal. See *KS*, 16 March 1837; *SCJ*, 26 April 1841. In 1837, John Smith announced in Toronto that he intended to launch *The British North American, Religious, Agricultural, Literary and Scientific Monthly Magazine*, See *Constitution*, 4 January 1837; *CA*, 4 January 1837. In 1839, J.H. Sears of St. Catharines printed at least one issue of his *Canadian Cultivator*. See *KCG*, 18 September 1839. For Sears's address to the public, see *SCJ*, 21 November 1839.

61 Initially formed as the England Agricultural Society, the institution received its Royal Charter from Queen Victoria in 1840. Goddard, *Harvests of Change*, 1. Powell, *History of the Smithfield Club*, 6–7, especially the note on p. 7. Watson, *The History of the Royal Society of England, 1838–1939*; Hudson, *Patriotism with Profit*, 57–9.

62 *BAC*, April 1842.

63 *BAC*, April, May, June 1842.

64 *BAC*, April, June 1842.

65 Perhaps this had to do with age and background. Evans was in his late fifties. He had been born in Ireland and was already a devoted agricultural improver before his departure for Lower Canada in 1819. In contrast, Edmundson was only in his late twenties, had likely been born in the province and had been raised near Newmarket, north of Toronto. Robert, "Evans, Robert"; MacKenzie, "Edmundson, William Graham."

66 *BAC*, April 1842; Marti, "Early Agricultural Societies," 324, 326; Hedrick, 120–2.

67 *BAC*, April 1843.

68 Jones, "Fergusson, Adam." Also see *UCH*, 9 October 1833; Board of Agriculture of Upper Canada, *The Canada Herd Book*, vol. 1, xiv.

69 *BAC*, May 1843.

70 See McCalla, "The Commercial Politics of the Toronto Board of Trade, 1850–1860," 51–67.

71 *BAC*, June 1843. This issue also featured a letter of support for a board from J.W. Rose of Williamsburg West.

72 *BAC*, April 1842.

73 Ian Radforth, "Sydenham and Utilitarian Reform," 82–5; Careless, *Union of the Canadas*, 53–4.
74 Radforth, "Sydenham and Utilitarian Reform," 84.
75 *BAC*, July 1843. For Edmundson's initial discussions of a provincial association see various articles and letters published in issues of the *BAC* from March to July and October to December 1843.
76 AO, *Minutes of the Home District Municipal Council*, August 1843, Appendix #21 "Of Special Committee on Mr. Edmundson's Communication on the subject of Agricultural Societies."
77 See his outline on the front page of the *BAC*, October 1843 and the entire plan, *BAC*, November 1843.
78 MacKenzie, "Edmundson, William Graham"; and TRL, L5, Robert Baldwin Papers, A43, no. 72, "Edmundson to Baldwin, 25 October 1843." *BAC*, January 1842; LAC, RG5, C1, Canada West; Provincial Secretary's numbered correspondence files, vol. 83, PSO/CW file 147/2924 of 1842 on reel C-13568, "William Edmundson to Hon J.B. Harrison, 19 February 1842"; Brock University, Special Collections, Niagara District Agricultural Society – Letters and Papers, "List of Subscribers to B.A. Cultivator," n.d.
79 *BAC*, November 1843.
80 In Edmundson's article, both W.B. Jarvis and E.W. Thomson are listed as vice-presidents. Perhaps there was a typographical error in listing one of these individual's roles. *BAC*, December 1843. An October 1843 report of the recent cattle show identified Jarvis as a vice-president. It made no mention of Thomson, but one might presume he was president. There is no indication that any other person assumed the role of president of the HDAS during this period. *BAC*, October 1843.
81 *BAC*, December 1843, January 1844.
82 *BAC*, January 1844; February 1844; April 1844.
83 *BAC*, March 1844; May1844. There was also a positive response received from the Gananoque Agricultural Society. *BAC*, February 1844. On 16 April and 14 May 1844, the MDAS discussed "a plan (set forth in the *British American Cultivator*) for forming Township Agricultural Societies in connection with an Institution to be established under the name and title of the "Canada Agricultural Association." *BW*, 23 April 1844,17 May 1844; *KCG*, 24 April 1844, 22 May 1844; *BC*, 28 May 1844. Also see supportive letter from James S. Wetenhall, giving notice his intentions to propose that the Gore District Agricultural Society adopt the plan at its next annual meeting. *BAC*, April 1844.
84 *BAC*, January 1845.
85 It may have been a petition from the Fourth Riding of York Agricultural Society, requesting an "alteration in the mode of distributing money" that spawned the formation of this committee. LAC, RG5 C1, Canada West: Provincial Secretary's numbered correspondence files, PSO/CW file 73/6513

of 1844, "Petition, Fourth Riding Agricultural Society of the County of York"; Province of Canada, Legislative Assembly, *Journals*, 13 October 1843. The Agricultural Society of the County of Huntington in Lower Canada had earlier submitted a request that agricultural societies there be "placed on the same footing" as those in Upper Canada. Province of Canada, Legislative Assembly, *Journals*, 17 September 1842. For a list of committee members see: Province of Canada, Legislative Assembly, *Journals*, 27 October, 14 November 1843.

86 Province of Canada, Legislative Assembly, *Journals*, 17, 20, 21, 28 February and 5, 6, 7, 13, 14, 17, 18, 19, 25 March 1845. Province of Canada, *Statutes*, 1845, 8 Vic, c. 54.

87 Province of Canada, *Journals of the Legislative Council of the Province of Canada*, 2, 3, 5, 6, 9 August 1850; Province of Canada, *Journals of the Legislative Assembly*, 6, 9, 10 August 1850; Province of Canada, *Statutes*, 1850, 13&14 Vic., c. 73.

88 Armstrong, *Handbook of Upper Canadian Chronology*, 149,160; "An Act for abolishing territorial divisions in Upper Canada into districts," Upper Canada, Statutes, 1849, 12 Vic., c.78.

89 Province of Canada, Legislative Assembly, *Journals*, 3 July, 1, 25, 27, 30 August 1851; Legislative Council, *Journals*, 28 August 1851; *Statutes*, 1851, 14&15 Vic, c. 127.

90 Spragge, "The Districts of Upper Canada 1788–1849," 34–42; Armstrong, *Handbook of Upper Canadian Chronology*, 149–52. Province of Canada, *Statutes*, 1851, 14&15 Vic, c. 127. There may have been other county agricultural societies that did not submit their financial report to the Board of Agricultural in 1852. Province of Canada, Board of Agriculture, *Journal and Transactions of the Board of Agriculture of Upper Canada*, vol. 1, 239.

91 Province of Canada, *Statutes*, 1851, 14&15 Vic, c. 127. A similar bill adjusting agricultural societies to the county level was passed in November 1852 for Lower Canada. *Statutes* 1852, 16 Vic., c.18. In October 1852, the Provincial Agricultural Association (discussed in chapter seven) recommended county agricultural societies erect permanent buildings for their exhibitions and promised that it would lobby the government for matching grants of £250 for counties interested in doing so. *CdnAg*, October 1852. Henry Youle Hind et al. noted in 1863 that some permanent buildings for this purpose had been erected in Toronto, Hamilton, London, and Kingston. Hind et al., *Eighty Years Progress of British North America*, 47.

**5 The Farming Compact: York, 1830**

1 Romney, "A Struggle for Authority," 14.

2 An exceptional incident was the 1835 annual meeting of the Bathurst District Agricultural Society during which a group of Ottawa Valley timberers

staged a successful coup and maintained control of the organization for two years. Cross, "The Shiners" War," 1–26.

3 Romney, "A Struggle for Authority," 34.
4 Firth, *Town of York, 1815–1834*, lxxxii.
5 Romney, "A struggle for authority," 14; LeSueur, *William Lyon Mackenzie*, 137–8.
6 Romney, "From the Types Riot to the Rebellion,"113–44.
7 Firth, *Town of York 1815–1834*, xl–xlii.
8 Firth, *Town of York, 1815–1834*, xxviii, xli, lxxxi; Romney, "A Struggle for Authority," 9–14.
9 Mackenzie and Ketchum represented the two-member rural riding of York, and Cawthra was the elected member for the riding of Simcoe. Armstrong, 85.
10 Johnson, *Becoming Prominent*, 155–6.
11 Dominated by reform supporters, the list suggests that Mackenzie approached those whom he believed would be supportive of his proposed institution. It included two of Mackenzie's reform associates in the House of Assembly, Robert Baldwin, and Jesse Ketchum, along with eleven merchants, nine farmers, two butchers, two physicians, two lawyers, a brewer, the House of Assembly clerk, a watchmaker, a tanner, a shipwright, and a British author writing an immigrant guide to Upper Canada. See Firth, *The Town of York, 1815–1834*, vol. 2, 126–9. The travel writer was William Cattermole. *CA*, 8 July 1830.
12 *CA*, 11 March 1830; *CG*, 20 March 1830.
13 *CA*, 11 March 1830, 8 July 1830. It is unclear if Mackenzie attended this meeting, for he argued a few months later that he neither called nor attended it. See *CA*, 1 July 1830. The six newspapers were the *Upper Canada Gazette, Observer, Colonial Advocate, Christian Guardian, Courier of Upper Canada*, and the *Canadian Freeman*.
14 *CA*, 1 April 1830.
15 *CA*, 1 April 1830.
16 AO, F 529 Mary Sophia O'Brien fonds, Journal #32, 8 April 1830.
17 These proceedings were sent to the *Upper Canada Gazette*, but it did not publish them. See *CA*, 1 April 1830; *CG*, 3 April 1830.
18 Mackenzie disliked this decision, as he later stated that Doyle had "considered himself authorized to call an adjourned meeting." *CA*, 8 July 1830.
19 In the by-election for the vacant House of Assembly seat, Small had unsuccessfully presented himself as an independent candidate after John Beverley Robinson's promotion to the post of chief justice. He was infuriated that during the campaign Mackenzie had called him a tory government supporter. Armstrong, "Small, James Edward"; Lindsey, *The Life and Times of William Lyon Mackenzie*, vol. 1, 173–7; Miller, "Yonge Street Politics," 105–7; *CA*, 8 July 1830.

20 Miller, *The Journals of Mary O'Brien 1828–1838*, xvii; Macpherson, "O'Brien, Edward George." "Dr. John Strachan and Christopher Hagerman were York residents on whom the family frequently called when they visited the capital ... they also called on Sir John and Lady Colborne." Miller, "Yonge Street Politics," 101.
21 AO, F 529 Mary Sophia O'Brien fonds, Journal #32, 8 April 1830; Miller, "Yonge Street Politics," 107.
22 AO, F 529 Mary Sophia O'Brien fonds, Journal #32, 8 April 1830; Miller, "Yonge Street Politics," 107; *CA*, 22 April 1830.
23 Perhaps this is why Mackenzie's proposed constitution was so similar to that of the UCAS – the very society he had criticized in his newspaper when he expressed his hopes that the "institution now in embryo" would have a longer life than its predecessor. *CA*, 1 April 1830.
24 Chassé, Girard-Wallot, and Wallot, "Neilson, John"; Fairley, ed., *The Selected Writings of William Lyon Mackenzie 1824–1837*, 294; AO, F 37 Mackenzie – Lindsey family fonds, W.L. Mackenzie Correspondence, "Mackenzie to Mr. John Neilson, Esq.," Quebec, 25 April 1830.
25 Mackenzie copied the item from the *Courier* for the benefit of his readers. *CA*, 13 May 1830.
26 Lownsbrough, "Boulton, D'Arcy (1759–1834)."
27 Burns, "Jarvis, William Botsford"; Crawford, *Rosedale*, 13–14.
28 *CA*, 8 July 1830.
29 *CA*, 13 May 1830.
30 This was Mary O'Brien's description. AO, F 529 Mary Sophia O'Brien fonds, Journal # 32, 8 April 1830.
31 Raible, *Muddy York Mud*, 21, 25. For the account of the meeting see *CA*, 20 May 1830, 24 June 1830. In a later issue, Clod informed readers of the *Colonial Advocate* that there was "no *lampooning* at all" about the courthouse meeting. His report "was as correct as it was possible for a reporter to make it." *CA*, 1 July 1830.
32 *CA*, 24 June 1830.
33 Isaac, *The Transformation of Virginia 1740–1790*, 93–4; Gidney and Millar, *Professional Gentlemen*, 127; Romney, *Mr Attorney*, 37.
34 See Romney, "From the Types Riot to the Rebellion," 113–44.
35 Henri Pilon, "Elmsley, John (1801–63)."
36 Martyn, *Aristocratic Toronto*, 41–6.
37 In collaboration, "Allan, William."
38 McKenzie, "FitzGibbon, James"; Neary, "Stanton, Robert"; Dyster, "Gamble, John William."
39 MacKenzie, "Thomson, Edward William."
40 *CA*, 24 June 1830.
41 *CA*, 24 June 1830, 1 July 1830.

42 AO, F 529 Mary Sophia O'Brien fonds, Journal #36, 15 May 1830.
43 *CA*, 20 May 1830; *UCG*, 27 May 1830; *CUC*, [?] June 1830.
44 Armstrong, "Crookshank, George."
45 Firth, "Wood, Alexander."
46 LAC, RG5 A1, Upper Canada Sundries, "Petition to His Excellency Sir John Colborne, 15 May 1830," 56403–4.
47 *CA*, 1 July 1830.
48 *CA*, 20 May 1830.
49 *CA*, 10 June 1830.
50 AO, F 529 Mary Sophia O'Brien fonds, Journal #38, 4 June 1830; Journal # 36, 15 May 1830; Journal #39, "Letter from Edward O'Brien to Anthony Gapper," 6 July 1830.
51 Anthony was a "gentlemen amateur" who had toured Upper Canada to observe and record mammals that he encountered. He published his findings upon his return to England. Miller, *Journals of Mary O'Brien*, xiii–xv.
52 *CA*, 8 July 1830, 15 July 1830.
53 AO, F 529 Mary Sophia O'Brien fonds, Journal #39, "Letter from Edward O'Brien to Anthony Gapper," 6 July 1830.
54 AO, F 529 Mary Sophia O'Brien fonds, #39, "Letter from Edward O'Brien to Anthony Gapper," 6 July 1830.
55 Mackenzie also noted two other individuals: a "_____ Crookshank", and "a _____ Young." *CA*, 8 July 1830.
56 *CA*, 15 July 1830.
57 *CA*, 8 July 1830.
58 *CA*, 8 July 1830. Edward O'Brien appended to his letter to Anthony Gapper the versions of the story as recorded by Mackenzie in the *Colonial Advocate* and by Francis Collins in his *Canadian Freeman*. Unfortunately, no copy of this issue of the *Canadian Freeman* exists. Edward noted that Mackenzie's "account [was] so far false that no one attempted to strike him." See AO, F 529 Mary Sophia O'Brien fonds, Journal #39, "Letter from Edward O'Brien to Anthony Gapper," 6 July 1830.
59 A fragment of the letter by "Stockholder" is found in *CUC*, [?] June 1830. See McNairn, *The Capacity to Judge*, 96, footnote 91, for reference to this letter. McNairn, however, ascribes "Stockholder's" account to the HDAS meeting.
60 Isaac, *Transformation of Virginia*, 94.
61 *CA*, 15 July 1830.
62 *CA*, 15 July, 28 October 1830. On 7 November 1830, Mary O'Brien recorded that the news at church had been that Mackenzie's election victory satisfied him with public opinion; thus, he did not plan to "carry on the prosecution for the assault or rather turn out he got at the Agricultural meeting." AO, Mary Sophia O'Brien fonds, Journal #45, 7 November 1830.

## 6 The Home, Midland, and Niagara District Agricultural Societies, 1830–1850

1 *NG*, 16 July 1831; *HFP*, 2 August 1831.
2 Spragge, "The Districts of Upper Canada 1788–1849," 34–42. Armstrong, *Handbook of Upper Canadian Chronology*, 275.
3 See Noel, *Patrons. Clients, Brokers*, especially Chapter 4.
4 That this event was hosted in conjunction with the York Fair at Market Square is taken from Francis Collins's editorial written after the event, in which he discusses the two events in tandem – and complains that a monthly one-day fair would be better than a six-day fair. *CF*, 28 September 1830; 7 October 1830. Notice of the public fair is found in *Observer* 6 September 1830.
5 *CF*, 19 May 1831; *CG*, 21 May 1831; *NG*, 4 June 1831.
6 LAC, RG5 A1, Upper Canada Sundries, "Petition of the President and Directors of the Home District Agricultural Society, 12 October 1831," 62155–8.
7 Admittedly, Ryerson's criticism may have been nothing more than a veiled complaint that the HDAS did not provide him with the prize lists of the October cattle show. Generally, the HDAS did send its advertisements to the reform newspapers such as Ryerson's *Christian Guardian*, Mackenzie's *Colonial Advocate*, and Francis Collins' *Canadian Freeman*. *CG*, 22 October 1831.
8 *CF*, 24 May 1832. Also see notice for the cattle show in *CA*, 10 May 1832.
9 Collins's death from cholera in a second epidemic to hit York in 1834 ended his *Canadian Freeman*, a valuable source of critical commentary on the activities of the HDAS. *CG*, 17 April 1833; *CA*, 18 April 1833; *TP*, 19 April 1833; *CUC*, 20 April 1833; *CA*, 19 September 1833; *CG*, 25 September 1833; *CUC*, 28 September 1833.
10 With no copy of the HDAS constitution, it is unknown if presidents were limited to a one-year term.
11 Pilon, "Elmsley, John (1801–63).
12 *TP*, 3 May 1833, 27 September 1833; *CA*, 19 September 1833, 1 May 1834; *CG*, 25 September 1833; *CUC*, 28 September 1833.
13 The Town of York was incorporated into the City of Toronto in March 1834. *TP* 19 September 1834; *CG* 24 September 1834; *Toronto Recorder and General Advertiser*, 1 October 1834.
14 *TP*, 25 August 1835.
15 Richard Gapper would be elected a vice-president in 1836 and 1837. For a list of HDAS executives, 1830–46, see my "Gentlemen Famers," Appendix 11, 336–8.
16 Burns, "Jarvis, William Botsford."
17 Craig, *Upper Canada*, 235–41.

18 MacKenzie, "Thomson, Edward William"; *CA*, 1 April 1830; *CG*, 3 April 1830; Johnson, *Becoming Prominent*, 15–16.
19 Craig, "Wells, Joseph."
20 Walton, *York Commercial Directory, Street Guide, and Register 1833–4*, 23; Romney, "Struggle for Authority," 11.
21 Armstrong, *Handbook of Upper Canadian Chronology*, 272; Careless, *Toronto to 1918*, 54.
22 Upper Canada, House of Assembly, *Journals*, Appendix, "Population Returns for 1830," 1 Wm. 4, 1st Sess., 11th Parl., 1831; Province of Canada. Legislative Assembly, *Journals*, Appendix T, "Population Returns, 1841." 4–5 Vic., 1st Sess., 1st Parl., 1841; *Journals*, Appendix L "Return of the Counties Cities and Towns in Upper Canada … by the Census of 1850" 14–15 Vic., 4th Sess., 3rd Parl., 1851. Armstrong, *Handbook*, 272.
23 Stagg, "Gibson, David"; AO, David Gibson Fonds, Document 96, "William Morrison, Forfar, Scotland to David Gibson," York, 17 June 1834; Walton, *The City of Toronto and the Home District Commercial Directory and Register … for 1837*, 191.
24 Stagg, "Lount, Samuel"; Hunter, *The History of Simcoe County, Volume II*, 102.
25 Walton, *City of Toronto*, 191.
26 *TP*, 14 July 1835. It appears that the agricultural society founded in 1835 for the "East Riding of the County of York" became known as "The Agricultural Society of the Townships of Pickering and Whitby" by the following year. *TP*, 21 March 1835, 23 September 1836, 22 September 1837. Likewise, it appears that "[t]he Agricultural Society for the Second Riding of the County of York might also have become known as the Township of Toronto Agricultural Society. *CG*, 20 January 1836; *CA*, 21 January 1836; *Constitution*, 25 January 1837.
27 It is unclear how many cattle shows the HDAS hosted in 1836 and 1837, although each was held in Toronto. There is a notice of planning for a May 1836 cattle show but no reports of the show. *CA*, 28 March 1836, Extra. It did host and October 1836 cattle show and another in October 1837. *RS*, 9 November 1836; *TP*, 7 July 1837; *UCG*, 13 July 1837; *Constitution*, 25 October 1837; 1 November 1837.
28 Read and Stagg, eds., *The Rebellion of 1837 in Upper Canada*, xxvii–xxxi; Craig, *Upper Canada*, 226–47.
29 Stagg, "Gibson, David."
30 Quoted in Ryerson, *The Story of My Life*, 183–4. Stagg, "Lount, Samuel"; Burns, "Jarvis, William Botsford."
31 It is unclear how quickly the HDAS recovered following the rebellion. There is a March 1838 notice about planning the cattle show for May 1838, but no prize list or report of the show. *TP*, 2 March 1838. The society held its fall cattle show that year and returned to a pattern of hosting

spring and fall events in 1839. 1838: 25 September, 5 October; *BC*, 4 and 11 October. 1839: *TP*, 3 and 14 May; *BC*, 24 April, 15 May, 7 August, 9 October.

32 Oddly, the first notices of Sir Arthur as patron of the HDAS appear after his departure from the province. *MSTT*, 24 April 1941; *BAC*, August 1842, August 1843; *TP*, 11 October 1844, 7 October 1845; *Globe*, 12 May 1849. Buckner, "Arthur, Sir George." For evidence of Bond Head's patronage, see *RS*, 9 November 1836, 18 January 1837; *TP*, 13 January 1837; *CG*, 18 January 1837; *Constitution*, 25 January 1837.

33 LAC, RG5 A1, Upper Canada Sundries, "William Atkinson, Treasurer in a/c current ... 31 August 1839," 124597–9.

34 For the Newmarket Agricultural Society, see *BC*, 9 October 1839, 21 October 1840, 9 June 1841; *Mirror*, 15 May 1840, 12 November 1841; *TE*, 10 November 1841; *TP*, 12 November 1841; *TS*, 13 November 1841. For the Etobicoke Agricultural Society see: *TS*, 13 November 1841. *TP*, 8 May 1840; *BC*, 24 June 1840. For Vaughan, York, Scarborough, and Whitby township societies, see: *BAC*, February 1844, April 1844.

35 *BAC*, February 1844, April 1844, June 1844

36 *BAC*, March 1845.

37 *BAC*, February 1844; Dyster, "Gamble, John William."

38 Brown, *Brown's Toronto City and Home District Directory, 1846–7*, 32; Senior, "Boulton, William Henry"; Gates, "Price, James Hervey"; R. Lynn Ogden, "Connor, George Skeffington"; Gagan, *The Denison Family of Toronto, 1792–1925*, 20–1; (George Miller), *Globe*, 29 April 1880. The list of directors also included several unidentified individuals: Robert Cooper, Alexander Shaw, John Scarlett, and a Dr. Hamilton.

39 Starting in October 1840, the HDAS had established a regular home for its exhibitions in the space in front of the jail and courthouse on King Street, along a central stretch of the city's high street. By 1841, the dates for the spring and fall cattle shows had become regular, the second Wednesday in May and in October. 1840: *TP*, 10 March, 21, 24 and 28 April; *BC*, 11 March, 30 September; *SCJ*, 5 November. 1841: *MSTT*, 24 April, 16 September, 21 October; *TE*, 28 April; *TP*, 4 May; *BC*, 19 May, 22 September, 20 October; *TH*, 19 October.

40 *BAC*, February, April, June, July, October, November 1844; *TP*, 27 September 1844, 8 October 1844, 11 October 1844; *TS*, 16 October 1844; TRL, L5 Robert Baldwin Papers, A43 no. 73, "William Edmundson to Robert Baldwin, Newmarket, 16 January 1845"; A 44 no. 26, "M.P. Empey to Robert Baldwin, M.P. Montreal, Newmarket, 22 February 1845"; *BC*, 20 May 1845; *BAC*, April 1845.

41 *BC*, 18 March 1845; *BAC*, April 1845; *TH*, 3 April 1845.

42 *TP*, 13 May 1845.

43 *TE*, 21 May 1845; Careless, "Lesslie, James."

44 *TE*, 28 May 1845.
45 *TP*, 22 and 30 May 1845.
46 *TE*, 28 May 1845.
47 In December 1837, Edmundson was nineteen years old and living with his parents near Newmarket. On 2 December, two days before the rebellion, he met David Gibson who apparently convinced him to join the gathering of rebels at Montgomery's Tavern. Edmundson followed Gibson for an unknown distance down Yonge Street but later claimed that he did reach the tavern. Instead, fearing for his fate, he began an escape to the United States. After travelling some seventy miles, he decided it would be safe to return home. It was not. After two days at home, he was captured and delivered to the Toronto jail. There, he remained until July 1838. He was first sentenced to transportation for seven years, then a prison term of similar length, and finally he was pardoned and released. LAC, RG5 A1, Upper Canada Sundries, "Jane Emily FitzGerald to Sir George Arthur, n.d.," 105014–17; "John Strachan to J. Joseph, Esq., 16 March 1838," 105018–19; "Adam Graham and William Graham Edmundson to Sir Francis Bond Head, 15 March 1838," 105020–2." Also see AO, F 37 Mackenzie – Lindsey family fonds, "Silas Fletcher to William Lyon Mackenzie, Fredonia, 19 July 1840." Concerning Toronto's threatened tory elite in the aftermath of the rebellion and political union, see Way, "Political Process and Social Conflict."
48 *TP*, 7 October 1845.
49 *CdnAg*, 1 March 1849.
50 *CdnAg*, 1 March, 1 September, 1 October, and 1 November 1849; *BC*, 11 May 1849; 5 October 1849.
51 *Globe*, 19 February 1850, 11 May 1850, 10 October 1850; *CG*, 15 May 1850; *Mirror*, 10 May 1850.
52 *UCH*, 17 March 1830, 14 April 1830; *KC*, 17 April 1830.
53 *UCH*, 21 April 1830; *KC*, 8 May 1830.
54 Hugh C. Thomson and John Macaulay led several projects in the Midland District's development in the early 1830s. For example, the provincial government commissioned them to report on the founding of a provincial penitentiary in Kingston. *KC*, 8 May 1830; Fraser, "Macaulay, John"; Bindon, "Kingston," 454, 591; Wise, "John Macaulay: Tory for all Seasons," 185–202; Gundy, "Thomson, Hugh Christopher"; Gundy, "Hugh C. Thomson: Editor, Publisher and Politician, 1791–1834," 203–22.
55 Johnson, *Becoming Prominent*, 214; *Kingston Daily News*, 8 March 1872; LAC, MG24 I43 John Bennett Marks fonds.
56 Herrington, *History of the County of Lennox and Addington*, 375; Lennox and Addington Centennial Brochure Committee, *Historical Glimpses of Lennox and Addington County*, 54–5; Prince Edward Historical Society, *Historic*

*Prince Edward*, 19; Armstrong, *Handbook of Upper Canadian Chronology*, 73; Johnson, *Becoming Prominent*, 234; Lennox and Addington Museum and Archives, William Bell Papers, "Correspondence, 1779–1836".

57 LAC, RG5 A1, Upper Canada Sundries, "Midland District Agricultural Society Petition, 16 November 1830," 558489–91; *KC*, 8 May 1830.

58 Upper Canada, House of Assembly, *Journals*, Appendix, "Population Returns for 1830," 1 Wm. 4, 1st Sess., 11th Parl., 1831; Province of Canada. Legislative Assembly, *Journals*, Appendix T, "Population Returns, 1841." 4–5 Vic., 1st Sess., 1st Parl., 1841; Appendix F. F. "Return of Enumeration ..." 7 Vic., 3rd Sess., 1st Parl., 1843; *Journals*, Appendix L "Return of the Counties Cities and Towns in Upper Canada ... by the Census of 1850" 14–15 Vic., 4th Sess., 3rd Parl., 1851.

59 *KC*, 31 July 1830; *UCH*, 28 July 1830.

60 LAC, RG5 A1, Upper Canada Sundries, "Midland District Agricultural Society Petition, 16 November 1830," 558489–91; "H.C. Thompson to Z. Mudge, Kingston, 17 November 1830," 58500–2; *UCH*, 10 November 1830, 17 November 1830, 18 May 1831; *KC*, 13 November 1830, 20 November 1830.

61 *UCH*, 16 June 1830, 18 May 1831. For information concerning the Prince Edward Agricultural Society see *HFP*, 31 May, 14 June 1831, 3 May 1832, 5 March, 28 May, 21 October 1833, 12 May, 23 June, 29 September 1834; LAC, RG5 A1, Upper Canada Sundries, "L.P. MacPherson to Sir John Colborne, Hallowell, 6 July 1831," 61305–18.

62 Upper Canada, *Statutes*, 1831, 1 Wm 4. c.7. See Spragge, "Districts of Upper Canada", 34–42, 40; Armstrong, *Handbook*, 190.

63 LAC, RG5 A1, Upper Canada Sundries, "L.P. MacPherson to Sir John Colborne, Hallowell, 6 July 1831," 61305–18; "H.J. Boulton to Edward McMahon, Attorney General's Office, York, 1 October 1831," 62114–15; *UCH*, 2 May 1832.

64 *UCH*, 18 May 1831.

65 Upper Canada, *Statutes*, 1830 11 Geo. 4 c. 10.

66 LAC, RG5 A1, Upper Canada Sundries, "John Macaulay to Edward McMahon, Kingston, 23 December 1831," 63021–4; "John Macaulay to Colonel Rowan, 24 December 1834," 81033–4; "John Macaulay to Colonel Rowan, 4 September 1835," 85969–71. Spragge, "Districts of Upper Canada," 40.

67 LAC, RG5 A1, Upper Canada Sundries, "John Macaulay to Edward MacMahon, Kingston, 23 December 1831," 63021–4; "John Macaulay to Edward MacMahon, Kingston, 31 January 1832," 63973–5. *UCH*, 2 May 1832.

68 As discussed in chapter four. *KC*, 5 March 1831.

69 *UCH*, 18 May 1831, 24 August 1831, 14 September 1831, 12 October 1831, 26 October 1831; *KC*, 27 August 1831, 17 September 1831, 15 October 1831.

70 *UCH*, 26 October 1831; 2 May 1832.
71 *UCH*, 1 February 1832, 15 February 1832, 22 February 1832, 2 May 1832; *KC*, 4 February 1832.
72 *UCH*, 3 October 1832, 17 October 1832. For a Scottish officer's report of this event, see Alexander, *Transatlantic Sketches*, 33. Also see Guillet, *The Pioneer Farmer and Backwoodsman*, vol. 2, 133–4.
73 *UCH*, 1 May 1833.
74 *KC*, 4 February 1832, 24 March 1832; *UCH*, 1 February 1832, 24 March 1832, 25 April 1832.
75 *UCH*, 1 May 1833.
76 The lack of a newspaper published in the Napanee area may skew conclusions about the relationship between the fair and agricultural society. Frontenac County: *UCH*, 2 October 1833, 9 October 1833, 20 October 1835; *KCG*, 5 October 1833, 12 October 1833, 13 October 1835, 20 October 1835; *BW*, 28 July 1835, 14 October 1834, 7 October 1835, 14 October 1835; *KCG*, 18 October 1834. Lennox and Addington Counties: *UCH*, 25 June 1834, 21 July 1835, 17 September 1834, 8 September 1835, 31 August 1838; *KCG*, 20 September 1834, 22 July 1835, 7 September 1835, 18 August 1838, 22 August 1838; *BW*, 30 August 1838.
77 *UCH*, 1 May 1833.
78 Again, the lack of a newspaper published in the Napanee area may skew this assessment.
79 *KCG*, 5 October 1833, 12 October 1833, 28 December 1833, 26 April 1834; *UCH*, 2 October 1833, 9 October 1833, 30 April 1834.
80 *KCG*, 26 April 1834; *UCH*, 30 April 1834.
81 *KCG*, 26 April 1834; Gundy, "Thomson, Hugh Christopher."
82 *KCG*, 13 September 1834, 18 October 1834; *BW*, 14 October 1834, 21 October 1834; *TP*, 21 October 1834; *KS*, 28 August 1834.
83 LAC, RG5 A1, Upper Canada Sundries, "John Macaulay to Lieutenant Colonel Rowan, Kingston, 24 December 1834," 81033–4; "Macaulay to Rowan, Kingston, 31 December 1834," 81425–9. RG7 G16, Civil Secretary's Letter books, "Rowan to Macaulay, 3 January 1835" and "Rowan to Macaulay, 10 January 1835."
84 *KCG*, 18 April 1835, 9 May 1835, 13 June 1835, 22 July 1835; *BW*, 19 May 1835, 28 July 1835; *UCH*, 21 July 1835.
85 Thomas Rice was elected as the new secretary. *KCG*, 22 July 1835; *UCH*, 21 July 1835; *BW*, 28 July 1835.
86 *KCG*, 30 April 1836; *UCH*, 3 May 1836.
87 *KCG*, 13 August 1836; 3 September 1836; 8 October 1836; *UCH*, 17 August 1836; 11 October 1836; *UCH*, 18 October 1836.
88 Fraser, "Macaulay, John."
89 *BW*, 2 May 1845; Johnson, *Becoming Prominent*, 214.

90 Members from Lennox and Addington were to choose whether they wished to form one or two county societies.
91 *KCG*, 10 May 1837, 3 August 1837; *BW*, 12 May 1837; *UCH*, 22 August 1837.
92 The MDAS did publish its new constitution. Midland District Agricultural Society, *Constitution of the Midland District Agricultural Society* ([Kingston?]: s.n., 1837?).
93 Frontenac County: *KCG*, 27 September 1837, 14 October 1837, 8 September 1838, 19 September 1838; 31 October 1838, 18 September 1839, 9 October 1839, 12 October 1839, 5 September 1840, 14 October 1840, 17 October 1840; *UCH*, 3 October 1837, 10 October 1837, 25 September 1838, 2 October 1838 15 October 1839; *BW*, 20 October 1837, 20 October 1840; *BC*, 21 October 1840. Lennox and Addington Counties: *KCG*, 18 August 1838, 22 August 1838, 20 February 1839; *UCH*, 31 August 1838; *BW*, 30 August 1838.
94 *KCG*, 29 June 1839, 6 July 1839; *UCH*, 25 June 1839, 2 July 1839; LAC, RG5 A1, Upper Canada Sundries, "J. Marks to S. B. Harrison, Kingston, 18 October 1839," 126503–8.
95 *KCG*, 20 June 1840, 27 June 1840, 18 July 1840; *UCH*, 16 June 1840, 30 June 1840.
96 The Lennox and Addington cattle show of October 1842 was held once again in conjunction with the Napanee Fair but held at Bath the following year. *UCH*, 4 May 1841, 28 September 1841, 2 November 1841, 30 August 1842; *KCG*, 29 September 1841, 9 October 1841, 23 October 1841, 3 September 1842; *CFM*, 15 September 1841; *TS*, 4 November 1843.
97 *KCG*, 30 April 1842, 14 May 1842, 25 May 1842, 29 June 1842; *UCH*, 5 July 1842. Likewise, the Frontenac County cattle show held its last cattle show in Kingston in 1842, relocating to Waterloo for its October 1843 exhibition. *KCG*, 12 October 1842, 15 October 1842, 12 July 1843; *UCH*, 11 July 1843; 19 September 1843, 10 October 1843.
98 *KCG*, 10 June 1840; *UCH*, 16 June 1840.
99 There is no evidence that any such reports were ever produced.
100 *KCG*, 11 July 1840, 25 July 1840.
101 *UCH*, 11 July 1843; *KCG*,12 July 1843.
102 *BW*, 23 April 1844, 17 May 1844; *KCG*, 24 April 1844, 22 May 1844; *BC*, 28 May 1844; *UCH*, 30 April 1844. The eight townships included: Adolphustown, Camden, Ernestown, Fredericksburgh, Kingston, Pittsburgh, Richmond, and Sheffield. For their activities in the first year, see *BW*, 17 May 1844, 3 September 1844, 27 September 1844, 4 October 1844; *KCG*, 22 May 1844; *BC*, 28 May 1844; *UCH*, 11 June 1844, 13 August 1844, 22 October 1844; *KN*, 24 October 1844.
103 *BW*, 11 July 1845, 15 July 1845, 15 August 1845; *UCH*, 26 August 1845.
104 A detailed thirty-page report of active township branch agricultural societies in the Midland District during 1845 was prepared by the

MDAS Secretary and Treasurer in May 1846. LAC, RG5 C1, Canada West: Provincial Secretary's numbered correspondence files, v. 182., Register entry for file 115/1380 of 1846 "Thos. Glassup to Dominick Daly, Kingston, 6 May 1846."

105 The constitutional change responded to a November letter the inspector general had sent to all district agricultural societies, drawing their attention to the administrative requirements set forth by the revised agricultural societies legislation of 1845. It is not entirely clear how the 1845 constitution of the MDAS did not adhere to that act and why the MDAS believed it must create yet another constitution. *KN*, 18 May 1846; *UCH*, 26 May 1846; *BW*, 12 February 1847, 21 August 1847, 9 February 1848.

106 *BW*, 19 August 1848, 29 August 1849.

107 *CdnAg* March 1850, March 1851, August 1852.

108 The committee also included James Clendinning who remains unidentified. *SOT*, 24 June 1830; *NG*, 26 June 1830; Wilson, *Marriage Bonds of Ontario 1803–1834*, 87, 106, 215, 218, 337; "A Walk around Town! Q," Junius [Oliver Seymour Phelps], *St. Catharines A to Z*.

109 Information concerning this clause comes from its removal in 1834. *BAJ*, 24 June 1834.

110 Parker, "Street, Samuel (1775–1844)"; *SOT*, 24 June 1830; *NG*, 26 June 1830. Another list of thirteen directors appears in a petition requesting government funds. Presumably, these directors were elected as proper officials of the NDAS following the drafting of a constitution sometime between June and December 1830. LAC, RG5 A1, Upper Canada Sundries, "Petition to Sir John Colborne, December 1830," 59142–4.

111 Upper Canada, House of Assembly, *Journals*, Appendix, "Population Returns for 1830," 1 Wm. 4, 1st Sess., 11th Parl., 1831; Province of Canada. Legislative Assembly, *Journals*, Appendix T, "Population Returns, 1841." 4–5 Vic., 1st Sess., 1st Parl., 1841; Ibid., Appendix F. F. "Return of Enumeration ..." 7 Vic., 3rd Sess., 1st Parl., 1843; *Journals*, Appendix L "Return of the Counties Cities and Towns in Upper Canada ... by the Census of 1850" 14–15 Vic., 4th Sess., 3rd Parl., 1851.

112 Jackson, *St. Catharines*, 33, 35–47.

113 *SCJ*, 16 August 1844; "A Walk around Town! O," *St. Catharines A to Z*; *UCG*, 10 June 1795; Romney, "Randal, Robert"; *NH*, 21 March 1804; *UCG*, 2 December 1797; Jackson, *St. Catharines*, 33–4.

114 Robertson, *History of Freemasonry in Canada*, vol. 1, 453, 502.

115 Gray, *Soldiers of the King*, 72, 92, 250.

116 *SCJ*, 21 July 1842, 16 August 1844; "A Walk about Town! O," *St. Catharines A to Z*; Johnson, *Becoming Prominent*, 62. For notices about Adam's farming, see *SCJ*, 26 January 1837, 23 September 1841; *NC*, 13 May 1841.

117 *FJWI*, [?] January 1831; 12 October 1831; 16 November 1831; *NG*, 19 February 1831; 22 October 1831; 26 November 1831.
118 *FJWI*, 18 January 1832, 17 May 1832.
119 *NG*, 12 May 1832; *FJWI*, 17 May 1832.
120 "A Walk Around Town! P," *St. Catharines A to Z*; *FJWI*, 1 February 1826.
121 *FJWI*, 18 January 1832; 18 April 1833.
122 *NG*, 18 May 1833; *FJWI*, 18 April 1833; [12 July] 1833; *BAJ*, 24 June 1834.
123 *BAJ*, 7 May 1835, 25 June 1835; *SCJ*, 12 November 1835.
124 LAC, RG5 A1, Upper Canada Sundries, "To His Excellency Sir Francis Bond Head … Toronto, 4 May 1836," 88931–3.
125 *SCJ*, 5 May 1836; 16 June 1836. *BAJ*, 7 May 1835, 25 June 1835; *SCJ*, 12 November 1835; *SCJ*, 5 May 1836; 16 June 1836, 20 October 1836; 8 December 1836.
126 *SCJ*, 20 October 1836.
127 *SCJ*, 17 May 1838; 31 May 1838.
128 Read, "The Short Hills Raid of June 1838, and its Aftermath," 93–4, 98–9; Read, *The Rising in Western Upper Canada 1837–8*, 137; *SCJ*, 14 June 1838, 21 June 1838.
129 *SCJ*, 5 July 1838.
130 *SCJ*, 5 July 1838.
131 *NR*, 17 May 1839, 10 May 1839, 14 June 1839.
132 The society settled on a standard date for its cattle shows from 1844–6. Its spring event, uncoupled from the St. Catharines Fair, was held the second week of May and its autumn show in the third week of October. These are the locations of the NDAS cattle shows from 1839 to 1847. The first location named hosted the spring cattle show, and the second location named hosted the fall cattle show. 1839: Niagara. *NR*, 14 June 1839. 1840: St. Catharines, Stamford. *SCJ*, 23 July 1840; 17 September 1840. 1841: Drummondville, St. Catharines. *SCJ*, 27 May 1841; 24 June 1841; 28 October 1841; 4 and 18 November 1841. 1842: Drummondville, St. Catharines. *SCJ*, 21 and 28 April 1842; 26 May 1842; *BAC*, June 1842; *SCJ*, 3 and 17 November 1842. 1843: St. Davids, Niagara. *SCJ*, 4 May 1843; 1 June 1843; *TS*, 7 June 1843; *SCJ*, 12 October 1843; *BAC*, December 1843. 1844: St. Catharines, Drummondville. *SCJ*, 17, 24 and 31 May 1844; 14 November 1844. 1845: St. Davids, St. Catharines. *SCJ*, 1 and 29 May 1845; 18 September 1845; 16 and 23 October 1845; 6 November 1845. 1846: Drummondville, Beaverdams. *SCJ*, 16 April 1846; 14 and 28 May 1846; 17 September 1846; 22 October 1846; *NC*, 23 October 1846. 1847: St. Catharines, Stamford. Special Collections and Archives, James A. Gibson Library, Brock University, St. Catharines, Ontario. Niagara District Agricultural Society – Letters and Papers, "John Gibson in acct. with the Niagara District Agricultural Society, [1848]."

133 *SCJ*, 26 May 1842, 9 June 1842.
134 *SCJ*, 10 October 1844.
135 *SCJ*, 1 May 1845. Forman, ed., *Legislators and Legislatures of Ontario*, vol. 1, 103.
136 "A Walk around Town! U," *St. Catharines A to Z*. For mentions of Samuel Wood's role as coroner and his involvement in other in other local institutions, see: *FJWCI*, 17 February 1827, 7 March 1827, 2 May 1827; *SCJ*, 15 October 1835, 26 November 1835, 15 December 1836, 4 May 1843.
137 *SCJ*, 6 November 1845. This letter was published as having been written by the Secretary of the society A.K. Boomer. See *SCJ*, 13 November 1845 for a correction of the author; *SCJ*, 22 October 1846; Special Collections and Archives, James A. Gibson Library, Brock University, St. Catharines, Niagara District Agricultural Society – Letters and Papers, "Circular," St. Catharines, Grantham, 28 October 1846."
138 The county society was Haldimand, and the ten township societies were Clinton, Grimsby, Crowland, Louth, Gainsboro, Stamford, Niagara, Pelham, Thorold, and Bertie. *SCJ*, 6 November 1845; *Globe*, 4 November 1846; *BC*, 10 November 1846; LAC, RG 5 C1, Provincial Secretary's numbered correspondence files, v. 203, PSO/CW File 170/17020 of 1847, "A.K. Boomer to Dominick Daly," St. Catharines, 15 June 1847; Niagara Historical Society Museum, 978.502.1, "Niagara Township Agricultural Society Membership List," 23 August 1847; Special Collections and Archives, James A. Gibson Library, Brock University, St. Catharines, Niagara District Agricultural Society – Letters and Papers, "John Gibson in acct with the Niagara District Agricultural Society, [November] 1847"; AO, F 380, Percy C. Band collection, "Richard Graham, Secretary, Bertie Agricultural Society to A.K. Boomer, Secretary, NDAS, Fort Erie, 22 December 1845."
139 *CdnAg*, June 1850.
140 The NDAS re-elected John Lemon and William McMicking as president and secretary, respectively. Vice presidents were Edward Jones, Samuel Parker, and James Williams. *CdnAg*, April 1851.

## 7 District Agricultural Societies and Their Improvements, 1830–1850

1 Smith, *Smith's Canadian Gazetteer*, 246–7.
2 Taylor, *Narrative of a Voyage to, and Travels in Upper Canada*, 71–2.
3 Upper Canada, *Statutes*, 1830, 11 Geo.4, c. 10.
4 For an overview of the theory of exhibition see Heaman, *Inglorious Arts of Peace*, 10–28.
5 Canniff, *History of the Province of Ontario*, 590–1.
6 See Sean Gouglas, "The Influences of Local Environmental Factors on Settlement and Agriculture in Saltfleet Township, Ontario, 1790–1890."

7 *FJWCI*, 16 November 1831.
8 Derry, *Ontario's Cattle Kingdom*, 3, 7.
9 The livestock was purchased from an L. Jenkins. *FJWCI*, [?] January 1831, 18 January 1832; *NG*, 19 February 1831, 17 March 1832; *CG*, 21 May 1831; *NC*, 13 May 1841. For an advertisement about the sale of Jenkins's livestock see *UCH*, 2 June 1830.
10 *NG*, 17 March 1832; *NC* 13 May 1841.
11 *NG*, 5 May 1832; 18 May 1833; *FJWI*, 31 May 1832.
12 *TS*, 7 June 1843.
13 *SCJ*, 24 February 1842.
14 *KC*, 5 March 1831, 27 August 1831; *UCH*, 18 May 1831, 20 July 1831, 24 August 1831, 2 May 1832.
15 Barrie had previously offered his bull for breeding purposes at the rate of a dollar per cow. The *Kingston Chronicle and Gazette* would later remark that a bull and cow that Barrie had imported had offspring spread throughout the Kingston area. Following the closure of the Kingston Dockyard, Barrie returned to England in 1834, and afterwards, his herd was dispersed. Barrie had also imported a pedigreed bay horse "Daghee," which he sold in 1834 to a company of gentlemen led by two officers of the MDAS, J.B. Marks and George Yarker. *KCG*, 14 October 1837; LAC, RG5 A1, Upper Canada Sundries, "J. Marks to J. Macaulay, Kingston, 20 September 1838," 113748–51. Brock, "Barrie, Sir Robert"; *KC*, 26 March 1831, 19 February 1831; *TP*, 29 March 1832; *KCG*, 10 May 1834. See notice of the sale of his animals, *KCG*, 17 October 1835. For a description of "Daghee" see *KC*, 19 February 1831. Also see notice of the sale of John B. Marks' "Heart of Oak", descended from Barrie's Durham bull. *KCG*, 7 October 1843.
16 *KCG*, 17 January 1835; *UCH*, 7 June 1836.
17 *KCG*, 20 June 1840; *UCH*, 16 June 1840.
18 *UCH*, 20 October 1840.
19 *KCG*, 29 June 1842, 12 July 1843; *UCH*, 5 July 1842, 11 July 1843.
20 *KCG*, 29 June 1842; *UCH*, 5 July 1842.
21 *UCH*, 4 May 1841, 5 July 1842; *KCG*, 9 October 1841, 2 April 1842, 29 June 1842. For details of Somonocodrom's pedigree and his previous availability for breeding in Ancaster and Toronto, see *TP*, 14 March 1837, 3 May 1839; *KCG*, 2 April 1842.
22 *CG*, 17 April 1833; *CA*, 18 April 1833; *TP*, 19 April 1833; *CUC*, 20 April 1833; "Extracts from Memorandums of William Helliwell," Firth, *Town of York, 1815–1834*, 338.
23 Board of Agriculture of Upper Canada, *The Canada Herd Book*, xiv; Dominion Short-Horn Breeders' Association, *History of Short-Horn Cattle Imported into the Present Dominion of Canada from Britain and United States*, xi.

24 *RS*, 9 November 1836.

25 *TP*, 25 September 1838.

26 *BC*, 24 April 1839, 22 September 1841, 4 May 1842, 18 May 1842; *TP*, 3 May 1839, 2 May 1843, 19 September 1843; *SCJ*, 5 November 1840; *TS*, 16 September 1841; 13 May 1843; *BAC*, August 1843, April 1845.

27 Two Toronto butchers also joined the ranks of HDAS directors in 1846. *BC*, 24 April 1839; *TS*, 16 September 1841; Brown, *Brown's Toronto City and Home District Directory, 1846–7*, 32.

28 *BC*, 19 May 1841; *RS*, 9 November 1836.

29 *KC*, 5 March 1831, 27 August 1831, 17 September 1831; *UCH*, 18 May 1831, 24 August 1831, 14 September 1831, 2 May 1832.

30 *UCG*, 7 September 1831, 22 February 1832, 2 May 1832; *KC*, 10 September 1831, 14 September 1831.

31 *UCH*, 1 February 1832; *KC*, 4 February 1832.

32 *UCG*, 2 May 1832.

33 *KCG*, 5 October 1833, 12 October 1833, 28 December 1833, 26 April 1834; *UCH*, 2 October 1833, 9 October 1833, 30 April 1834.

34 *KCG*, 31 October 1838; LAC, RG5 A1, Upper Canada Sundries, "J. Marks to S. B. Harrison, Kingston, 11 October 1839," 126503–8.

35 *UCH*, 22 February 1832.

36 *KCG*, 22 July 1835, 9 September 1835; *UCH*, 21 July 1835, 8 September 1835, 20 October 1835, 16 May 1843; *BW*, 14 October 1835; *UCH*, 16 May 1843.

37 *UCH*, 2 October 1833, 15 October 1839, 16 June 1840, 30 June 1840; *KCG*, 18 August 1838, 20 February 1839, 18 September 1839, 12 October 1839, 20 June 1840, 27 June 1840.

38 *UCH*, 26 October 1831, 20 October 1835, 10 October 1837; *BW*, 14 October 1835; *KCG*, 14 October 1837.

39 *CF*, 28 September 1830; *CA*, 18 April 1833, 25 July 1833.

40 No prize list was included in advertisements for the October cattle show. *CG*, 17 April 1833, 25 September 1833; *CA*, 18 April 1833, 19 September 1833; *TP*, 19 April 1833; *CUC*, 20 April 1833; *CG*, 17 July 1833; *CA*, 25 July 1833.

41 *TP*, 19 September 1834; *CG*, 24 September 1834.

42 *RS*, 9 November 1836; *Constitution*, 25 October, 1 November 1837.

43 *BAC*, August 1843; *TP*, 19 September 1843.

44 *SCJ*, 14 November 1844, 23 October 1845, 22 October 1846; *NC* 23 October 1846; LAC, RG5 C1, Canada West: Provincial Secretary's numbered correspondence files, v. 176, PSO/CW file 66/13,207 of 1846, "A.K. Boomer To His Excellency Earl Cathcart, 4 April 1846"

45 *NG*, 18 May 1833; *BAC*, December 1843; *SCJ*, 5 July 1838, 22 October 1846.

46 *FJWCI*, [?] January 1831, 17 May 1832; *NG*, 12 May 1836; *SCJ*, 16 June 1836, 8 December 1836, 5 July 1838, 24 May 1844, 16 October 1845, 6 November 1845.

47 A young Miss Cecilia Moore exhibited a "highly credible" sample of worsted work at the fall 1846 cattle show, but "being foreign to Agriculture" the judges were unable to award a prize. *SCJ*, 16 June 1836, 8 December 1836, 22 October 1846. The participation by these women in cattle show competitions is the earliest era of women in agricultural fairs that Jodey Nurse explores in *Cultivating Community*.

48 *CF*, 5 May 1831; *CA*, 5 May 1831, Supplement; *NG*, 4 June 1831

49 *CG*, 22 October 1831; *CA*, 10 May 1832

50 *BC*, 18 March 1845, 20 May 1845, 15 May 1846; *TH*, 3 April 1845; *TS*, 22 May 1845; 11 October 1845, 16 May 1846; *Globe*, 15 May 1847, 13 May 1848, 12 May 1849, 23 October 1849, 11 May 1850. The HDAS's aborted grand exhibition of October 1844 had intended to offer prizes in categories of Farming Implements (scotch plough, subsoil plough, fanning mill, cultivator, frill barrow, portable threshing machine, straw cutter, clover machine, flax and hemp dressing machine, horse rake, ribbing plough), Dairy (fifty pounds of butter, one hundred pounds of cheese), and Domestic Manufacture (pair of woollen blankets, ten yards fulled cloth, fifty yards woollen carpet), but none of these categories was offered in the scaled-down cattle show that the HDAS struggled to host instead. *BAC*, June 1844; *TP*, 27 September 1844.

51 *KC*, 4 February 1832; *UCH*, 26 October 1831, 1 February 1832, 22 February 1832.

52 *KCG*, 12 October 1833, 18 October 1834, 18 August 1838; *CFM*, 15 September 1841.

53 *BW*, 14 October 1835; *UCH*, 20 October 1835. Women were also awarded prizes by the MDAS for their domestic manufacturers in 1837 and 1839. *KCG*, 14 October 1837, 12 October 1839; *UCH*, 15 October 1839. At the HDAS spring cattle show, Richard Gapper's wife won prizes for showing draught mares in April 1840 and May 1841. *TP*, 28 April 1840; *BC*, 19 May 1841.

54 *KCG*, 9 September 1835; *UCH*, 8 September 1835, 20 October 1835; *BW*, 14 October 1835.

55 *UCH*, 17 October 1832. Also see his speech, *KCG*, 12 October 1833 and report, *KN*, 18 May 1846.

56 *CF*, 28 September 1830, 7 October 1830, 5 May 1831; *CA*, 5 May 1831, Supplement; *NG*, 4 June 1831; *BAC*, August 1842, May 1843, June 1844, April 1845; *BC*, 21 September 1842, 2 October 1842; *TS*, 29 September 1842, 13 October 1842; *TP*, 7 May 1844.

57 *FJWI* 16 November 1831; *NG* 26 November 1831; *SCJ* 4 November 1841, 18 November 1841, 17 November 1842, 31 May 1844, 22 October 1846; *NC* 23 October 1846.

58 *Canadian Magazine*, January, April 1833; *CG*, 24 April 1833; *TP*, 3 May 1833; LAC, RG5 A1, Upper Canada Sundries "William Sibbald to Colborne,

York, 6 May 1833," 71047–50; LAC, RG 11, C.O. 42 v. 424 "Sibbald to Stanley Toronto, 7 May 1834." See advertisements for Sibbald's lectures in *TP*, 15 November 1833 and issues of the *Canadian Correspondent*, 23 November 1833–8 March 1834, and commentary about his lectures, 7 December 1833.

59 *KCG*, 8 May 1830, 3 August 1837; *UCH*, 21 April 1830, 22 August 1837.

60 *KCG*, 4 February 1832, 25 July 1840; *UCH*, 1 February 1832; *BW*, 19 May 1835.

61 In 1845, Edmundson highlighted the "unexampled liberal support to the agricultural press" offered by the "old and popular Association" of the Midland District. *UCH*, 16 May 1843; 26 August 1845; *KCG*, 11 September 1844; *BAC*, September 1845.

62 *BW*, 12 February 1847, 9 February 1848.

63 *FJWI*, 12 October 1831; *CG*, 13 August 1831; LAC, RG5 C1, Canada West: Provincial Secretary's numbered correspondence files, v. 176, PSO/CW file 66/13,207 of 1846, "A.K. Boomer To His Excellency Earl Cathcart, 4 April 1846"; RG5 C1 v. 203, PSO/CW file 170/17020 of 1847 "A.K. Boomer to Dominick Daly, St. Catharines, 15 June 1847"; Special Collections and Archives, James A. Gibson Library, Brock University, St. Catharines, Niagara District Agricultural Society – Letters and Papers, "List of Subscribers to B.A. Cultivator," J. Eastwood, [n.d.].

64 [Reverend Thomas Brock Fuller], *The Canadian Agricultural*; *SCJ*, 6 November 1845.

65 Ruggle, "Fuller, Thomas Brock."

66 *NC*, 17 December 1845, 24 April 1846; *BC*, 13 March 1846, 23 June 1846; *SCJ*, 19 March 1846; *KN*, 18 May 1846; *UCH*, 26 May 1846.

67 *UCH*, 16 May 1843, 26 May 1846; *BW*, 3 September 1844.

68 Edmundson responded to this market for his newspaper by promising that, in future, he would devote two pages of his monthly journal to horticultural matters. *BAC*, December 1843.

69 *BAC*, January 1844; *BC*, 12 January 1844; *TS*, 14 February 1844.

70 Crawford, "The Roots of the Toronto Horticultural Society," 125–9. Crawford misses the fact that there were two distinct horticultural societies in Toronto. *CA*, 3 May 1832; *CF*, 3 May 1832; *TP*, 23 May 1834; AO, F 44 John Beverley Robinson family fonds, "Regulations & Bye Laws of the Toronto Horticultural Society ... The 1st day of May, 1834"; The last published notice of the first Toronto Horticultural Society is found in *Constitution*, 21 September 1836.

71 Thornton, "The Moral Dimensions of Horticulture in Antebellum America," 5, 17.

72 *BC*, 18 July 1845.

73 Other executives included its vice-presidents: William H. Boulton, who in 1844 was a Toronto alderman; F.T. Billings, the Home District Treasurer;

and George W. Allan, a prominent York gentleman and resident of his father William's "Moss Park" estate. Robert Maitland, the owner of the Church Street wharf was recording secretary, and William Edmundson was elected corresponding secretary. Several individuals who operated nurseries in the city were also appointed to a committee of management. *BAC*, January and June 1844; *TS*, 14 February 1844; *BC*, 26 March 1844.

74 *BC*, 13 September 1844; *TS*, 14 September 1844; Brown, *Brown's Directory*, 32.

75 *BC*, 16 March 1849, 22 March 1853.

76 *TP*, 28 April 1840.

77 Stagg, "Thorne, Benjamin."

78 *TP*, 28 April 1840, 11 October 1844

79 Stagg, "Thorne, Benjamin."

80 *BC*, 19 May 1841; McCalla, "Ridout, George Percival."

81 See notices of the Agricultural Central Committee. *TS*, 20 November 1841, 2 December 1841, 30 April 1842; *BC*, 1 December 1841, 8 December 1841, 15 December 1841. Jarvis also led the effort to gather thousands of signatures on another agricultural petition in response to legislation related to commerce being considered by the Canadian parliament. *BAC*, October 1843; *UCH*, 11 July 1843; *KCG*, 12 July 1843; *TS*, 16 September 1843. *UCH*, 16 May 1843. *BAC*, August 1846; *BC*, 21 August and 25 September 1846; *Globe*, 6 January 1847. In the summer of 1842, NDAS president Adams, received a letter from MPP Merritt informing farmers of an anti–free trade petition circulated by the North American Colonial Committee. *SCJ*, 21 July 1842.

82 *KA*, 13 February 1846, 6 March 1846.

83 The Ontario Plowmen's Association also hosts the annual International Plowing Match and Rural Expo. See https://www.plowingmatch.org. For a study of ploughing matches in Upper Canada/Ontario, see Catharine Anne Wilson, "A Manly Art: Plowing, Plowing Matches, and Rural Masculinity in Ontario, 1800–1930."

## 8 The Agricultural Association of Upper Canada, 1846–1852

1 *BC*, 11 September 1846.

2 *BAC*, February 1844; April 1844.

3 *BAC*, March 1844.

4 *BAC*, February 1845. No mention of a provincial agricultural association appears in Edmundson's report of the HDAS meeting in February. *BAC*, March 1845.

5 *BAC*, January 1845.

6 *KN*, 27 November 1845; *BAC*, January 1846. The committee members were President Jarvis, Vice-President Thomson, Secretary Wells, Edmundson,

and Peter Perry of Whitby. Edmundson did not receive the reports of this November meeting in time to publicize the meeting, scheduled for the second Wednesday of February, in his December *British American Cultivator*. He did so in his January 1846 issue.

7 The committee was composed of Wells, Edmundson, and William B. Crew, an HDAS member and councillor for Toronto Township. *BAC*, June 1846; *Journal and Transactions*, 20–21; "A History of the Agricultural and Arts Association," 139–40.

8 *BAC*, August 1846.

9 *BAC*, September 1846. *TS*, 29 July 1846; *NC*, 31 July 1846; *BC*, 21 August 1846; *SCJ*, 27 August 1846; *BAC*, August 1846; September 1846; *NF*, 1 September 1846; *Journal and Transactions*, 21.

10 *TP*, 25 August 1846; *Banner*, 4 September 1846. Edmundson and Joseph Hartman, a young farmer and superintendent of education for Whitchurch Township, had travelled to the New York State Fair at Utica, New York, in 1845 on behalf of the Fourth Riding of York Agricultural Society to report upon what they saw that might be "calculated to benefit Canadian Agriculturists." *BAC*, November and December 1845; Jarvis, "Hartman, Joseph."

11 *TP*, 25 August 1846; *Banner*, 4 September 1846; *BAC*, September 1846; *BC*, 21 August 1846; *Journal and Transactions*, 21–5.

12 Ruttan's name does not appear on the list of individuals present at the meeting. *BAC*, April 1844. John Wetenhall of Nelson Township was "one of the principal shorthorned breeders in the province and an extensive farmer." LAC, R 3831-0-5-E, R. Kay Diary [1845], 39; *Journal and Transactions*, 21–5; *BAC*, September 1846.

13 Jones, "Fergusson, Adam."

14 *TP*, 25 August 1846; *Banner*, 4 September 1846; *SCJ*, 27 August 1846. The Canada Company also offered £50 to the Association, £25 of which was to be offered as a premium for growing wheat for the next season. *BC*, 20 October 1846.

15 *BC*, 11 September 1846.

16 For example, see: *BC* 11 September 1846; *TP*, 15 September 1846; *KN*, 14 September 1846.

17 *Globe*, 13 October 1846; *CG*, 30 September 1846; *TE*, 16 September 1846.

18 *BAC*, October 1846.

19 A "day or two previous to the exhibition," handbills were distributed notifying the change to the new location. *TE*, 28 October 1846. For people from across the province to attend, steamboat operators and stagecoach proprietors offered special services for the exhibition. *BC*, 11 September 1846.

20 *BC*, 11 September 1846; *TE*, 14 October 1846. For the complete list of categories see *BAC*, October 1846.

21 The reported number of guests varied widely among newspaper accounts. *BC*, 11 September 1846 and 27 October 1846; *BAC*, November 1846; *Banner*, 23 October 1846; *CG*, 28 October 1846.

22 *BC*, 23 October 1846. The *Examiner* noted a cold wind blew on each day, and attendance was "as great as could, under the circumstances, be expected." *TE*, 28 October 1846. The *Niagara Chronicle* reported that the exhibition was attended by a "respectable number of Farmers and others, interested in agricultural pursuits." *NC*, 23 October 1846. *BC*, 27 October 1846; *BAC*, October 1846.

23 *Banner*, 4 September 1846.

24 BAC, November 1846; *Journal and Transactions*, 41–4.

25 Mills, *Idea of Loyalty*, 136. In Lower Canada, "The Canadian Agricultural Society" would choose its incorporated name to be "The Lower Canada Agricultural Society." Province of Canada, *Statutes*, 1847, 10 & 11 Vic., c. 60.

26 *BAC*, August 1846; September 1846.

27 *BAC*, November 1846; *Journal and Transactions*, 41–2.

28 *BAC*, November 1846; *Journal and Transaction*, 44.

29 *BAC*, October 1847; *Journal and Transactions*, 43–4.

30 Province of Canada, *Journals*, 21 June, 9, 22, 23, 26, 28 July 1847; *Statutes*, 1847, 10 & 11 Vic., c. 61.

31 *BAC*, December 1846; February 1847.

32 *Canada Farmer*, 26 February 1847.

33 *BAC*, June 1847. The Home and Gore Districts offered £50 each, Northumberland County £25, and the Simcoe and London Districts £10 each. *BAC*, July 1847.

34 *BAC*, September 1847.

35 *BAC*, October 1847; *Globe*, 9 October 1847; *Canada Farmer*, 9 October 1847; *Journal and Transactions*, 47–8, 65. The destruction of exhibits was noted in *NF*, 1 October 1848, to showcase the waterproof buildings of the exhibition at Cobourg that year.

36 He had served as the first vice president, and the promotion from first vice president to president would become standard practice for the AAUC. *Journal and Transactions*, 64.

37 *Journal and Transactions*, 69. The amount equalled six grants of the maximum amount of £250 available to district agricultural societies under the existing legislation. Province of Canada, *Statutes*, 1845, 8 Vic, c. 54.

38 *ACJ*, 1 March 1848; *Globe*, 15 March 1848. For the circular, see *NF*, 1 April 1848; *Globe*, 29 April 1848; *ACJ*, 15 May 1848.

39 *NF*, 1 October 1848; *Journal and Transactions*, 75.

40 *NF*, 1 October 1848; *FM*, November 1848; *Globe*, 7 October 1848; *ACJ*, 16 October 1848. For a full report on the AAUC's exhibition at Cobourg, see *FM*, November 1848.

41 *NF*, October 1848. The competitions in Niagara and Toronto were much smaller, with only three awards provided for a few examples of moccasins. *CdnAg* October 1850, November 1852.
42 The date had already been shifted from the last week to the first week of October, and the decision for the third week of September was a change from an earlier decision to hold the exhibition in the first week of September. *Journal and Transactions*, 77.
43 *FM*, 1 March 1849; *CdnAg*, 1 March, 2 July, 1 September 1849; *Journal and Transactions*, 76–8.
44 Province of Canada, *Journals of the Legislative Assembly*, 29 January, 8, 19, 20 February, 16 March 1849. The petition of the AAUC was read on 5 February.
45 Careless, *Union of the Canadas*, 122–6.
46 Province of Canada, *Journals of the Legislative Assembly*, 18 April, 3 May 1849.
47 *BAC*, April 1847; "An Act to incorporate *The Lower Canada Agriculture Society*" received royal assent on the same day as the act incorporating the Agricultural Association of Upper Canada. Province of Canada, *Statutes*, 1847, 10 & 11 Vic., c. 60.
48 Province of Canada, *Journals of the Legislative Assembly*, 4 April. *Journal and Transactions*, 77. Receipt of these grants was not at all a certainty before the end of the session. See *CdnAg*, 1 June 1849.
49 *CdnAg*, January, February 1850; *Journal and Transactions*, 123–4.
50 MacKenzie, "Edmundson, William Graham." A short obituary appears in the *Globe*, 25 November 1852.
51 For information about the university bill, see Friedland, *The University of Toronto: A History*, 26–9.
52 MacKenzie, "Buckland, George"; *Globe*, 2 March 1885.
53 *CdnAg*, 1 June 1849.
54 The NYSAS's 1848 invitation brought Johnston to North America for most of 1849. Wynn, "Johnston, James Finlay Weir."
55 *CdnAg*, 1 September 1849. *Globe*, 22 September 1849. Johnston's lecture can be found in: *CdnAg*, 1 October 1849; *Journal and Transactions*, 82–9.
56 It is unknown whether attendance numbers for any provincial exhibition discussed in this chapter were based on gate receipts or simple guesswork.
57 The "no, nos" were quickly drowned out by cheers of support. *CdnAg*, 1 October, 1 November 1849. *Globe*, 22 September 1849; *Journal and Transactions*, 79–117.
58 *CdnAg*, 1 October, 1 December 1849, February 1850; *Journal and Transactions*, 113–15, 118–22.
59 *CdnAg*, 1 October, 1 December 1849, February 1850; *Journal and Transactions*, 113–15, 118–22.

60 *CdnAg*, February 1850. See the report of the Kingston committee in *CdnAg*, June 1850. Donations were also received from two districts (a municipal category defunct after 1849) as well as the Town of London. *Journal and Transactions*, 148.
61 Coleman, "Cameron, Malcolm"; Cross, "Hopkins, Caleb"; Clarke, *Sixty Years in Upper Canada*, 78–9; *Globe*, 26 February 1850, 7 March 1850; *CdnAg*, July 1850. During the rest of 1850, AAUC executives expressed their sadness over the loss of one of the AAUC's most zealous supporters. *CdnAg*, July and October 1850; *Journal and Transactions*, 135.
62 Province of Canada, Statutes, 1850, 13&14 Vic., c. 73.
63 *CdnAg*, February, June, July, August 1850; *Globe*, 26 February 1850; *Journal and Transactions*, 127, 148.
64 *CdnAg*, September 1850. For information on the exhibition, see *CdnAg*, April, August, September, October 1850; *Journal and Transactions*, 130–48.
65 *CdnAg*, February, August, October 1850; *Journal and Transactions*, 130–1, 146; Craig, "Croft, Henry Holmes." Inspired by his message, the AAUC appointed a committee at its executive meeting following the exhibition to draft a petition to the legislature requesting convicts at the provincial penitentiary in Kingston be used as labour for burning lime for fertilizer and other purposes. *CdnAg*, October 1850. At its July 1851 meeting, the AAUC adopted the memorial about convicts burning lime and sent it to the governor in council. There is no evidence of government support. *CdnAg*, July 1851.
66 *Journal and Transactions*, 148.
67 *CdnAg*, July 1851.
68 *CdnAg*, 1 October, 1 November 1849. *Globe*, 22 September 1849; *Journal and Transactions*, 113.
69 *ACJ*, 16 October 1848.
70 *CdnAg*, September 1849, October 1849.
71 *CdnAg*, February 1850; *Journal and Transactions*, 122.
72 *CdnAg*, October 1850.
73 Careless, *Union of the Canadas*, 127–31; *CdnAg*, 1 September 1849; February 1850; *Globe*, 14 and 26 February 1850; *Journal of Education for Upper Canada*, June 1850; Keefer, *The Canals of Canada*. Nelles, "Keefer, Thomas Coltrin."
74 *CdnAg*, January 1851.
75 *CdnAg*, April 1851.
76 *CdnAg*, May, July, August 1851.
77 *CdnAg*, November 1850, January 1851.
78 "First Annual Report of the Board of Agriculture of Upper Canada for 1851–2," Appendix (S), Province of Canada, *Appendix to the eleventh volume of the journals of the Legislative Assembly of the Province of Canada, 1st session, 4th Parliament*, 1851–2.

79 *CdnAg*, October 1851.
80 See address of J.B. Marks, *CdnAg*, January 1851 and address of the Brockville exhibition's local committee, *CdnAg*, April 1851. See rules about female competitors and admission in prize report: *CdnAg*, July 1851. See report of the Floral Hall in *CdnAg*, October 1851. *CFH*¸18 September 1852.
81 *Journal and Transactions*, 210; *CdnAg*, January, February, March, May, September 1852.
82 *CdnAg*, November 1849; *Journal and Transactions*, 89–112.
83 *CdnAg*, October 1851; *Journal and Transactions*, 170–79.
84 *CdnAg*, October 1851; Swainson, "Street, Thomas Clark."
85 Mackey, *Steamboat Connections*, 168–9, 182, 337.
86 Legget, "Tredwell (Treadwell), Nathaniel Hazard"; Treadwell, *Arguments in Favor of the Ottawa and Georgian Bay Ship Canal*.
87 Fleming had been appointed the AAUC's seedsman at its February 1849 directors meeting. He operated a nursery on Yonge Street in Toronto. An April 1850 notice suggested Fleming's role to be importing rare and other seeds for members upon request and his interest in purchasing "timothy, clover, and other agricultural seed" from farmers of the province. *FM*, 1 March 1849; *CdnAg*, 1 March, 2 July, 1 September 1849; *Journal and Transactions*, 76–8; *CdnAg*, February, April 1850; *Globe* 19 March 1850; *Journal and Transactions*, 122.
88 Denison, *A History of the Denison Family in Canada*, 16–7; *Journal and Transactions*, 211.
89 Careless, *Toronto to 1918*, 201.
90 *CdnAg*, March, June, August, September 1852; *CFH*¸18 September 1852. *Journal and Transactions*, 207, 211. During the previous October, executives of the County of York Agricultural Society had marched in the procession to a sod-turning ceremony witnessed by a crowd of twenty thousand that initiated the construction of the railway. The Ontario, Simcoe, and Huron Railway would begin operations in 1853, causing Board of Agriculture president E.W. Thomson to remark in that year on the potential of railways to bring more people and articles of competition to the provincial exhibitions and for the possibility of holding the exhibition in different parts of the province. "Second Annual Report of the Board of Agriculture of Upper Canada, 1853–4," Appendix (I.I.), Province of Canada, *Appendix to the Thirteenth Volume of the Journals of the Legislative Assembly of the Province of Canada, 1st session, 5th Parliament*, 1854–5. Careless, *Toronto to 1918*, 77.
91 *CFH*, 2 October 1852.
92 A new Normal School had been opened in Toronto during the previous year. Two of its eight acres were turned into a botanical garden, with three acres reserved for agricultural experiments. These were to provide a

practical illustration of related courses taught within the Normal School's classrooms. *TE*, 29 September 1852. *Journal of Education for Upper Canada*, November 1853; *Canada Farmer*, 16 July 1866.

93 *CFH*, 2 October 1852.

94 *BC*, 1 October 1852. Also see: *BC*, 24 September 1852; *TE*, 29 September 1852; *Globe*, 22 September 1852. For extensive reports on the 1852 exhibition, see *CFH*, 18 and 25 September, 9 October 1852; *CdnAg*, October, November, December 1852; *The Canadian Journal: A Repertory of Industry, Science, and Art*, 1:3 (October 1852): 51–68.

95 Moodie, *Life in the Clearings versus the Bush*, 313–14, 319–22, 324–27; Ballstadt, "Strickland, Susanna."

**9 A Board and a Bureau of Agriculture, 1850–1852**

1 *TE*, 29 September 1852; *Globe* 28 September 1852.

2 *CdnAg*, February, June 1850; *Globe*, 26 February, 14 May 1850.

3 Hodgetts, *Pioneer Public Service*, 35.

4 Careless, *Union of the Canadas*, 166–84. Ormsby, "Sir Francis Hincks," 148–93; Ormsby, "Hincks, Sir Francis"; Cross, *A Biography of Robert Baldwin*, 284–95. As several historians note, a George Brown version of history has dominated our understanding of Francis Hincks and the Hincks–Morin administration. Being a vocal political foe of Hincks, his version emphasizes the man who opened the door for widespread corruption to take over Canadian politics. Davison, "Francis Hincks and the Politics of Interest," 21; Curtis, *The Politics of Population*, 140, 340–1, note 16.

5 *CdnAg*, February 1850. See also, *CdnAg*, June 1850; *Globe*, 14 May 1850; *Journal and Transactions*, 128–9. The draft bill is published in *CdnAg*, June 1850.

6 Careless, *Union of the Canadas*, 166–7; Cross, *Robert Baldwin*, 284.

7 Cross, *Robert Baldwin*, 294.

8 Careless, "Christie, David."

9 Careless, "Christie, David"; Turner, "Perry, Peter"; Coleman, "Cameron, Malcolm"; Cross, *Robert Baldwin*, 287–9.

10 *CdnAg*, September 1850.

11 *CdnAg*, June, September 1850. Careless, "Robert Baldwin," *Pre-Confederation Premiers*, 138–40; Cross, *Robert Baldwin*, 295–309.

12 The bill faced immediate opposition from Adam Johnston Fergusson, the Reform MPP for Waterloo and son of the Honourable Adam Fergusson, whose concern turned out to be a misunderstanding as to the proposed financial relationship between county and township agricultural societies. Not until 2 August was the younger Fergusson satisfied. *CdnAg*, June, September 1850; Province of Canada, *Journals of the Legislative Council of the*

*Province of Canada*, 2, 3, 5, 6, 9 August 1850; Province of Canada, *Journals of the Legislative Assembly*, 6, 9, 10 August 1850; Province of Canada, *Statutes*, 1850, 13&14 Vic., c. 73. *Journal and Transactions*, 158–60.

13 During the brief debate on the bill the idea of appointing a minister of agriculture had been raised but rejected outright. See George Brown's statement to parliament on 5 October 1852 in Gibbs, ed., *Debates of the Legislative Assembly of United Canada*, v. 11, part 2 (1852–3), 849.

14 Province of Canada, *Statutes*, 1850, 13&14 Vic., c. 73. Compare with draft bill. *CdnAg*, June 1850. The board would report £70 in expenses during its first year of operation. "Receipts and Expenditure of the Board of Agriculture of Upper Canada for the year 1851–2" in "First Annual Report of the Board of Agriculture of Upper Canada, for 1851–2," Appendix (S.), Province of Canada, *Appendix to the Eleventh Volume of the Journals of the Legislative Assembly of the Province of Canada, 1st session, 4th Parliament*, 1851–2; *Journal and Transactions*, 247–8.

15 *CdnAg*, September 1850.

16 The Board of Agriculture can be considered semi-public because it received no public funding, and unlike the Board of Registration and Statistics, its ministerial appointment was *ex officio*. Province of Canada, *Statutes*, 1850, 13&14 Vic., c. 73. *CdnAg*, June 1850. Legislation establishing a board of registration and statistics for the Canadas had been passed in 1847, requiring the receiver general, the secretary general, and the provincial secretary to form the board, which was to issue a "Report on the Statistics of the Province," relating information on "Trade, Manufacturers, Agriculture and Population." Province of Canada, *Statutes*, 1847, 9&11 Vic., c. 14. Also see Curtis, 66–7.

17 Province of Canada, *Statutes*, 1850, 13&14 Vic., c. 73; *CdnAg*, February 1850.

18 *CdnAg*, January 1851.

19 In 1847, Harland had been a member of the local committee to organize the AAUC's exhibition at Hamilton. "The Harland Family", *Historical Atlas of the County of Wellington, Ontario*, 32. *BAC*, November 1846.

20 *CdnAg*, July, November 1851; *Journal and Transactions*, 164–6.

21 Careless, "Robert Baldwin," *Pre-Confederation Premiers*, 140–1; Cross, *Robert Baldwin*, 343–7.

22 Province of Canada, *Journals of the Legislative Assembly*, 3 July, 1, 25, 27, 28, 30 August 1851; Province of Canada, *Statutes*, 1851, 14&15 Vic., c. 127.

23 Province of Canada, *Statutes*, 1851, 14&15 Vic., c. 127.

24 *CdnAg*, July, November 1851. *Journal and Transactions*, 165–6.

25 *CF*, 12 March 1847; *BAC*, April 1847.

26 Edmundson pointed out advances in agricultural education in the United States, such as professors of agriculture at Harvard and Yale, plus the state of Georgia that had provided recent funding for a chair in the subject

at its state university. He also pointed to Ireland, where plans had been adopted to establish agricultural colleges in each county. *BAC*, June 1847. The subject was discussed by the province's agricultural press in several editorials. *CF*, 9 April, 19 June, 3 July, 14 August 1847.

27 *ACJ*, 15 February, 15 March, 1 April, 15 April, 1 May 1848.

28 McDougall made it clear that Buckland was not the author of his February 1850 editorial, because Buckland had come to Upper Canada "with a view of becoming a candidate for the Chair, which he was told would be established." Clearly, Buckland did not wish to ruin his chances by being viewed as a harsh critic of a decision to not fill the chair of agriculture. *CdnAg*, January, February, April, May, July 1850. Several of these articles highlighted examples of public funds being used to build a model farm in conjunction with the chair in agriculture at the University of Cork, Ireland, as well as the New York State Agricultural College and Farm, funded by the state government. Within Lower Canada, it noted, the College of Chambly taught agricultural chemistry and operated a fifty-four-acre model farm.

29 *CdnAg*, November 1850. An additional notice is found in *CdnAg*, January 1851.

30 Ontario, Department of Education, *Documentary History of Education*, v. 9, 134, 268, 282–3; *CdnAg*, January 1851. For the university senate bill see *CdnAg*, February 1851.

31 Conners, "Fife, David."

32 Hind would spend only a short time at the Normal School, becoming a professor of chemistry at Toronto's Trinity College. Jarrell, "Hind, Henry Youle"; *Documentary History of Education*, vol. 9, 282–3; vol. 10, 5–6, 286; *Journal of Education for Upper Canada*, November 1853; *CF*, 16 July 1866. Hind's first book was *Two Lectures on Agricultural Chemistry* (1850).

33 *Documentary History of Education*, v. 9, 282–3; v. 10, 208; *Journal and Transactions*, 205; *CdnAg*, February 1851, February 1852. William McDougall, however, announced the news in his *North American* as a "most excellent appointment." *NA*, 13 February 1852.

34 *CdnAg*, December 1850.

35 See David Bain, "William Mundie, landscape gardener," 298–308; Court, "An Erosion of Imagination," 166–91.

36 *Documentary History of Education*, v. 9, 278–9; *CdnAg*, February 1851.

37 *CdnAg*, March 1853, January 1855. The board spent some £46 on this task. "Receipts and Expenditure of the Board of Agriculture of Upper Canada for the year 1851–2" in "First Annual Report of the Board of Agriculture of Upper Canada, for 1851–2," Appendix (S.), Province of Canada, *Appendix to the Eleventh Volume of the Journals of the Legislative Assembly of the Province of Canada, 1st Session, 4th Parliament*, 1851–2.

38 *CdnAg*, January, November 1851, January, May 1852; *Journal and Transactions*, 214.
39 *CdnAg*, July, November, December 1851, January, May 1852. Hind's book was first advertised in December 1850, and by July 1851, the AAUC had ordered one hundred copies for distribution to county agricultural societies. At the same time, it also ordered fifty copies of a forthcoming pamphlet by Major Robert Lachlan. This was likely a pamphlet on "The Agriculture and Agricultural Societies of Canada," which was not published due to a lack of subscribers. *CdnAg*, December 1850, July 1851, February 1852.
40 Careless, *Union of the Canadas*, 172; Ormsby, "Sir Francis Hincks," 162–3. Ormsby, "Hincks, Sir Francis."
41 Careless, *Union of the Canadas*, 178–9.
42 Ormsby, "Hincks, Sir Francis."
43 Gordon, rev. David Huddleston, "Thomas Dix Hincks."
44 Davison, 11; Cross, *Robert Baldwin*, 284; Curtis, 140–1.
45 Davison, 2, Davison calculates that when the Hincks–Morin ministry met parliament in Quebec in August 1852, twenty lawyers represented a predominant rump within the government benches, though thirteen of this group had other business interests. There were also eight merchants, six lumber merchants, fifteen railway promoters, six doctors, five notaries, three seigneurs, three large-scale farmers from Lower Canada, and two mill owners. Davison, 238–9.
46 Davison, 2, 4–6; Curtis, 140–1. "Hon. F. Hinks, Governor Windward Isles," appears in an 1858 list of AAUC Life Members, though he had left the province by then. *Journal and Transactions*, 230. Ormsby, "Hincks, Sir Francis."
47 *NA*, 13 and 16 January 1852.
48 *NA*, 23 January 1852. Hincks, *Reminiscences*, 255–6.
49 *NA*, 23 January 1852. For the £800 salary figure, see *NA*, 30 January 1852.
50 Parliamentary rules of the day forced any new appointee to a cabinet position to receive approval from the voters in his riding by way of a by-election. *BC*, 23 and 27 January 1852; *Globe* 24 and 27 January 1852; *TE*, 21 January 1852. Unfortunately, no copy exists of Brown's 22 January issue in which he made his initial attack on Cameron's deal. Also see reports copied from other Canadian newspapers reprinted in *NA*, 27 January 1852.
51 Careless, *Union of the Canadas*, 177–8; Ormsby, "Sir Francis Hincks," 162–3. Ormsby, "Hincks, Sir Francis."
52 *Globe*, 24 and 27 January 1852.
53 *NA*, 30 January 1852. Likewise, the *Examiner* relayed the accepted belief that the Board of Registration and Statistics was currently "of little if any practical utility." *TE*, 4 February 1852.

54 *CdnAg*, February 1852.
55 *NA*, 3 February 1852; *BC*, 6 February.
56 *NA*, 3, 6, 10, 13, 17, 24, 27 February 1852; *Globe*, 3 February 1852; *BC*, 13 February 1852; *TE*, 11, 18, 25 February, 3 March 1852.
57 Cameron was returned as elected on 12 May 1852. Forman, ed. *Legislators and Legislatures of Ontario*, vol. 1, 128.
58 *CdnAg*, May 1852; *Journal and Transactions*, 212–3.
59 *CdnAg*, September 1852.
60 In his letter to the Board of Agriculture, Cameron had inquired about the incorporation of township agricultural societies and a lowering of the amount they needed to raise to be eligible for the government grant. The board replied with its approval; accordingly, he amended the bill. *CdnAg*, September 1852. *Journal and Transactions*, 241–3.
61 *Globe*, 28 September 1852; *NA*, 24 September 1852. Also see the report from the *Quebec Gazette* of 6 October, copied in *NA*, 19 October 1852, which claimed it was one individual "who did not know how to behave himself."
62 *NA*, 24 September 1852. Also see short reports on this event in *CdnAg*, November 1852; *CFH*, 25 September 1852.
63 Province of Canada, Legislative Assembly, *Journals*, 24 September, 26 October, 4, 6, 10 November 1852; Legislative Council, *Journals*, 9 November 1852; *Statutes*, 1852, 16 Vic. c.18.
64 Province of Canada, Legislative Assembly, *Journals*, 24 September 1852. For McDougall's reporting on the bill while under debate, see *NA*, 28 September, 1, 19, 26 October 1852; *CdnAg*, October, November 1852.
65 Gibbs, ed., *Debates*, v. 11, part 2 (1852–3), 842.
66 Gibbs, ed., *Debates*, v. 11, part 2 (1852–3), 843–6.
67 Christie had been a schoolmate of Brown's in Edinburgh. Careless, *Brown of the Globe*, v.1, 5.
68 Gibbs, ed., *Debates*, v. 11, part 2 (1852–3), 847–57.
69 Gibbs, ed., *Debates*, v. 11, part 2 (1852–3), 1187–96; Province of Canada, Legislative Assembly, *Journals*, 22 October 1852. The division on second reading was fifty-one members in favour and seventeen opposed. The Tory W.B. Robinson of Simcoe attempted an amendment requesting that the bill be referred to a select committee to consider and report "whether or not it is expedient to establish a Bureau of Agriculture." His motion was not approved.
70 The most significant amendments concerned the fourth and eighth clauses. The fourth clause was amended to remove from the governor in council the role of selecting a vice-president for each board of agriculture, and restored authority for selecting both a president and vice-president to each board. The eighth clause dealt with the collection of statistical information and any response to queries that might sent forth by the minister of

agriculture. As amended, the minister would retain the authority to collect information from "All Boards of Agriculture, Agricultural Societies, Associations, Municipal Councils, Mechanics' Institutes, Public Institutions, and Public Offices" in the Canadas. Indicative of the increasing importance of statistical information to inform government initiatives, penalties for ignoring the minister's queries were set at a fine of £10. Province of Canada, Legislative Assembly, *Journals*, 26 October, 5, 6 November 1852. Compare the clauses of bill printed in *CdnAg*, October 1852 with Province of Canada, *Statutes*, 1852, 16 Vic. c.11. Also see McDougall's explanation of the bill in his *CdnAg*, December 1852 in which he marked the amended clauses and itemized the minor changes and their implications.

71 The vote on third reading was twenty-eight for and nine against, with the latter number being less due to those who were absent, and not due to any change in opinion from the vote on second reading. Province of Canada, Legislative Assembly, *Journals*, 6 and 10 November 1852; Legislative Council, *Journals*, 9 November 1852; *Statutes*, 1852, 16 Vic. c.11.Concerning the cholera outbreak see: Legislative Assembly, *Journals*, 3 November 1852; *NA*, 5 November 1852.

72 Curtis, *Politics of Population*, 142, 145.

73 See minutes of the August Board of Agriculture meeting, *CdnAg*, September 1852; and Street's comments during debate on the bill, Gibbs, ed., *Debates*, v. 11, part 2 (1852–3), 844–5.

74 See McDougall's discussion of the draft bill and the act as passed. *CdnAg*, December 1852.

75 *CdnAg*, October 1852.

76 Bruce Curtis uncovers these details in his study of the Board of Registration and Statistics. Curtis, *Politics of Population*, 142.

77 Province of Canada, *Statutes*, 1852, 16 Vic. c.11 and 18. The Lower Canadian legislation repealed all previous legislation regarding Lower Canadian agricultural societies while retaining some of the characteristics unique to the Lower Canadian system rather than impose entirely the Upper Canadian system upon it (e.g., no provision was made for township societies in Lower Canada). These reformed county agricultural societies maintained their connection to government by way of an annual grant equal to three times the amount a county agricultural society had collected in annual subscriptions from thirty or more members.

78 Curtis, *Politics of Population*, 142; "First Annual Report of the Board of Agriculture of Upper Canada, for 1851–2," Appendix (S), 1851–2; *Journal and Transactions*, 243–8.

79 As some indication of the relative state of agricultural improvement in Upper Canada, McDougall reported proudly on his visit to the New

York exhibition that all machines or implements "of real utility are either already known to Canadian farmers," or soon would be on display at the AAUC's annual provincial exhibition. He could make only a few recommendations to Canadian farmers, such as growing flax and using clay tile for field drainage. "Mr. William McDougall's Report on American Implements, Seeds, &c." in "Documents submitted by the Bureau of Agriculture to the Legislature of Canada," Appendix (I.I), 1854–5.

80 The German-language pamphlet also included a map of the province's infant railway system. "Hon. Mr. Cameron's Report," in "Documents submitted by the Bureau of Agriculture to the Legislature of Canada," Appendix (I.I), 1854–5.

81 Craig, "Rolph, John." Although E.W. Thomson addressed his Board of Agriculture report to Rolph, it was Cameron who wrote the minister's report. "Report from the Upper Canada Board of Agriculture," Appendix (I.I), 1854–5.

82 Careless, *Union of the Canadas*, 191–4; Ormsby, "Sir Francis Hincks," *Pre-Confederation Premiers*, 180–3.

83 Careless, *Union of the Canadas*, 186; Davison, "Francis Hincks," 242; Ormsby, "Hincks, Sir Francis."

**Conclusion**

1 "Hon. Mr. Cameron's Report," Appendix (I.I), Province of Canada, *Appendix to the thirteenth volume of the journals of the Legislative Assembly of the Province of Canada, 1st session, 5th Parliament*, 1854–5.

2 MacKenzie, "Thomson, Edward William."

3 "First Annual Report of the Board of Agriculture of Upper Canada, for 1851–2," Appendix (S), Province of Canada, *Appendix to the Eleventh Volume of the Journals of the Legislative Assembly of the Province of Canada*, 1st session, 4th Parliament, 1851–2; *Journal and Transactions*, 246–7, 279, 281, 283, 314–5; *CdnAg*, March 1853, January 1855, February 1856. Crowley, *The College on the Hill*, 14; Lawr, "Agricultural Education in Nineteenth-Century Ontario: An Idea in Search of an Institution."

4 Curtis, *Politics of Population*, 159.

5 Hodgetts, *Pioneer Public Service*, 226–7.

6 Curtis, *Politics of Population*, 145–62; Boyce, *Hutton of Hastings*, 178, 194–5; Turner, "Hutton, William." Also see Worton, *The Dominion Bureau of Statistics*, 5; Hodgetts, *Pioneer Public Service*, 40.

7 Hodgetts, *From Arm's Length to Hands-On*, 25–6.

8 Hodgetts, *From Arm's Length to Hands-On*, xi–xii, 5–6, 9, 18–20, 47–8, 57, 97–8; Ontario Agricultural Commission, *Report of the Commissioners*, 8–9. MacKenzie, "Buckland, George."

9 One can view these atlases at The Canadian County Atlas Digital Project, by McGill University Libraries. https://digital.library.mcgill.ca/countyatlas/

10 Studies that analyse elements of the Canadian state sprouted from the "potting shed" of the Province of Canada's Bureau of Agriculture include Curtis, *Politics of Population* and Worton, *The Dominion Bureau of Statistics*. Wilson's *Thomas D'Arcy McGee, Volume 2* also provides valuable insights, as McGee was appointed minister of the bureau in 1863.

# Bibliography

**Abbreviations**

| | |
|---|---|
| *ACJ* | *Agriculturist and Canadian Journal* |
| *BAC* | *British American Cultivator* |
| *BAJ* | *British American Journal* |
| *BC* | *British Colonist* |
| *BW* | *British Whig* |
| *CC* | *Canadian Constellation* |
| *CdnAg* | *Canadian Agriculturist* |
| *CFH* | *Canadian Family Herald* |
| *CFM* | *Canadian Farmer and Mechanic* |
| *CF* | *Canadian Freeman* |
| *CG* | *Christian Guardian* |
| *CA* | *Colonial Advocate* |
| *CUC* | *Courier of Upper Canada* |
| *DCB* | *Dictionary of Canadian Biography* |
| *DNB* | *Dictionary of National Biography* |
| *FM* | *Farmer and Mechanic* |
| *FJWCI* | *Farmers' Journal and Welland Canal Intelligencer* |
| *GG* | *Gore Gazette* |
| *HFP* | *Hallowell Free Press* |
| *KA* | *Kingston Argus* |
| *KC* | *Kingston Chronicle* |
| *KCG* | *Kingston Chronicle and Gazette* |
| *KG* | *Kingston Gazette* |
| *KN* | *Kingston News* |
| *KS* | *Kingston Spectator* |
| *MSTT* | *Morning Star and Toronto Transcript* |
| *NF* | *Newcastle Farmer* |

| | |
|---|---|
| *NC* | *Niagara Chronicle* |
| *NG* | *Niagara Gleaner* |
| *NH* | *Niagara Herald* |
| *NR* | *Niagara Reporter* |
| *NA* | *North American* |
| *QG* | *Quebec Gazette* |
| *RS* | *Royal Standard* |
| *SCJ* | *St. Catharines Journal* |
| *SOT* | *Spirit of the Times* |
| *TE* | *Toronto Examiner* |
| *TP* | *Toronto Patriot* |
| *TS* | *Toronto Star* |
| *UCG* | *Upper Canada Gazette* |
| *UCH* | *Upper Canada Herald* |

## Archival Sources

### *Archives of Ontario*

F 32 Macaulay family fonds
F 37 Mackenzie–Lindsey family fonds
F 44 John Beverley Robinson family fonds
F 46 Peter Russell fonds
F 47 Simcoe Family fonds
F 378 Hiram Walker Historical Museum collection
F 380 Percy C. Band collection
F 474 John Askin fonds
F 535 John Steele fonds
F 592 Mary Sophia O'Brien fonds
F 775 Miscellaneous Collection, 1825, #2
F 775 Miscellaneous Collection, 1832–1835
F 775 Miscellaneous Collection, 1906–1909
F 1078 Riddell Family fonds
F 1961 City of Toronto fonds
F 4354 Lois Darroch fonds
Mun Doc Minutes of the Home District Municipal Council, from 1842 to 1849, inclusive

### *Lennox and Addington County Museum and Archives*

William Bell Papers

*Library and Archives Canada*

MG11 CO 42 Colonial Office [Great Britain] fonds. Canada, formerly British North America, Original Correspondence
MG23-GIII7 John Porteus Papers
MG23-HII7 Hugh Hovell Farmar fonds
MG24-A40 John Colborne, 1st Baron Seaton fonds
MG24-E1 William Hamilton Merritt collection
MG24-H71 R. Kay Diary
MG24-I43 John Bennett Marks fonds
RG1 E1 Upper Canada State Book C and State Book D
RG1 E3 Upper Canada: State Submissions to Executive Council
RG5 A1 Upper Canada Sundries. Civil Secretary's correspondence. Upper Canada and Canada West
RG5 B21 Upper Canada and Canada West: Civil Secretary, Correspondence on Immigration
RG5 C1 Canada West: Provincial Secretary, Numbered Correspondence Files
RG7 G16, Civil Secretary's Letter books [of Upper Canada]

*Marilyn and Charles Baillie Special Collections Centre, Toronto Reference Library*

1801. Society of arts. Account of certain premiums offered in 1801 by the Society instituted at London for the encouragement of Arts, Manufactures and Commerce, to encourage the cultivation of hemp in Upper and Lower Canada. Adelphi, London, April 9, 1801.
B15 David W. Smith Papers
*First Report of the Select Committee appointed to take into consideration the Internal Resources of the Province of Upper Canada in its agriculture and exports and the practicability and means of enlarging them, also to consider the expediency of granting encouragement to domestic manufacture.* York: Upper Canada Gazette Office, 1821.
L5 Robert Baldwin Papers A34 correspondence
S106 John Graves Simcoe and Wolford Papers

*Provincial Archives of New Brunswick*

MC211 MS4/7/8 Gorham, R.P. "The Development of Agricultural Administration in Upper Canada during the Period before Confederation." Unpublished manuscript. December 1932.

*Special Collections and Archives, James A. Gibson Library, Brock University, St. Catharines, Ontario*

Niagara District Agricultural Society – Letters and Papers

*Government Documents Upper Canada/Lower Canada/Province of Canada/Ontario/Canada*

Board of Agriculture of Upper Canada. *The Canada Herd Book*. Toronto: W.C. Chewett and Co., 1867.

Forman, Debra, ed. *Legislators and Legislatures of Ontario*, vol. 1. Ontario: Legislative Library Research and Information Services, 1984.

"A History of the Agricultural and Arts Association." in Ontario. Sessional Papers (No. 28), Appendix D. *Fiftieth Annual Report of the Agriculture and Arts Association of Ontario, 1895*.

James, C.C. "The Development of Agriculture in Ontario," in Ontario, "Appendix to the Report of the Bureau of Industries, 1896" (1898).

– "The First Agricultural Societies," Appendix in Ontario: Department of Agriculture. *Annual Report of the Bureau of Industries for the Province of Ontario, 1901* (1902), Appendix No. 26, 111–35.

Lower Canada. *Statutes*.

Ontario Agricultural Commission. *Report of the Commissioners*. Toronto: C. Blackett Robinson, 1881.

Ontario. Agricultural and Horticultural Organizations Act, *R.S.O.* 1990, c. A.9.

Ontario. Department of Education. *Documentary History of Education*. vols. 9 and 10. Toronto: L.K. Cameron, 1902.

Ontario Heritage Trust. *Online Plaque Guide*. The Niagara Agricultural Society. https://www.heritagetrust.on.ca/en/index.php/plaques/niagara-agricultural-society

Ontario. Legislative Library. *The Upper Canada Gazette and its Printers*. Toronto: Legislative Library, 1993.

"Political State of Upper Canada in 1806." Note D. in Public Archives of Canada. *Report on Canadian Archives*. Ottawa: S.E. Dawson, 1893.

Province of Canada. Board of Agriculture. *Journal and Transactions of the Board of Agriculture of Upper Canada*. vol. 1. 1856.

Province of Canada. House of Assembly. *Journal*.

Province of Canada. Legislative Council. *Journal*.

Province of Canada. *Statutes*.

Upper Canada. *Statutes*.

Upper Canada. House of Assembly. *Journal*.

Upper Canada. Legislative Council. *Journal*.

*Nova Scotia*

Martell, J.S. "The Achievements of Agricola and the Agricultural Societies 1818–25," in *Bulletin of the Public Archives of Nova Scotia*, Public Archives of Nova Scotia, vol. 2, no. 2, 1940.

*Great Britain*

Board of Agriculture (Great Britain). *Premiums offered by the Board of Agriculture, 1804*. [London]: B. M'Millan, [1804].
– *Premiums offered by the Board of Agriculture, 1808*. London: B. McMillan, 1808.
Great Britain. Parliament. House of Commons. *The Parliamentary History of England from the earliest period to the year 1803*. vol. 30, 1817.

*United States*

True, Alfred Charles. *A History of Agricultural Education in the United States 1785–1925*. Washington: United States Government Printing Office, 1929.
United States, Department of Agriculture, Bureau of Animal Industry, *Special Report of the History and Present Condition of the Sheep Industry of the United States* (Washington, DC: Government Printing Office, 1892), 218–20.

## Newspapers and Periodicals

*Agriculturist and Canadian Journal*
*Banner*
*British American Cultivator*
*British American Journal*
*British Colonial Argus*
*British Colonist*
*British Whig*
*Canadian Agriculturalist or Journal and Transactions of the Board of Agriculture of Upper Canada*
*Canada Constellation*
*Canadian Correspondent*
*Canadian Family Herald*
*Canada Farmer*
*Canadian Farmer and Mechanic*
*Canadian Freeman*
*The Canadian Journal: A Repertory of Industry, Science and Art*
*Canadian Magazine*

*Christian Guardian*
*Colonial Advocate*
*Constitution*
*Correspondent and Advocate*
*Courier of Upper Canada*
*Examiner*
*Farmers' Journal and Welland Canal Intelligencer*
*Globe*
*Gore Gazette*
*Hallowell Free Press*
*Journal of Education for Upper Canada*
*Kingston Argus*
*Kingston Chronicle*
*Kingston Chronicle and Gazette*
*Kingston Daily News*
*Kingston Gazette*
*Kingston News*
*Kingston Spectator*
*Mirror*
*Morning Star and Toronto Transcript*
*Newcastle Farmer*
*Niagara Chronicle*
*Niagara Gleaner*
*Niagara Herald*
*Niagara Reporter*
*Niagara Spectator*
*North American*
*Observer*
*Patriot*
*Royal Standard*
*Spirit of the Times*
*St. Catharines Journal*
*Toronto Herald*
*Toronto Recorder and General Advertiser*
*Upper Canada Gazette*
*Upper Canada Guardian*
*Upper Canada Herald*

## Contemporary Printed Sources

*An Abstract of the Proceedings which occurred at the two meetings of the Provincial Agricultural Society during the Session of 1823.* Halifax, NS: Holland and Co., April 1823.

The Agricultural Society in Canada. *Papers & Letters on Agriculture, Recommended to the Attention of the Canadian Farmers.* Quebec: Samuel Neilson, 1790.

Alexander, Captain J.E. *Transatlantic Sketches*. Philadelphia: Key and Biddle, 1833.

Bell, Reverend William. *Hints to Emigrants; in a Series of Letters from Upper Canada*. Edinburgh: Waugh and Innes, 1824.

Bouchette, Joseph. *The British Dominions in North America.* vol 1. London: Longman, Rees, Orme, Brown, Green and Longman, 1832.

Boulton, D'Arcy. *Sketch of His Majesty's Province of Upper Canada.* London: C. Rickaby, 1805.

Boulton, Henry John. *A Short Sketch of the Province of Upper Canada*. London: John Murray, 1826.

Brown, George. *Toronto City and Home District Directory 1846–7*. Toronto: George Brown, 1846.

Burton, Reverend J.E. *Essay on Comparative Agriculture or a brief examination into the state of agriculture as it now exists in Great Britain and Canada*. Montreal: Montreal Gazette, 1828.

[Cameron, John]. *The Upper Canada Almanac for the Year of Our Lord 1810*. York: Printed by John Cameron, [1809].

[Cartwright, Richard]. *Letters from an American Loyalist in Upper Canada to his friend in England: on a pamphlet published by John Mills Jackson, Esquire, entitled A view of the province of Upper Canada.* Halifax, [1810].

Cattermole, William. *Emigration: The Advantages of Emigration to Canada*. London: Simpkin and Marshall, 1831.

Conder, Josiah. [*United States of America and Canada*]. 2 vols. London: J. Duncan, 1830.

Evans, William. *A Treatise on the Theory and Practice of Agriculture*. Montreal: Fabre, Perrault and Co., 1835.

– *Supplementary Volume to a Treatise on the Theory and Practice of Agriculture.* Montreal: L. Perrault, 1836.

Fidler, Reverend Isaac. *Observations on Professions, Literature, Manners and Emigration in the United States and Canada made during a residence there in 1832*. New York: J & J Harper, 1833.

Fothergill, Charles. *The York Almanac and Royal Calendar of Upper Canada for the Year 1825 Being the First after Bissextile or Leap Year.* York, UC: Charles Fothergill, 1824.

[Fuller, Reverend Thomas Brock]. *The Canadian Agricultural Reader*. Niagara: John Simpson, 1845.

Galt, John. *The Autobiography of John Galt*. 2 vols. London: Cochrane and M'Crone, 1833.

– "A statistical account of Upper Canada," *Philosophical Magazine* (London), [1st ser.], 29 (October 1807–January 1808): 3–10.

Goldie, John. *Diary of a Journey Through Upper Canada And Some of the New England States – 1819*. Toronto: Wm. Tyrrell and Co., 1897.

Gourlay, Robert. *General introduction to statistical account of Upper Canada, compiled with a view to a grand system of emigration, in connexion with a reform of the Poor Laws*. London, 1822; repr. [East Ardsley, Eng.], 1966.

– *Statistical account of Upper Canada, compiled with a view to a grand system of emigration*. 2 vols. London, Simpkin and Marshall, 1822.

– *Statistical Account of Upper Canada*. Reprint with introduction by S.R. Mealing. Toronto: McClelland and Stewart Ltd., 1974.

Grece, Charles F. *Facts and Observations Respecting Canada and the United States of America*. London: J. Harding, 1819.

Haw, Reverend William. *Fifteen Years in Canada; Being a Series of Letters on its Early History and Settlement*. Edinburgh: Charles Ziegler, 1850.

Hincks, K.C.M.G, C.B., Sir Francis. *Reminiscences of His Public Life*. Montreal: William Drysdale & Co., 1884.

Hind, Henry Youle, *Two Lectures on Agricultural Chemistry*. Toronto: H. Scobie, 1850.

Howison, John. *Sketches of Upper Canada*. 1821. Reprint. Toronto: Coles Publishing Company, 1970.

Inches, James. *Letters on Emigration to Canada*. Perth: C.G. Sidney, 1836.

Jackson, John Mills. *A View of the Political Situation of the Province of Upper Canada in North America*. London: W. Earle, 1809.

Jameson, Anna Brownell. *Winter Studies and Summer Rambles in Canada*. 1838. Reprint. Toronto: McClelland and Stewart Inc., 1990.

Kames, (Henry Home) Lord. *The Gentleman Farmer, Being an attempt to improve agriculture, by subjecting it to the test of rational principles*. 3rd ed. Edinburgh: Printed for John Bell and G.G.J. and J. Robinson, London, 1788.

Keefer, T.C. *The Canals of Canada: Their Prospects and Influence*. Toronto: Andrew H. Armour and Co., 1850.

La Rochefoucauld-Liancourt, François-Alexandre-Frédéric, duc de. *Travels through the United States of North America and the Country of the Iroquois and Upper Canada, in the years 1795, 1796, and 1797*. London: R. Phillips, 1799.

*Letters and Pape[rs] on Agriculture Extracted from the Correspondence of a Society Instituted at Halifax for Promoting Agriculture in the Province of Nova Scotia*. vol 1. Halifax: John Howe, 1791.

Mackenzie, William Lyon. *Sketches of Canada and the United States*. London: Effingham Wilson, 1833.

Midland District Agricultural Society. *Constitution of the Midland District Agricultural Society*. [Kingston: s.n., 1837?].

Moodie, Susanna. *Life in the Clearings versus the Bush*. London: Richard Bentley, 1853.

Murray, Hugh. *An Historical and Descriptive Account of British America*. vol. 1. Edinburgh: Oliver and Boyd, 1839.

*Report of the Loyal and Patriotic Society of Upper Canada*. Montreal, LC: William Gray, 1817.

*Reports of the Provincial Agricultural Society of Nova Scotia*. Halifax, N.S: Holland and Co., 1821.

Rolph, Thomas. *A Descriptive and Statistical Account of Canada*. London: Smith, Elder and Co., 1841.

Royal Society of Arts. *Transactions of the Society Instituted at London for the Encouragement of Arts, Manufactures, and Commerce.* vols. 19–45 (1801–1827).

Shirreff, Patrick. *A Tour of North America; together with a Comprehensive View of the Canadas and United States*. Edinburgh: Oliver and Boyd, 1835.

Sinclair, Sir John. *The code of agriculture*. London: Printed for Sherwood, Neely and Jones, 1821.

– *The Correspondence of The Right Honourable Sir John Sinclair, Bart*. 2 vols. London: Henry Colburn and Richard Bentley, 1831.

Smith, William Henry. *Smith's Canadian Gazetteer*. Toronto: H. & W. Rowsell, 1846.

*Societe d'Agriculture Etablie en Canada*. [1789].

Talbot, Edward Allen. *Five Years" Residence in the Canadas*. 2 vols. 1824. Reprint. New York: Johnson Reprint Co., 1968.

Taylor, Charles. *Remarks on the Culture and Preparation of Hemp in Canada*. [1805].

Taylor, James. *Narrative Of A Voyage To, And Travels In Upper Canada*. Hull: John Nicholson, 1846.

Treadwell, C.P. *Arguments in Favor of the Ottawa and Georgian Bay Ship Canal.* Ottawa, Ottawa Citizen, 1856.

Watson, Elkanah. *History of the rise, progress, and existing condition of the western canals in the state of New York: from September 1788, to the completion of the middle section of the Grand Canal, in 1819: together with the rise, progress, and existing state of modern agricultural societies, on the Berkshire system …* Albany, NY: D. Steele, 1820.

Walton, George. *The City of Toronto and the Home District Commercial Directory and Register with Almanack and Calendar for 1837*. Toronto: T. Dalton and W.J. Coates, 1837.

– *York Commercial Directory, Street Guide, and Register 1833–4*. York, UC: Thomas Dalton, [1833].

Washington, George. *Letters from his excellency George Washington to Arthur Young … and Sir John Sinclair.* Alexandria [VA.]: Cottom and Stewart, 1803.

Young, Arthur. *Annals of Agriculture and other Useful Arts*. London: Richard Phillips, 1804.

Young, John. *The Letters of Agricola on the Principles of Vegetation and Tillage.* Halifax, NS: Holland and Co., 1822.

## Secondary Sources

Abbott, Richard H. "The Agricultural Press Views the Yeoman: 1819–1859." *Agricultural History* 42 (1968): 35–48.

Addison, William. *English Fairs and Markets*. London: B.T. Batsford Ltd., 1953.

Anderson, Virginia DeJohn. *How Domestic Animals Transformed Early America*. Oxford: Oxford University Press, 2002.

Angus, Margaret. "The Macaulay Family of Kingston." *Historic Kingston* 5 (1956): 3–12.

Armstrong, Frederick H. *A City in the Making: Progress, People & Perils in Victorian Toronto*. Toronto: Dundurn Press, 1988.

– "Crookshank, George." In *Dictionary of Canadian Biography*. http://www.biographi.ca/en/bio/crookshank_george_8E.html

– *Handbook of Upper Canadian Chronology*. Rev. ed. Toronto: Dundurn Press, 1985.

– "Small, James Edward." In *Dictionary of Canadian Biography*. http://www.biographi.ca/en/bio/small_james_edward_9E.html

–, and Ronald J. Stagg, "Mackenzie, William Lyon." In *Dictionary of Canadian Biography*. http://www.biographi.ca/en/bio/mackenzie_william_lyon_9E.html.

Baatz, Simon. *"Venerate the Plough" A History of the Philadelphia Society for Promoting Agriculture. 1785–1985*. Philadelphia: Philadelphia Society for Promoting Agriculture, 1985.

Baillie, Jr., James L. "Charles Fothergill 1782–1840." *Canadian Historical Review* 25 (1944): 376–96.

Bain, David. "William Mundie, landscape gardener." *Journal of Garden History* 5:3 (1985): 298–308.

Ballstadt, Carl P.A. "Strickland, Susanna." In *Dictionary of Canadian Biography*. http://www.biographi.ca/en/bio/strickland_susanna_11E.html.

Barker, George Fisher Russell. "Thomas Coke," *Dictionary of National Biography*, vol. 4, 705–7.

Berman, Morris. "'Hegemony' and the Amateur Tradition in British Science." *Journal of Social History* 8 (1975): 30–50.

– *Social Change and Scientific Organization: The Royal Institution, 1799–1844*. Ithaca, NY: Cornell University Press, 1978.

Bowler, R. Arthur and Bruce G. Wilson. "Butler, John." In *Dictionary of Canadian Biography*. http://www.biographi.ca/en/bio/butler_john_1796_4E.html.

Bowsfield, Hartwell. "Maitland, Sir Peregrine." In *Dictionary of Canadian Biography*. http://www.biographi.ca/en/bio/maitland_peregrine_8E.html.

Boyce, Gerald E. *Hutton of Hastings: The Life and Letters of William Hutton, 1801–61*. Belleville, ON: Hastings County Council, 1972.

Brock, Thomas L. "Barrie, Sir Robert." In *Dictionary of Canadian Biography*. http://www.biographi.ca/en/bio/barrie_robert_7E.html
Brown, Howard Morton. *Lanark Legacy: Nineteenth-Century Glimpses of an Ontario County*. Perth, ON: Corporation of the County of Lanark, 1984.
Buckner, Phillip. "Arthur, Sir George." In *Dictionary of Canadian Biography*. http://www.biographi.ca/en/bio/arthur_george_8E.html.
Burns, Robert J. "Jarvis, William." In *Dictionary of Canadian Biography*. http://www.biographi.ca/en/bio/jarvis_william_5E.html.
– "Jarvis, William Botsford." In *Dictionary of Canadian Biography*. http://www.biographi.ca/en/bio/jarvis_william_botsford_9E.html.
– "Markland, George Herchmer." In *Dictionary of Canadian Biography*. http://www.biographi.ca/en/bio/markland_george_herchmer_9E.html.
Burroughs, Peter. "Ramsay, George, 9th Earl of Dalhousie." In *Dictionary of Canadian Biography*. http://www.biographi.ca/en/bio/ramsay_george_7E.html.
Bushman, Richard L. *The American Farmer in the Eighteenth Century: A Social and Cultural History*. New Haven, CT: Yale University Press, 2018.
– *The Refinement of America: Persons, Houses, Cities*. New York: Vantage Books, 1993.
Cameron, Wendy. "Robinson, Peter." In *Dictionary of Canadian Biography*. http://www.biographi.ca/en/bio/robinson_peter_7E.html.
"Canadian Letters: Description of a Tour thro" the Provinces of Lower and Upper Canada" *Canadian Antiquarian and Numismatic Journal*. 3rd ser., 9 (July–October, 1912).
Canniff, William. *History of the Province of Ontario*. Toronto: A.H. Hovey, 1872.
Canny, Nicholas, and Anthony Pagden, eds. *Colonial Identity in the Atlantic World, 1500–1800*. Princeton, NJ: Princeton University Press, 1989.
Careless, J.M.S. *Brown of the Globe*, vol. 1. Toronto: Macmillan Co. of Canada Ltd., 1959.
– "Christie, David." In *Dictionary of Canadian Biography*. http://www.biographi.ca/en/bio/christie_david_10E.html.
– "Lesslie, James." In *Dictionary of Canadian Biography*. http://www.biographi.ca/en/bio/lesslie_james_11E.html.
– *The Pre-Confederation Premiers: Ontario Government Leaders, 1841–1867*. Toronto: University of Toronto Press, 1980.
– *Toronto to 1918: An Illustrated History*. Toronto: James Lorimer & Company, Publishers, 1984.
– *The Union of the Canadas*. Toronto: McClelland and Stewart, 1967.
Carnochan, Janet. *History of Niagara*. 1914. Reprint. Belleville: Mika Publishing, 1973.
– "Members of the Agricultural Society, 1792–1805." *Niagara Historical Society* [Publications] 27 (1914–15): 17.

– "Niagara Library, 1800 to 1820" *Niagara Historical Society* [Publications] 6 (1900): 1–30.
Carter, Harold B. *His Majesty"s Spanish Flock.* [Sydney]: Angus & Robertson, [1964].
Cartwright, Reverend C.E. *Life and Letters of the Late Hon. Richard Cartwright.* Toronto: Belford Brothers, 1876.
Chassé, Sonia, Rita Girard-Wallot and Jean-Pierre Wallot. "Neilson, John." In *Dictionary of Canadian Biography*. http://www.biographi.ca/en/bio/neilson_john_7E.html.
Clark, Colonel John. "Memoirs of Colonel John Clark of Port Dalhousie, C.W." *Ontario Historical Society Papers and Records* 7 (1906): 157–90.
Clarke, Charles. *Sixty Years in Upper Canada, with Autobiographical Recollections.* Toronto: William Briggs, 1908.
Clarke, John. "Baby, James." In *Dictionary of Canadian Biography*. http://www.biographi.ca/en/bio/baby_james_6E.html
– *Land, Power, and Economics on the Frontier of Upper Canada*. Montreal & Kingston: McGill-Queen's University Press, 2001.
– *The Ordinary People of Essex: Environment, Culture, and Economy on the Frontier of Upper Canada*. Montreal & Kingston: McGill-Queen's University Press, 2010.
Clarke, Sir Ernest. "The Board of Agriculture 1793–1822." *Journal of the Royal Agricultural Society of England* 9 (1898): 1–41.
Coad, Jonathan G. *The Royal Dockyards 1690–1850: Architecture and Engineering Works of the Sailing Navy.* Aldershot, UK: Scholar Press, 1989.
Cochrane, D.J. "Agricultural Societies, Cattle Fairs, Agricultural Shows and Exhibitions of Upper Canada prior to 1867." Unpublished manuscript. McLaughlin Library, University of Guelph.
Coleman, Margaret. "Cameron, Malcolm." In *Dictionary of Canadian Biography*. http://www.biographi.ca/en/bio/cameron_malcolm_10E.html.
Colgate, William. "The Diary of John White." *Ontario History* 47 (1955): 147–70.
*Commemorative Biographical Record of the County of York Ontario.* Toronto: J.H. Beers, 1907.
Conners, I.L. "Fife, David." In *Dictionary of Canadian Biography*. http://www.biographi.ca/en/bio/fife_david_10E.html.
Court, John P.M. "An Erosion of Imagination: Unfulfilled Plans for a University Botanical Gardens and Taddle Creek, 1850 to 1884." *Ontario History* 95:2 (2003): 166–91.
Craig, Gerald M. "Croft, Henry Holmes." In *Dictionary of Canadian Biography*. http://www.biographi.ca/en/bio/croft_henry_holmes_11E.html.
– "Rolph, John." In *Dictionary of Canadian Biography*. http://www.biographi.ca/en/bio/rolph_john_9E.html.

– *Upper Canada: The Formative Years 1784–1841*. Toronto: McClelland and Stewart Ltd., 1963.

– "Wells, Joseph." In *Dictionary of Canadian Biography*. http://www.biographi.ca/en/bio/wells_joseph_8E.html.

Crawford, Bess Hillery. *Rosedale*. Erin, ON: Boston Mills Press, 2000.

Crawford, Pleasance. "The Roots of the Toronto Horticultural Society." *Ontario History* 89 (1997): 125–39.

Crosby, Jr., Alfred W. *America, Russia, Hemp and Napoleon: American Trade with Russia and the Baltic, 1783–1812*. Columbus: Ohio State University Press, 1965.

Cross, Michael S. "'The Age of Gentility': The Formation of an Aristocracy in the Ottawa Valley." in Canadian Historical Association *Historical Papers* (1967): 105–17.

– *A Biography of Robert Baldwin: The Morning Star of Memory*. Toronto: Oxford University Press, 2012.

– "Duncombe, Charles." In *Dictionary of Canadian Biography*. http://www.biographi.ca/en/bio/duncombe_charles_9E.html.

– "Hopkins, Caleb." In *Dictionary of Canadian Biography*. http://www.biographi.ca/en/bio/hopkins_caleb_10E.html

– "The Shiners" War: Social Violence in the Ottawa Valley in the 1830s." *Canadian Historical Review* 54 (1973): 1–26.

Crowley, Terry. *The College on the Hill: A New History of the Ontario Agricultural College, 1874–1999*. Toronto: Dundurn Press, 1999.

Cruikshank, E. A., ed. *The Correspondence of Lieutenant Governor John Graves Simcoe*. vols. 1–5. Toronto: Ontario Historical Society, 1923–1931.

–, ed. *The Correspondence of the Honourable Peter Russell*, vol. 1. Toronto: Ontario Historical Society, 1932.

–, ed. "Petitions for Grants of Land in Upper Canada, Second Series, 1796–99." *Ontario Historical Society Papers and Records* 26 (1930): 97–379.

– "Records of Niagara in the Days of Commodore Grant and Lieut.-Governor Gore. 1805–1811." *Niagara Historical Society* [Publications] 42 (1931).

– *The Story of Butler's Rangers and the Settlement of Niagara*. 1982. Reprint. Welland, ON: Tribute Printing House, 1893.

– "Ten Years of the Colony of Niagara," *Niagara Historical Society* [Publications] 17 (1908).

Curtis, Bruce. *The Politics of Population: State Formation, Statistics, and the Census of Canada, 1840–1875*. Toronto: University of Toronto Press, 2001.

– "Representation and State Formation in the Canadas, 1790–1850." *Studies in Political Economy*, 28 (Spring 1989): 59–87.

– *True Government by Choice Men? Inspection, Education, and State Formation in Canada West*. Toronto: University of Toronto Press, 1992.

Dalton, Ian Robert. *Thomas Dalton and the "pretended" Bank*. Toronto: Ian R. Dalton, 2002.

Dehli, Kari. "Creating A Dense and Intelligent Community: Local State Formation in Early 19th Century Upper Canada." *Journal of Historical Sociology* 3:2 (June 1990): 109–32.

Denison, Robert Evelyn. *A History of the Denison Family in Canada, 1792–1910.* Grimsby, Ont., [n.p], 1910.

Derry, Margaret E. *Art and Science in Breeding: Creating Better Chickens.* Toronto: University of Toronto Press, 2012.

– *Masterminding Nature: The Breeding of Animals, 1750–2010.* Toronto: University of Toronto Press, 2015.

– *Ontario's Cattle Kingdom: Purebred Breeders and their World, 1870–1920.* Toronto: University of Toronto Press, 2001.

Dodds, Philip F., and H.E. Markle. *The Story of Ontario Horticultural Societies, 1854–1873.* Picton: Picton Gazette Publishing Company (1971) Ltd., 1973.

Dominion Short-Horn Breeders' Association. *History of Short-Horn Cattle Imported into the Present Dominion of Canada From Britain and United States, Chronologically Arranged.* n. p., 1894[?].

Douglas, R. Alan. *Uppermost Canada: The Western District and the Detroit Frontier, 1800–1850.* Detroit, MI: Wayne State University Press, 2001.

Drayton, Richard. *Nature's Government: Science, Imperial Britain, and the 'Improvement' of the World.* New Haven & London: Yale University Press, 2000.

Dunham, Aileen, *Political Unrest in Upper Canada 1815–1836,* Toronto: McClelland and Stewart Ltd., 1963.

Dunquemin, Colin K, comp. *A Lodge of Friendship: The History of Niagara Lodge, No. 2 A.F & A.M. G.R.C., Niagara-on-the-Lake, Ontario, Canada, 1792–1992.* Niagara-on-the-Lake, ON: Niagara Lodge No. 2, A.F. & A.M., G.R.C., 1991.

Dyster, Barrie. "Gamble, John William." In *Dictionary of Canadian Biography.* http://www.biographi.ca/en/bio/gamble_john_william_10E.html.

Ennals, Peter. "Burnham, Zacheus." In *Dictionary of Canadian Biography.* http://www.biographi.ca/en/bio/burnham_zacheus_8E.html.

– "Spilsbury, Francis Brockell." In *Dictionary of Canadian Biography.* http://www.biographi.ca/en/bio/spilsbury_francis_brockell_6E.html.

Errington, Jane. *The Lion. The Eagle and Upper Canada: A Developing Colonial Ideology.* Montreal & Kingston: McGill-Queen's University Press, 1987.

– "Markland, Thomas." In *Dictionary of Canadian Biography.* http://www.biographi.ca/en/bio/markland_thomas_7E.html.

Fair, Ross. "The Kingston Royal Naval Dockyard and Canadian hemp supply, 1822–34." *Transactions of the Naval Dockyards Society* 15 (September 2021): 69–82.

– "A Most Favourable Soil and Climate: Hemp Cultivation in Upper Canada, 1800–1813." *Ontario History* 96 (Spring 2004): 41–61.

Fairley, Margaret., ed. *The Selected Writings of William Lyon Mackenzie 1824–1837.* Toronto: Oxford University Press, 1960.
Ferry, Darren. *Uniting in Measures of Common Good: The Construction of Liberal Identities in Central Canada.* Montreal and Kingston: McGill-Queen's University Press, 2008.
Firth, Edith, ed. *Profiles of A Province.* Toronto: Ontario Historical Society, 1967.
– *The Town of York, 1783–1815: A Collection of Documents of Early Toronto.* Toronto: The Champlain Society, 1962.
–, ed. *The Town of York, 1815–1834: A Further Collection of Documents of Early Toronto.* Toronto: The Champlain Society, 1966.
– "Wood, Alexander." In *Dictionary of Canadian Biography.* http://www.biographi.ca/en/bio/wood_alexander_7E.html.; *UCG*, 27 May 1830.
–, and Curtis Fahey. "Scadding, Henry." In *Dictionary of Canadian Biography.* http://www.biographi.ca/en/bio/scadding_henry_13E.html.
Fleming, Patricia Lockhart. *Upper Canadian Imprints, 1801–1841.* Toronto: University of Toronto Press, 1988.
Flindall, R.D. *J.M. Flindall: The Uncommon Man.* Cobourg, ON: Ron and Linda Flindall, 1988.
Fowke, Vernon C. *Canadian Agricultural Policy: The Historical Pattern.* 1946. Reprint. Toronto: University of Toronto Press, 1978.
Fraser, Robert L. "'All the privileges which Englishmen possess': Order, Rights, and Constitutionalism in Upper Canada," in Fraser, ed. *Provincial Justice: Upper Canadian Legal Portraits from the Dictionary of Canadian Biography.* Toronto: University of Toronto Press, 1992.
– "Dorland, Thomas." In *Dictionary of Canadian Biography.* http://www.biographi.ca/en/bio/dorland_thomas_6E.html.
– "Jackson, John Mills." In *Dictionary of Canadian Biography.* http://www.biographi.ca/en/bio/jackson_john_mills_7E.html.
– "Macaulay, John." In *Dictionary of Canadian Biography.* http://www.biographi.ca/en/bio/macaulay_john_8E.html
– "Nichol, Robert." In *Dictionary of Canadian Biography.* http://www.biographi.ca/en/bio/nichol_robert_6E.html
– "Washburn, Ebenezer." In *Dictionary of Canadian Biography.* http://www.biographi.ca/en/bio/washburn_ebenezer_6E.html.
Friedland, Martin. *The University of Toronto: A History.* Toronto: University of Toronto Press, 2002.
Fryer, Mary Beacock. *Elizabeth Posthuma Simcoe 1762–1850, A Biography.* Toronto: Dundurn Press, 1989.
–, and Christopher Dracott. *John Graves Simcoe, 1752–1806: A Biography.* Toronto: Dundurn Press, 1998.
Fussell, G.E. *More Old English Farming Books From Tull to the Board of Agriculture 1731–1793.* London: Crosby Lockwood, 1950.

Gagan, David. *The Denison Family of Toronto, 1792–1925*. Toronto: University of Toronto Press, 1973.
Garcia, Robert. "Fort Amherstburg in the War of 1812," http://www.warof1812.ca/fortambg.htm.
Gascoigne, John. *The Enlightenment and the Origins of European Australia*. Cambridge: Cambridge University Press, 2002.
– *Joseph Banks and the English Enlightenment: Useful Knowledge and Polite Culture*. Cambridge: Cambridge University Press, 1994.
– *Science in the Service of Empire: Joseph Banks, the British State and the Uses of Science in the Age of Revolution*. Cambridge: Cambridge University Press, 1998.
Gates, Lillian F. *Land Policies in Upper Canada*. Toronto: University of Toronto Press, 1968.
– "Price, James Hervey." In *Dictionary of Canadian Biography*. http://www.biographi.ca/en/bio/price_james_hervey_11E.html
Gates, Paul W. *The Farmer's Age: Agriculture 1815–1860* New York: Holt, Rinehart and Winston, 1960.
Gazley, John G. *The Life of Arthur Young 1741–1820*. vol. 97. Philadelphia: American Philosophical Society, 1973.
Gibbs, Elizabeth, ed., *Debates of the Legislative Assembly of United Canada*, vol. 11, part 2 (1852–3). Montreal: Presses de l'Ecole des Hautes études commerciales, 1970–).
Gidney, R.D. and W.P.J. Millar. *Professional Gentlemen: The Professions in Nineteenth Century Ontario*. Toronto: University of Toronto Press, 1994.
Gilchrist, J. Brian, ed. *Inventory of Ontario Newspapers*, Toronto: Micromedia Ltd., 1987.
Glazebrook, G.P. de T. *The Story of Toronto*. Toronto: University of Toronto Press, 1971.
Goddard, Nicholas. *Harvests of Change: The Royal Agricultural Society of England, 1838–1988*. London: Quiller Press, 1988.
Gordon, Alexander, rev. David Huddleston. "Thomas Dix Hincks." *Dictionary of National Biography* vol. 27, 261–62.
Gouriévidis, Laurence. *The Dynamics of Heritage: History, Memory and the Highland Clearances*. Farnham: Ashgate Publishing Limited, 2010.
Graham, Gerald S. *British Policy and Canada 1774–1791* [1930]. Reprint, Westport, CT: Greenwood Press, 1974.
Gray, William. *Soldiers of the King: The Upper Canadian Militia 1812–1815*. Erin, ON: Boston Mills Press, 1995.
Graymont, Barbara. "Thayendanegea." In *Dictionary of Canadian Biography*. http://www.biographi.ca/en/bio/thayendanegea_5E.html.
Greer, Allan, "Davison, George." In *Dictionary of Canadian Biography*. http://www.biographi.ca/en/bio/davison_george_4E.html.

–, and Ian Radforth, eds. *Colonial Leviathan, State Formation in Mid-Nineteenth Century Canada*. Toronto: University of Toronto Press, 1992.
Guillet, Edwin C. *Early Life in Upper Canada*. Toronto: The Ontario Publishing Co., Ltd., 1933.
– *The Pioneer Farmer and Backwoodsman*. 2 vols. Toronto: The Ontario Publishing Co. Ltd., 1963.
Gundy, H.P. "Thomson, Hugh Christopher." In *Dictionary of Canadian Biography*. http://www.biographi.ca/en/bio/thomson_hugh_christopher_6E.html.
Hall, Roger and Nick Whistler, "Galt, John." In *Dictionary of Canadian Biography*. http://www.biographi.ca/en/bio/galt_john_7E.html
Hamil, Fred Coyne. *The Valley of the Lower Thames 1640 to 1850*. Toronto: University of Toronto Press, 1951.
Hancock, David. *Citizens of the World: London merchants and the integration of the British Atlantic community, 1735–1785*. Cambridge: Cambridge University Press, 1995.
Harland-Jacobs, Jessica L. *Builders of Empire: Freemasonry and British Imperialism, 1717–1927*. Chapel Hill, N.C.: The University of North Carolina Press, 2007.
Harris, Cole. *The Reluctant Land: Society, Space, and Environment in Canada before Confederation*. Vancouver: UBC Press, 2008.
Hawes, Clement. "*Gulliver's Travels*: Colonial Modernity Satirized," in Hawes, ed. *Gulliver's Travels and Other Writings*. Boston: Houghton Mifflin, 2004. 1–31.
Heaman, E.A. *Civilization: From Enlightenment Philosophy to Canadian History*. Kingston and Montreal: McGill-Queen's University Press, 2022.
– *The Inglorious Arts of Peace: Exhibitions in Canadian Society during the Nineteenth Century*. Toronto: University of Toronto Press, 1999.
– *A Short History of the State in Canada*. Toronto: University of Toronto Press, 2015.
Hedrick, Ulysses Prentice. *A History of Agriculture in the State of New York*. 1933. Reprint. New York: Hill and Wang, 1966.
Herrington, Walter S. *History of the County of Lennox and Addington*. Toronto: The MacMillan Company of Canada, 1913.
Hillman, Thomas A. *A Statutory Chronology of Ontario: Counties and Municipalities*. Gananoque, ON: Langdale Press, 1988.
Hind, Henry Youle, et al. *Eighty Years Progress of British North America*. Toronto: L. Stebbins, 1864.
*Historical Atlas of the County of Wellington, Ontario*. Toronto: Historical Atlas Publishing Co., 1906.
Hobsbawm, Eric, and Terrence Ranger, ed. *The Invention of Tradition*. Cambridge: Cambridge University Press, 1983.

Hodgetts, J.E. *From Arm's Length to Hands-On: The Formative Years of Ontario's Public Service, 1867–1940.* Toronto: University of Toronto Press, 1995.
– *Pioneer Public Service: An Administrative History of the United Canadas, 1841–1867.* Toronto: University of Toronto Press, 1955.
Hofstadter, Richard. *The Age of Reform.* New York: Alfred A. Knopf, 1955.
Horn, Pamela. "The Contribution of the Propagandist to Eighteenth-Century Agricultural Improvement." *The Historical Journal* 25 (1982): 313–29.
Hudson, Derek and Kenneth W. Luckhurst. *The Royal Society of Arts 1754–1954.* London: John Murray, 1954.
Hudson, Kenneth. *Patriotism with Profit: British Agricultural Societies in the Eighteenth and Nineteenth Centuries.* London: Hugh Evelyn Ltd., 1972.
Hunter, Andrew F. *The History of Simcoe County, Volume II. – The Pioneers.* Barrie, ON: The County Council, 1909.
In collaboration. "Allan, William." In *Dictionary of Canadian Biography.* http://www.biographi.ca/en/bio/allan_william_8E.html.
In collaboration with Bruce A. Parker. "Street, Samuel." In *Dictionary of Canadian Biography.* http://www.biographi.ca/en/bio/street_samuel_1753_1815_5E.html.
– "Warren, John." In *Dictionary of Canadian Biography.* http://www.biographi.ca/en/bio/warren_john_5E.html.
Isaac, Rhys. *The Transformation of Virginia 1740–1790.* Chapel Hill: University of North Carolina Press, 1982.
Jackson, John N. *St. Catharines: Canada's Canal City.* St. Catharines, Ontario: Stonehouse Publications, 1993.
James, C.C. "The First Agricultural Societies." *Queen's Quarterly* 10 (1902): 218–23.
– "History of Farming." in Adam Shortt and Arthur G. Doughty eds. *Canada and its Provinces,* vol. 18. Toronto: Edinburgh University Press, 1914.
– "The Pioneer Agricultural Society of Ontario." *Farming World.* Special Fair Number, September 1902, 211–12.
Jarrell, Richard A. "Hind, Henry Youle." In *Dictionary of Canadian Biography.* http://www.biographi.ca/en/bio/hind_henry_youle_13E.html.
Jarvis, Eric. "Hartman, Joseph." In *Dictionary of Canadian Biography.* http://www.biographi.ca/en/bio/hartman_joseph_8E.html.
Johnson, J.K. *Becoming Prominent: Regional Leadership in Upper Canada, 1791–1841.* Montreal & Kingston: McGill-Queen's University Press, 1989.
– *In Duty Bound: Men, Women, and the State in Upper Canada, 1783–1841.* Montreal & Kingston: McGill-Queen's University Press, 2014.
Jones, Elwood H. "Fergusson, Adam." In *Dictionary of Canadian Biography.* http://www.biographi.ca/en/bio/fergusson_adam_9E.html.
– "Wyatt, Charles Burton." In *Dictionary of Canadian Biography.* http://www.biographi.ca/en/bio/wyatt_charles_burton_7E.html.

Jones, Robert Leslie. *A History of Agriculture in Ontario 1613–1880*. Toronto: University of Toronto Press, 1946.

Junius [Oliver Seymour Phelps]. *St. Catharines A to Z*. St. Catharines, ON: The St. Catharines and Lincoln Historical Society, 1967.

Keane, David and Colin Read, eds., *Old Ontario: Essays in Honour of J.M.S. Careless*. Toronto: Dundurn Press, 1990.

Kelly, Kenneth. "The Impact of Nineteenth Century Agricultural Settlement on the Land" in J. David Wood, ed. *Perspectives on Landscape and Settlement in Nineteenth Century Ontario*. Toronto: McClelland and Stewart Ltd., 1975.

– "Notes on a Type of Mixed Farming Practiced in Ontario During the Early Nineteenth Century." *Canadian Geographer* 17:3 (1973): 205–19.

– "The Transfer of British Ideas on Improved Farming to Ontario During the First Half of the Nineteenth Century." *Ontario History* 63 (1971): 103–11.

Kelsay, Isabel T. "Tekarihogen (1794–1832)." In *Dictionary of Canadian Biography*. http://www.biographi.ca/en/bio/tekarihogen_1794_1832_6E.html

Kirby, William. *Annals of Niagara*. 1896 Reprint. Toronto: Macmillan, 1927.

Landon, Fred. "The Agricultural Journals of Upper Canada (Ontario)." *Agricultural History* 9 (1935): 167–75.

LaSeur, William Dawson. *William Lyon Mackenzie: A Reinterpretation*. Toronto: The Macmillan Company of Canada Ltd., 1979.

Lawr, Douglas. "Agricultural Education in Nineteenth-Century Ontario: An Idea in Search of an Institution" in Michael B. Katz and Paul H. Mattingly, eds. *Education and Social Change: Themes from Ontario's Past*. New York: New York University Press, 1975.

Legget, Robert Ferguson. "Tredwell (Treadwell), Nathaniel Hazard." In *Dictionary of Canadian Biography*. http://www.biographi.ca/en/bio/tredwell_nathaniel_hazard_8E.html.

Lennox and Addington Historical Society Centennial Brochure Committee. *Historical Glimpses of Lennox and Addington County*. [Napanee, ON]: Lennox and Addington County Council, 1964.

Lindsey, Charles. *The Life and Times of Wm. Lyon Mackenzie*. Toronto: P.R. Randall, 1867.

Lord, Clifford. "Elkanah Watson and New York's First County Fair." *New York History* 28 (1942): 437–448.

Lownsbrough, John. "Boulton, D'Arcy (1759–1834)." In *Dictionary of Canadian Biography*. http://www.biographi.ca/en/bio/boulton_d_arcy_1759_1834_6E.html.

– *The privileged few: the Grange & its people in nineteenth century Toronto*. Toronto, Art Gallery of Ontario, 1980.

Macdonald, Norman. "Hemp and Imperial Defence," *Canadian Historical Review* 17:4 (December 1936): 385–98.

MacDonald, R.W. Bro. William W. "Colonel John Butler: Soldier, Loyalist, Freemason, Canada's Forgotten Patriot, 1725–1796." *Canadian Masonic Research Association*, 76 (1964): 1319–22.
MacKenzie, Ann. "Buckland, George." In *Dictionary of Canadian Biography*. http://www.biographi.ca/en/bio/buckland_george_11E.html.
– "Edmundson, William Graham." In *Dictionary of Canadian Biography*. http://www.biographi.ca/en/bio/edmundson_william_graham_8E.html.
– "Thomson, Edward William." In *Dictionary of Canadian Biography*. http://www.biographi.ca/en/bio/thomson_edward_william_9E.html.
Mackey, Frank. *Steamboat Connections: Montreal to Upper Canada, 1816–1843*. Montreal & Kingston: McGill-Queen's University Press, 2000.
MacLean, R.A. "Young, John (1773–1837)." In *Dictionary of Canadian Biography*. http://www.biographi.ca/en/bio/young_john_1773_1837_7E.html.
Macpherson, K.R. "O'Brien, Edward George." In *Dictionary of Canadian Biography*. http://www.biographi.ca/en/bio/o_brien_edward_george_10E.html.
Mancke, Elizabeth. "Early Modern Imperial Governance and the Origins of Canadian Political Culture." *Canadian Journal of Political Science* 32:1 (March 1999): 3–20.
Mann, Michael. *The Sources of Social Power. Volume 2. The Rise of Classes and Nation-States, 1760–1914*. Cambridge: Cambridge University Press, 1986.
Marti, Donald B. "Early Agricultural Societies in New York: The Foundations of Improvement. *New York History* 48 (1967): 313–31.
– "In Praise of Farming: An Aspect of The Movement for Agricultural Improvement in the Northeast, 1815–1840." *New York History* 51(1970): 351–75.
Martyn, Lucy Booth. *Aristocratic Toronto: 19th Century Grandeur*. Toronto: Gage Publishing Ltd., 1980.
Matthews, Geoffrey J., and R. Cole Harris. *Historical Atlas of Canada*. vol. 2. Toronto: University of Toronto Press, 1987–1993.
McCalla, Douglas "The Commercial Politics of the Toronto Board of Trade, 1850–1860." *Canadian Historical Review* 50:1 (March 1969): 51–67.
– "The 'Loyalist' Economy of Upper Canada, 1784–1806." *Histoire sociale – Social History* 26:32 (November 1983): 279–304.
– *Planting the Province: The Economic History of Upper Canada, 1784–1870*. Toronto: University of Toronto Press, 1993.
– "Ridout, George Percival." In *Dictionary of Canadian Biography*. http://www.biographi.ca/en/bio/ridout_george_percival_10E.html.
McCann, Phillip. "Culture, State Formation, and the Invention of Tradition: Newfoundland, 1832–1855." *Journal of Canadian Studies/Revue d'études canadiennes* 23:1&2 (Spring/Summer 1988): 86–103.

McClenachan, Charles T. *History of the Most Ancient and Honorable Fraternity of Free and Accepted Masons in New York from the Earliest Date.* New York: The Grand Lodge, 1888.

McGill University Libraries. Digital Initiatives. *The Canadian County Atlas Digital Project*. McGill University, 2001. https://digital.library.mcgill.ca/countyatlas/

McIlwraith, Thomas F. "Jones, Charles." In *Dictionary of Canadian Biography*. http://www.biographi.ca/en/bio/jones_charles_7E.html.

McInnis, R. Marvin. "Perspectives on Ontario Agriculture 1815–1930" in Donald H. Akenson ed. *Canadian Papers in Rural History*. vol. 8. Gananoque: Langdale Press, 1992.

McKenna, Katherine M.J. *A Life of Propriety: Anne Murray Powell and Her Family 1755–1849*. Montreal & Kingston: McGill-Queen's University Press, 1994.

McKenzie, Ruth. "FitzGibbon, James." In *Dictionary of Canadian Biography*. http://www.biographi.ca/en/bio/fitzgibbon_james_9E.html.

McKillop, A. B., and Paul Romney eds. *God's Peculiar Peoples: Essays on Political Culture in Nineteenth Century Canada*. Ottawa: Carleton University Press, 1993.

McNairn, Jeffrey L. *The Capacity to Judge: Public Opinion and Deliberative Democracy in Upper Canada, 1791–1854*. Toronto: University of Toronto Press, 2000.

Mealing, S.R. "The Enthusiasms of John Graves Simcoe." in J.K. Johnson ed. *Historical Essays on Upper Canada*. Toronto: McClelland and Stewart Ltd., 1975.

– "Powell, William Dummer." In *Dictionary of Canadian Biography*. http://www.biographi.ca/en/bio/powell_william_dummer_6E.html

– "Simcoe, John Graves." In *Dictionary of Canadian Biography*. http://www.biographi.ca/en/bio/simcoe_john_graves_5E.html.

– "Small, John." In *Dictionary of Canadian Biography*. http://www.biographi.ca/en/bio/small_john_6E.html.

Merritt, Richard, Nancy Butler and Michael Power, eds. *The Capital Years: Niagara-on-the-Lake, 1792–1796*, Toronto and Oxford: Dundurn Press, 1991.

Middleton, J. E., and Fred Landon. *The Province of Ontario: A History*. vol. 1. Toronto: The Dominion Publishing Co., Ltd., 1927.

Mika, Nick and Helma. *The Shaping of Ontario from Exploration to Confederation*. Belleville: Mika Publishing Company, 1985.

Milani, Lois Darroch. *Robert Gourlay, Gadfly*. Thornhill, ON: Ampersand Press, 1971.

Miller, Audrey Saunders. *The Journals of Mary O'Brien 1828–1838*. Toronto: Macmillan of Canada Ltd, 1968.

– "Yonge Street Politics, 1828 to 1832." *Ontario History* 62 (1970): 101–18.

Mills, David. *The Idea of Loyalty in Upper Canada 1784–1850*. Montreal and Kingston: McGill-Queen's University Press, 1988.

Minella, Timothy K. "A Pattern for Improvement: Pattern Farms and Scientific Authority in Early Nineteenth-Century America." *Agricultural History* 90:4 (Fall 2016): 434–58.

Miskell, Louise. "Putting on a show: the Royal Agricultural Society of England and the Victorian Town, c.1840–1876." *Agricultural History Review*, 60:1 (2012): 37–59.

Mitchison, Rosalind. "The Old Board of Agriculture (1793–1822)." *English Historical Review* 74 (1959): 41–69.

Mokyr, Joel. *The Enlightened Economy: An Economic History of Britain, 1700–1850.* New Haven, CT, and London: Yale University Press, 2009.

Moogk, Peter N "McNabb, Colin." In *Dictionary of Canadian Biography*. http://www.biographi.ca/en/bio/mcnabb_colin_5E.html

Morgan, Henry J. ed. *The Canadian Parliamentary Guide*. 6th ed. Montreal: Gazette Steam Printing House, 1871.

Morris, Martha. "Naval Cordage Procurement in Early Modern Europe," *International Journal of Maritime History*, 9 (June 1999): 81–99.

Morriss, Roger. *The Royal Dockyards during the Revolutionary and Napoleonic Wars.* Leicester: Leicester University Press, 1983.

Mosser, Christine ed. *York Upper Canada Minutes of Town Meetings and Lists of Inhabitants 1797–1823.* Toronto: Metropolitan Toronto Library Board, 1984.

Murray, J. McE. "A Recovered Letter: W.W. Baldwin to C.B. Wyatt, 6th April, 1813." *Ontario Historical Society Papers and Records* 35(1943): 49–55.

Neary, Hilary Bates. "Stanton, Robert." In *Dictionary of Canadian Biography*. http://www.biographi.ca/en/bio/stanton_robert_9E.html.

Neely, Wayne Cadwell. *The Agricultural Fair*. New York: Columbia University Press, 1935.

Nelles, H.V. "Keefer, Thomas Coltrin." In *Dictionary of Canadian Biography*. http://www.biographi.ca/en/bio/keefer_thomas_coltrin_14E.html.

Nelles, V.M. "Loyalism and Local Power: The District of Niagara 1792–1837." *Ontario History* 58 (June 1966): 99–114.

Noel, S.J.R. *Patrons, Clients, Brokers: Ontario Society and Politics 1791–1896.* Toronto: University of Toronto Press, 1990.

Nurse, Jodey. *Cultivating Community: Women and Agricultural Fairs in Ontario.* Montreal and Kingston: McGill-Queen's Press, 2022.

Ogden, R. Lynn. "Connor, George Skeffington." In *Dictionary of Canadian Biography*. http://www.biographi.ca/en/bio/connor_george_skeffington_9E.html.

Ontario Association of Agricultural Societies. *The Story of Agricultural Fairs and Exhibitions 1792–1967*. Picton: Picton Gazette Publishing Co., 1967.

Ormsby, William G. "Hincks, Sir Francis." In *Dictionary of Canadian Biography*. http://www.biographi.ca/en/bio/hincks_francis_11E.html

– "Sir Francis Hincks." in J.M.S. Careless, ed. *The Pre-Confederation Premiers: Ontario Government Leaders, 1841*–1867. Toronto: University of Toronto Press, 1980, 148–93.

Osborne, Brian S. "Trading on a Frontier: The Function of Peddlers, markets, and Fairs in Nineteenth-Century Ontario" in Donald H. Akenson ed. *Canadian Papers in Rural History*. vol. 1. Gananoque: Langdale Press, 1980.

–, and Donald Swainson. *Kingston: Building on the Past*. Westport, Ont.: Butternut Press, 1988.

Ouellette, David. "Crooks, James." In *Dictionary of Canadian Biography*. http://www.biographi.ca/en/bio/crooks_james_8E.html.

Parker, Bruce A. "Street, Samuel (1775–1844)." In *Dictionary of Canadian Biography*. http://www.biographi.ca/en/bio/street_samuel_1775_1844_7E.html

Patterson, G.H. "Thorpe, Robert." In *Dictionary of Canadian Biography*. http://www.biographi.ca/en/bio/thorpe_robert_7E.html.

– "Weekes, William." In *Dictionary of Canadian Biography*. http://www.biographi.ca/en/bio/weekes_william_5E.html.

Patterson, William J. *Lilacs and Limestone*. n.p.: Pitsburgh Historical Society, 1989.

Pawley, Emily. *The Nature of the Future: Agriculture, Science, and Capitalism in the Antebellum North*. Chicago and London: The University of Chicago Press, 2020.

Pemberton, Ian. "Sherwood, Levius Peters." In *Dictionary of Canadian Biography*. http://www.biographi.ca/en/bio/sherwood_levius_peters_7E.html.

Perkins, W. Frank, comp. *British and Irish Writers on Agriculture*, 2nd. ed. Lymington, UK: Chas. T. King, 1972.

Pilon, Henri. "Elmsley, John (1801–63)." In *Dictionary of Canadian Biography*. http://www.biographi.ca/en/bio/elmsley_john_1801_63_9E.html.

*Pioneer Life on the Bay of Quinte*. Toronto: Rolph and Clarke Ltd., [190–?].

Powell, E.J. *History of the Smithfield Club from 1798–1900*. London: The Smithfield Club, 1902.

Prince Edward Historical Society. *Historic Prince Edward*. [Picton, ON]: The Society, 1976.

Quaife, Milo M., ed. *The John Askin Papers*. Detroit: Detroit Library Commission, 1928.

Raible, Chris. *Muddy York Mud: Scandal and Scurrility in Upper Canada*. Creemore, ON: Curiosity House, 1992.

Raven, James. *London Booksellers and American Customers: Transatlantic Literary Community and the Charleston Library Society 1748–1811*. Columbia: University of South Carolina Press, 2002.

Rawlyk, George and Janice Potter. "Cartwright, Richard." In *Dictionary of Canadian Biography*. http://www.biographi.ca/en/bio/cartwright_richard_5E.html.

Read, Colin. *The Rising in Western Canada, 1837–8*. Toronto: University of Toronto Press, 1982.

–, and Ron Stagg, eds. *The Rebellion of 1837 in Upper Canada, a Collection of Documents*. Toronto: The Champlain Society, 1985.

– "The Short Hills Raid of June, 1838, and its Aftermath." *Ontario History* 68 (1976): 93–115.

Reaman, G. Elmore. *A History of Agriculture in Ontario*. 2 vols. Toronto: Saunders of Toronto Ltd, 1970.

Reid, William D. *Death Notices of Ontario*. Lambertville, NJ: Hunterdon House, 1980.

– *Marriage Notices of Ontario*. Lambertville, NJ: Hunterdon House, 1980.

Riddell, William Renwick. *The Life of John Graves Simcoe*. Toronto: McClelland and Stewart Limited, [1926].

Ritchie, Elizabeth. "Cows, Sheep & Scots: Livestock and Immigrant Strategies in Rural Upper Canada, 1814–1851." *Ontario History* 109:1 (Spring 2017): 1–26.

Ritvo, Harriet. *The Animal Estate: The English and Other Creatures in the Victorian Age*. Cambridge, MA: Harvard University Press, 1987.

Robert, Jean-Claude. "Evans, William." In *Dictionary of Canadian Biography*. http://www.biographi.ca/en/bio/evans_william_8E.html.

Roberts, Andrew. *George III: The Life and Reign of Britain's Most Misunderstood Monarch*. London: Allen Lane, 2021.

Robertson, H.H. "The First Agricultural Society Within the Limits of Wentworth–1806." *Journal and Transactions of the Wentworth Historical Society* 4 (1905): 93–5.

Robertson, John Ross. *The Diary of Mrs. John Graves Simcoe*. Toronto: William Briggs, 1911.

– *The History of Freemasonry in Canada*, vol. 1. Toronto: Hunter, Rose & Co, Ltd. 1899.

Roland, Charles G. "Kerr, Robert." In *Dictionary of Canadian Biography*. http://www.biographi.ca/en/bio/kerr_robert_6E.html.

Rollason, Bryan, ed. *County of a Thousand Lakes: The History of the County of Frontenac 1673–1973*. Kingston, ON: Frontenac County Council, 1982.

Romney, Paul. "A Conservative Reformer in Upper Canada: Charles Fothergill, Responsible Government and the 'British Party,' 1824–1840." Canadian Historical Association *Historical Papers* (1984): 42–61.

– "Fothergill, Charles." In *Dictionary of Canadian Biography*. http://www.biographi.ca/en/bio/fothergill_charles_7E.html

– "From the Types Riot to the Rebellion: Elite Ideology, Anti-legal Sentiment, Political Violence, and the Rule of Law in Upper Canada." *Ontario History* 79 (June 1987): 113–44.

– *Mr Attorney: The Attorney General for Ontario in Court, Cabinet, and Legislature, 1791–1899*. Toronto: University of Toronto Press, 1986.
– "Randal, Robert." In *Dictionary of Canadian Biography*. http://www.biographi.ca/en/bio/randal_robert_6E.html.
Ruggle, Richard E. "Fuller, Thomas Brock." In *Dictionary of Canadian Biography*. http://www.biographi.ca/en/bio/fuller_thomas_brock_11E.html.
Russell, Peter A. *Attitudes to Social Mobility in Upper Canada 1815–1840*. Queenston, ON: The Edwin Mellen Press, 1990.
– *How Agriculture Made Canada: Farming in the Nineteenth Century*. Montreal and Kingston: McGill-Queen's University Press, 2012.
Russell, Victor, ed. *Forging a Consensus: Historical Essays on Toronto*. Toronto: University of Toronto Press, 1984.
Ryerson, Egerton. *The Story of My Life: Being Reminiscences of Sixty Years' Public Service in Canada*. Toronto: W. Briggs, 1884.
Samson, Daniel. *Spirit of Industry and Improvement: Liberal Government and Rural-Industrial Society, Nova Scotia, 1790–1862*. Montreal and Kingston: McGill-Queen's University Press, 2008.
Sanderson, Charles R. ed. *The Arthur Papers*. Toronto: Toronto Public Libraries and University of Toronto Press, 1957.
Saunders, Robert E. "Robinson, Sir John Beverley." In *Dictionary of Canadian Biography*. http://www.biographi.ca/en/bio/robinson_john_beverley_9E.html.
Scadding, Henry Rev., ed. *Letter to Sir Joseph Banks, President of the Royal Society of Great Britain, written by Lieut. Governor Simcoe, in 1791*. Toronto: The Copp Clark Company, Ltd., Printers, 1890.
Scott, Guy. *A Fair Share: A History of Agricultural Societies and Fairs in Ontario 1792–1992*. Peterborough: Ontario Association of Agricultural Societies, 1992.
Seibel, George A. *The Niagara Portage Road: A History of the Portage on the West Bank of the Niagara River*. Niagara Falls, ON: City of Niagara Falls Canada, 1990.
Senior, Hereward. "Boulton, George Strange." In *Dictionary of Canadian Biography*. http://www.biographi.ca/en/bio/boulton_george_strange_9E.html.
– "Boulton, William Henry." In *Dictionary of Canadian Biography*. http://www.biographi.ca/en/bio/boulton_william_henry_10E.html
Sheppard, George. *Plunder, Profit, and Paroles: A Social History of the War of 1812 in Upper Canada*. Montreal -and Kingston: McGill-Queen's University Press, 1994.
Shipley, Robert J.M. "Meyers, John Walden." In *Dictionary of Canadian Biography*. http://www.biographi.ca/en/bio/meyers_john_walden_6E.html.

Smith, E.A. "Russell, Francis, fifth duke of Bedford (1765–1802)." *Oxford Dictionary of National Biography*. http://www.oxforddnb.com/view/article/24308.

Spadafora, David. *The Idea of Progress in Eighteenth Century Britain*. New Haven, CT: Yale University Press, 1990.

Spragge, George. *The John Strachan Letter Book, 1812–1834*. Toronto: The Ontario Historical Society, 1946.

Stagg, Ronald J. "Gibson, David." In *Dictionary of Canadian Biography*. http://www.biographi.ca/en/bio/gibson_david_9E.html.

– "Lount, Samuel." In *Dictionary of Canadian Biography*. http://www.biographi.ca/en/bio/lount_samuel_7E.html.

– "Thorne, Benjamin." In *Dictionary of Canadian Biography*. http://www.biographi.ca/en/bio/thorne_benjamin_7E.html.

Stoll, Steven. *Larding the Lean Earth: Soil and Society in Nineteenth Century America*. New York: Hill and Wang, 2002.

Stubbs, John. *Jonathan Swift: The Reluctant Rebel*. New York: W.W. Norton & Company, 2017.

Swainson, Donald. "Street, Thomas Clark." In *Dictionary of Canadian Biography*. http://www.biographi.ca/en/bio/street_thomas_clark_10E.html.

Talman, James J. "Agricultural Societies of Upper Canada." *Ontario Historical Society Papers and Records* 27 (1931): 545–52.

– *Historical Sketch to Commemorate the Sesqui-Centennial of Freemasonry in the Niagara District 1791–1942*. [n.p.]: Grand Lodge A.F. & A.M. of Canada in the Province of Ontario, 1942.

Thornton, Tamara Plakins. *Cultivating Gentlemen: The Meaning of Country Life among the Boston Elite 1785–1860*. New Haven, CT, and London: Yale University Press, 1989.

– "The Moral Dimensions of Horticulture in Antebellum America." *New England Quarterly* 57 (March 1984): 3–24.

True, Rodney H. "The Early Development of Agricultural Societies in the United States." *Annual Report of the American Historical Society for the year 1920* 1 (1920): 295–306.

Tulchinsky, Gerald, ed. *To Preserve and Defend: Essays on Kingston in the Nineteenth Century*. Montreal and Kingston: McGill-Queen's University Press, 1976.

Turner, H.E. "Addison, Robert." In *Dictionary of Canadian Biography*. http://www.biographi.ca/en/bio/addison_robert_6E.html.

– "Perry, Peter." In *Dictionary of Canadian Biography*. http://www.biographi.ca/en/bio/perry_peter_8E.html.

Turner, Wesley B. "Hutton, William." In *Dictionary of Canadian Biography*. http://www.biographi.ca/en/bio/hutton_william_9E.html

Van Wagenen, Jr., Jared. "Elkanah Watson–A Man of Affairs." *New York History* 13 (1932): 404–12.

Vass, Elinor. "The Agricultural Societies of Prince Edward Island." *The Island Magazine* 7 (Fall–Winter 1979): 31–7.

Walden, Keith. *Becoming Modern in Toronto: The Industrial Exhibition and the Shaping of a Late Victorian Culture*. Toronto: University of Toronto Press, 1997.

Wall, Cecil. "George Washington: Country Gentleman." *Agricultural History* 43 (1969): 5–6.

Watson, James A. Scott. *The History of the Royal Agricultural Society of England, 1839–1939*. London: Royal Agricultural Society, 1939.

Weaver, John C. *The Great Land Rush and the Making of the Modern World, 1650–1900*. Montreal and Kingston: McGill-Queen's University Press, 2003.

– "Hamilton, George (1788–1836)." In *Dictionary of Canadian Biography*. http://www.biographi.ca/en/bio/hamilton_george_1788_1836_7E.html

"Weddings at Niagara, 1792–1832." *Ontario Historical Society Papers and Records* 3 (1901): 53–65.

Williamstown Fair. History. http://williamstownfair.ca/about-us/history/

Wilmot, Sarah. "'The Business of Improvement': Agriculture and Scientific Culture in Britain, c.1700–c.1870." *Historical Geography Research Series*, 24 (November 1990).

Wilson, Alan. "Colborne, John, Baron Seaton." In *Dictionary of Canadian Biography*. http://www.biographi.ca/en/bio/colborne_john_9E.html.

Wilson, Bruce G. "Clench, Ralfe." In *Dictionary of Canadian Biography*. http://www.biographi.ca/en/bio/clench_ralfe_6E.html.

– "Dickson, William." In *Dictionary of Canadian Biography*. http://www.biographi.ca/en/bio/dickson_william_7E.html.

– *The Enterprises of Robert Hamilton: A Study of Wealth and Influence in Early Upper Canada 1776–1812*. Ottawa: Carleton University Press, 1983.

– "Hamilton, Robert." In *Dictionary of Canadian Biography*. http://www.biographi.ca/en/bio/hamilton_robert_5E.html.

Wilson, Catharine Anne. "A Manly Art: Plowing, Plowing Matches, and Rural Masculinity in Ontario, 1800–1930." *Canadian Historical Review* 95:2 (2014):157–186

– *Tenants in Time: Family Strategies, Land, and Liberalism in Upper Canada, 1799–1871*. Montreal and Kingston: McGill-Queen's University Press, 2009.

Wilson, David A. *Thomas D'Arcy McGee, Volume 2: The Extreme Moderate, 1857–1868*. Montreal: McGill-Queen's University Press, 2008.

Wilson, Thomas B. *Marriage Bonds of Ontario 1803–1834*. Lamberton, NJ: Huntendon House, 1985.

– *Ontario Marriage Notices*. Lambertville, NJ: Hunterdon House, 1982.

Wise, S.F. "Gourlay, Robert Fleming." In *Dictionary of Canadian Biography*. http://www.biographi.ca/en/bio/gourlay_robert_fleming_9E.html.

Withrow, John. "Born Out of Protest." in *Once Upon a Century: 100 year History of the "Ex."* Toronto: J.H. Robinson Publishing Ltd., 1978.

Wood, J. David. *Making Ontario: Agricultural Colonization and Landscape Re-creation before the Railway.* Montreal and Kingston, McGill-Queen's University Press, 2000.

Worton, David A. *The Dominion Bureau of Statistics: A History of Canada's Central Statistical Office and Its Antecedents, 1841–1972.* Montreal and Kingston: McGill-Queen's University Press, 1998.

Wulf, Andrea. *The Brother Gardeners: Botany, Empire and the Birth of an Obsession.* New York: Alfred A. Knopf, 2009.

– *Founding Gardeners: The Revolutionary Generation, Nature, and the Shaping of the American Nation.* New York: Alfred A. Knopf, 2012

Wynn, Graeme. "Exciting a Spirit of Emulation Among the 'Plodholes': Agricultural Reform in Pre-Confederation Nova Scotia." *Acadiensis* 20 (1990): 5–51.

– "Johnston, James Finlay Weir." In *Dictionary of Canadian Biography.* http://www.biographi.ca/en/bio/johnston_james_finlay_weir_8E.html.

Zeller, Suzanne. *Inventing Canada: Early Victorian Science and the Idea of a Transcontinental Nation.* Toronto: University of Toronto Press, 1987.

– *Land of Promise, Promised Land: The Culture of Victorian Science in Canada.* Ottawa: Canadian Historical Association Booklet No. 56, 1996.

## Theses

Averley, Gwendoline. "English Scientific Societies of the Eighteenth and Early Nineteenth Centuries." PhD diss., Council for National Academic Awards. [Teesside Polytechnic, University of Durham], 1989.

Bindon, Kathryn M. "Kingston: A Social History 1785–1830." PhD diss., Queen's University, 1979.

Blanton, Lynne. "The Agrarian Myth in Eighteenth and Nineteenth-Century American Magazines." PhD diss., University of Illinois at Urbana-Champaign, 1979.

Bowsfield, Hartwell. "Upper Canada in the 1820's: The Development of a Political Consciousness." PhD diss., University of Toronto, 1976.

Burns, Robert Joseph. "The First Elite of Toronto: An Examination of the Genesis, Consolidation and Duration of Power in an Emerging Colonial Society." PhD diss., University of Western Ontario, 1974.

Davison, George A. "Francis Hincks and the Politics of Interest." PhD diss., University of Alberta, 1989.

Fair, Ross D. "Gentlemen, Farmers, and Gentlemen Half-Farmers: The Development of Agricultural Societies in Upper Canada, 1792–1846." PhD diss., Queen's University, 1998.

Fraser, Robert L. ""Like Eden in Her Summer Dress": Gentry, Economy, and Society: Upper Canada, 1812–1840." PhD diss., University of Toronto, 1979.

Gouglas, Sean. "The Influences of Local Environmental Factors on Settlement and Agriculture in Saltfleet Township, Ontario, 1790–1890." PhD diss., McMaster University, 2001.

Keon, Daniel J. "The "New World" Idea in British North America: An Analysis of Some British Promotional, Travel and Settler Writings 1784 to 1860." PhD diss., Queen's University, 1984.

Marti, Donald B. "Agrarian Thought and Agricultural Progress: The Endeavor for Agricultural Development in New England and New York, 1815–1840." PhD diss., University of Wisconsin, 1966.

Nesmith, Tom. "The Philosophy of Agriculture: The Promise of the Intellect in Ontario Farming, 1835–1914." PhD diss., Carleton University, 1988.

Quealey, Francis M. "The Administration of Sir Peregrine Maitland Lieutenant-Governor of Upper Canada 1818–1828." PhD diss., University of Toronto, 1968.

Romney, Paul. "A Man Out of Place: The Life of Charles Fothergill; Naturalist, Businessman, Journalist, Politician, 1782–1840." PhD diss., University of Toronto, 1981.

Walton, John Bruce. "An End To All Order: A Study of Upper Canadian Conservative Response to Opposition, 1805–1810." MA thesis, Queen's University, 1977.

Way, Peter John. "Political Process and Social Conflict: Group Disorder in Tory Toronto of the 1840s." MA thesis, Queen's University, 1983.

White, John Douglas. "Speed the Plough: Agricultural Societies in PreConfederation New Brunswick." MA thesis, University of New Brunswick, 1976.

# Index

www.ingramcontent.com/pod-product-compliance
Ingram Content Group UK Ltd.
Pitfield, Milton Keynes, MK11 3LW, UK
UKHW042054240225
455503UK00003B/74/J